Developments in Statistics

VOLUME 1

CONTRIBUTORS

A. V. BALAKRISHNAN
DAVID R. BRILLINGER
P. R. KRISHNAIAH
M. M. RAO
PRANAB KUMAR SEN
J. N. SRIVASTAVA

Developments in Statistics

Edited by *PARUCHURI R. KRISHNAIAH*
DEPARTMENT OF MATHEMATICS AND STATISTICS
UNIVERSITY OF PITTSBURGH
PITTSBURGH, PENNSYLVANIA

Volume 1

ACADEMIC PRESS New York San Francisco London 1978
A Subsidiary of Harcourt Brace Jovanovich, Publishers

ACADEMIC PRESS, INC.
111 Fifth Avenue, New York, New York 10003

United Kingdom Edition published by
ACADEMIC PRESS, INC. (LONDON) LTD.
24/28 Oval Road, London NW1 7DX

LIBRARY OF CONGRESS CATALOG CARD NUMBER: 77–11215

ISBN 0–12–426601–0

PRINTED IN THE UNITED STATES OF AMERICA

Contents

List of Contributors

Numbers in parentheses indicate the pages on which the authors' contributions begin.

A. V. BALAKRISHNAN (1), System Science Department, University of California at Los Angeles, Los Angeles, California

DAVID R. BRILLINGER (33), Department of Statistics, University of California at Berkeley, Berkeley, California and Department of Mathematics, University of Auckland, Auckland, New Zealand

P. R. KRISHNAIAH (135), Department of Mathematics and Statistics, University of Pittsburgh, Pittsburgh, Pennsylvania 15260

M. M. RAO (171), Department of Mathematics, University of California at Riverside, Riverside, California

PRANAB KUMAR SEN (227), Department of Biostatistics, School of Public Health, University of North Carolina, Chapel Hill, North Carolina 27514

J. N. SRIVASTAVA (267), Department of Statistics, Colorado State University, Fort Collins, Colorado 80523

Preface

The series "Developments in Statistics" has been created to provide a central medium for the publication of long and important papers in various branches of statistics. The papers may be (i) expository papers, (ii) research papers, or (iii) papers that are partially expository in nature. The volumes in the series will appear at irregular intervals. The papers in these volumes are, in general, too long to be published in journals but too short to be published as separate monographs. The series will cover both theory and applications of statistics. The first volume consists of invited papers written by outstanding workers in the field. These papers give authoritative reviews of the present state of the art on various topics, including new material in many cases, in the general areas of stochastic control theory, point processes, multivariate distribution theory, time series, nonparametric methods, and factorial designs.

I wish to thank the Department of Mathematics and Statistics at the University of Pittsburgh and the Department of Statistics at Carnegie-Mellon University for providing the facilities to edit this volume. I also wish to thank Academic Press for its excellent cooperation.

Contents of Volume 2

Parameter Estimation in Stochastic Differential Systems: Theory and Application

A. V. BALAKRISHNAN†

SYSTEM SCIENCE DEPARTMENT
UNIVERSITY OF CALIFORNIA AT LOS ANGELES
LOS ANGELES, CALIFORNIA

1. INTRODUCTION

The estimation problem in essence is the following. We have an observed process $y(t)$ ($n \times 1$ matrix function) which has the form

$$y(t) = S(\theta, t) + N(t), \qquad 0 < t < T \tag{1.1}$$

where θ denotes a vector of unknown parameters which we want to estimate, $S(\theta, t)$ being a stochastic process (*signal*) which is completely specified once θ is specified (e.g., by means of a stochastic differential system) and $N(t)$ is a stochastic process which models the errors (that remain even after all

† Research was supported in part by Grant No. 73-2492, Applied Mathematics Division, AFOSR, USAF.

ISBN 0-12-426601-0

systematic errors, such as bias and calibration errors, have been accounted for). There is much evidence to suggest that the noise process may be well modeled as Gaussian, and independent of the signal process. This is a basic assumption throughout this chapter.

Under the title of "system identification" there is a large engineering literature dealing with such problems. This is well documented in the proceedings of three symposia [1] devoted exclusively thereto. In the bulk of this literature, the process $S(\theta, t)$ is taken to be deterministic, in which case the estimation is largely treated as a *least squares* problem of minimizing

$$\int_0^T \|y(t) - S(\theta, t)\|^2 \, dt$$

over a predetermined admissible set of parameters θ. Where the stochastic signal case is considered, it is reduced to the time-discrete version of (1.1):

$$y_n = S_n(\theta) + N_n \tag{1.2}$$

for the reason that the continuous time is mathematically too difficult to handle, and anyhow, in digital computer processing (as is the rule), it is so discretized in the analog-to-digital (A–D) conversion process. This is indeed true; but the authors invariably proceed to make the assumption that the noise samples $\{N_n\}$ are mutually independent. But this requires that the sampling rate (in the periodic sampling of the data) be not more than twice the postulated *bandwidth* of the noise, itself actually unknown. Indeed in most practical cases the sampling rate is far higher than twice the bandwidth. To meet this objection, one may then allow the $\{N_n\}$ to be correlated. But then the correlation function must be known, and anyone with experience in handling real data can easily appreciate that it is unrealistic to require that much knowledge of the noise process, even if the complication in the theory can be borne.

We maintain, in any event, that it is much better to work with the time-continuous model (1.1), allowing as high a sampling rate in the processing as the A–D converter is designed for. But in the time-continuous model we are faced with another problem. The basic tool in estimation is the likelihood functional (for fixed parameters) which is based on the Radon–Nikodym derivative of the probability measure induced by the process by $y(\cdot)$ to that induced by the noise process $N(t)$. But this derivative is too difficult to calculate even when the precise spectrum of $N(\cdot)$ is known, which it is not. What we can assert for sure is that the bandwidth of noise $N(t)$ is much larger than that of the process $S(\theta, t)$, which is essential in order that the measuring instrument does not *distort* the signal. At this point it was customary in the earlier engineering literature to introduce *white noise* in a formal way as a stationary stochastic process with constant spectral density to represent the *large bandwidth* nature of $N(t)$. With the advances in the theory of diffusion processes using the Ito integral, it became fashionable to use a Wiener process model as being *more rigorous* [2]. Thus we replace (1.1) by

$$Y(t) = \int_0^t S(\theta, \sigma)\, d\sigma + W(t) \tag{1.3}$$

where $W(t)$ is a Wiener process. We can then exploit the well-developed machinery of martingales and Ito integrals. In fact the likelihood function can then be expressed as (see Liptser and Shiryayev [2])

$$\exp\left\{-\tfrac{1}{2}\int_0^T \|\hat{S}(\theta, t)\|^2\, dt - 2\int_0^T [\hat{S}(\theta, t), dY(t)]\right\} \tag{1.4}$$

where $\hat{S}(\theta, t)$ is the best mean square estimate of $S(\theta, t)$ given the σ-algebra generated by $Y(s)$, $s \le t$. This formula can be justly considered as one of the triumphs of the Ito theory, the key to the success being the appearance of the Ito integral in the second form of (1.4). This integral is defined on the basis that $Y(t)$ is of unbounded variation with probability 1. Of course no physical instrument can produce such a waveform. To calculate it, given the actual observation (1.1), we can "retrace" our steps back from (1.3) and use $y(t)\, dt$ in place of $dY(t)$. But this is totally incorrect, unless $S(\theta, t)$ is deterministic, and any minimization procedure based on it leads to erroneous results. This point is not appreciated by authors using (1.3) as more rigorous, perhaps because they have not had occasion actually to calculate anything based on real data. In any data generated by digital computer simulation, which must perforce employ the discrete version (1.2), this point can be completely masked and hence never appreciated.

Faced with this difficulty we have to examine more precisely the model again, to see a physically more meaningful way of exploiting the fact that the noise bandwidth is large compared to the signal bandwidth. What is needed is the *asymptotic form* of the likelihood functional as the bandwidth expands to infinity in an arbitrary manner.

Such a theory has been developed by the author using a precise notion of white noise. This is explained in Section 2. Based on this theory we derive a likelihood functional in Section 3. It turns out that formula (1.4) is replaced by

$$\exp\left\{-\tfrac{1}{2}\int_0^T \|\hat{S}(\theta, t)\|^2\, dt - 2\int_0^T \hat{S}(\theta, t)y(t)\, dt + \int_0^T \left(\widehat{\|S(\theta, t)\|^2} - \|\hat{S}(\theta, t)\|^2\right) dt\right\} \tag{1.5}$$

where the caret denotes conditional expectation given the data up to time t. Note that a third term appears, which can also be expressed as

$$\int_0^T \|S(\theta, t) - \hat{S}(\theta, t)\|^2 \, dt$$

and in the case in which $S(\theta, t)$ is Gaussian, this reduces to

$$\int_0^T E[\|S(\theta, t) - \hat{S}(\theta, t)\|^2] \, dt$$

being thus the integral of the mean square error in estimation of the signal $S(\theta, t)$ from the observation up to time t. When the signal process can be described in terms of stochastic differential equations, whether finite or infinite dimensional, advantage can be taken of the fact that the mean square error can be evaluated by solving a Riccati equation. Section 4 is devoted to this specialization. Section 5 deals with the application to the problem of stability and control derivatives from flight test data taking turbulence into account. The algorithms used and results obtained on actual flight data are included.

2. WHITE NOISE: BASIC NOTIONS

Let H denote a real separable Hilbert space and let

$$W = L_2[0, T; H], \qquad 0 < T < \infty$$

denote the real Hilbert space of H-valued weakly measurable functions $u(\cdot)$ such that

$$\int_0^T [u(t), u(t)] \, dt < \infty$$

with the inner product defined by

$$[u, v] = \int_0^T [u(t), v(t)] \, dt$$

Let μ_G denote the Gaussian measure on W (on the cylinder sets with finite-dimensional Borel basis) with characteristic function

$$C_G(h) = \exp(-\tfrac{1}{2}[h, h]), \qquad h \in W$$

Elements of W under this (finitely additive) measure will be *white noise sample functions*, denoted by ω. This terminology appears to have the sanction of usage; see Skorokhod [3] for example. It is essential for us that W is an L_2-space over a finite interval.

Any function $f(\cdot)$ defined on W into another Hilbert space H_r such that the inverse images of Borel sets in H_r are cylinder sets with base in a finite-dimensional subspace will be called a *tame* function (see Gross [4]). As is readily seen, the class of tame functions is a linear class. Since the inverse image of the whole space H_r must be cylindrical, it is clear that any tame function has the form $f(P\omega)$ where P is a finite-dimensional projection.

To introduce the notion of a *random variable* let us first confine ourselves to the case in which H_r is finite dimensional: $H_r = R^n$ say. We introduce a metric into the linear space of tame functions by

$$|||f - g||| = \int_W \frac{\|f - g\|}{1 + \|f - g\|} \, d\mu_G$$

and then complete the space, the completion yielding a Frechet space. Every element of the completed space is called a random variable and if ζ denotes such an element and $f_n(\omega)$ a corresponding Cauchy sequence in probability, then we define the corresponding *distribution function* or probability measure on R^n to be that induced by the characteristic function

$$C_\zeta(h) - \lim_n E(\exp\{i[f_n(\omega), h]\}) \tag{2.0}$$

The latter limit exists (uniformly on bounded sets of $R^n = H_r$).

In the case in which H_r is no longer finite dimensional, we shall still identify Cauchy sequences in probability of tame functions as *weak random variables*. The limit in (2.0) still holds uniformly on bounded sets in H_r, but the limit may in general only define a *weak distribution* on H_r. We recall in this connection the Sazonov theorem [5] that the limit is the characteristic function of a probability measure if and only if it is continuous in the trace-norm topology (*S-topology*, see later). This is automatically the case if the sequence is Cauchy in the mean square sense, and we shall then drop the qualification weak.

Let $f(\omega)$ be any Borel measurable function mapping W into H_r. Then $f(P\omega)$ is tame for every finite-dimensional projection operator P. Let $\{P_n\}$ denote a sequence of finite-dimensional projections converging strongly to the identity; the sequence may be assumed to be monotone. If the sequence $f(P_n\omega)$ is Cauchy in probability, then we may associate a (weak, in general) random variable with $f(\cdot)$. Let us denote it by $f^\sim$ (a notation used by Gross). This limit of course can depend on the particular projection sequence chosen. Of primary interest to us are those function $f(\cdot)$ for which $\{f(P_n\omega)\}$ is Cauchy in probability for *every* such sequence of finite-dimensional projections, and moreover such that all such Cauchy sequences are equivalent so that the limit random variable $f^\sim$ is unique. In that case we say that $f(\omega)$ is a (weak) random variable. We shall use the term random variable if the corresponding measure is countably additive; with mean square convergence, this will be automatic.

The simplest function one can consider is perhaps the linear function

$$f(\omega) = L\omega$$

where L is a linear bounded transformation mapping W into H_r, where we now allow H_r to be infinite dimensional. Then it is easy to see that if L is Hilbert–Schmidt, then $\{LP_m\,\omega\}$ is Cauchy in the mean square sense, and $L\omega$ is a random variable. Conversely, L must be H–S if $L\omega$ is to be a random variable.

What is the class of functions which are random variables? To answer this question, at least in part, let us introduce the S-topology on W: This is the (locally convex) topology induced by seminorms of the form

$$\rho(\omega) = [S\omega, \omega]^{1/2} \tag{2.1}$$

where here (and hereinafter) S denotes a self-adjoint, nonnegative definite trace-class operator on W into W. For the case in which $H_r = R^1$, Gross [4] has given a sufficient condition: $f(\cdot)$ is a random variable if it is *uniformly* continuous in the S-topology. Uniform continuity means that given $\varepsilon > 0$, we can find $\rho(\cdot)$ such that

$$\|f(x) - f(y)\| < \varepsilon \qquad \text{for all} \quad x, y \quad \text{such that} \quad \rho(x - y) < 1$$

Unfortunately Gross does not seem to discuss nontrivial examples of functions satisfying this condition. Here we shall give a sufficient condition for a class of random variables with finite second moment.

Theorem 2.1. Let $p_n(\omega)$ denote a homogeneous polynomial of degree n mapping W into H_r. Suppose it is continuous at the origin in the S-topology. Let P denote any finite-dimensional projection. Then

$$\sup_P E(\|p_n(P\omega)\|^2) < \infty \tag{2.2}$$

where the supremum is taken over the class of all finite-dimensional projections. Conversely, if (2.2) holds, then $p_n(\cdot)$ is continuous at the origin in the S-topology.

Proof. We begin with a simple but useful lemma.

Lemma 2.1. Suppose $p_n(\cdot)$ is continuous in the S-topology at the origin. Then there exists a seminorm in the S-topology:

$$\rho(\omega) = [S\omega, \omega]^{1/2} \tag{2.3}$$

such that

$$\|p_n(\omega)\| \leq M\rho(\omega)^n \tag{2.4}$$

where M is a constant. Conversely, if (2.4) holds, then $p_n(\omega)$ is continuous in the S-topology at the origin.

Proof. Continuity in the S-topology at zero implies the following: Given $\varepsilon > 0$ we can find a seminorm of the form (2.3) such that

$$\|p_n(\omega)\| < \varepsilon \qquad \text{for all} \quad \omega \quad \text{such that} \quad \rho(\omega) \leq \delta \tag{2.5}$$

Hence for any ω for which $\rho(\omega) \neq 0$, we have that

$$\left\| p_n \left(\frac{\delta \omega}{\rho(\omega)} \right) \right\| < \varepsilon$$

or by the homogeneity of $p_n(\cdot)$,

$$\|p_n(\omega)\| < \left(\frac{\varepsilon}{\delta^n} \right) \rho(\omega)^n, \qquad \rho(\omega) \neq 0$$

If $\rho(\omega) = 0$, then for any positive number k,

$$\rho(k\omega) = 0$$

and hence from (2.5)

$$\|p_n(\omega)\| < \varepsilon | k^n \qquad \text{for all} \quad k > 0$$

and hence

$$p_n(\omega) = 0$$

Therefore (2.4) holds. The converse is obvious.

Proof of Theorem. Corresponding to a finite-dimensional projection P, we can find an orthonormal basis $\{\phi_i\}$ such that P is the projection operator corresponding to the space spanned by ϕ_i, $i = 1, 2, \ldots, m$. Let

$$p_n(\omega) = k_n(\omega, \ldots, \omega)$$

$k_n(\cdots)$ being the symmetric n-linear form, corresponding to $p_n(\cdot)$. Then

$$p_n(P\omega) = \sum_{i_1=1}^{m} \cdots \sum_{i_n=1}^{m} a_{i_1, \ldots, i_n} \zeta_{i_1}, \ldots, \zeta_{i_n} \tag{2.6}$$

where

$$a_{i_1, \ldots, i_n} = k_n(\phi_{i_1}, \ldots, \phi_{i_n}), \qquad \zeta_i = [\phi_i, \omega]$$

$\{\zeta_i\}$ is a sequence of independent zero-mean unit variance Gaussians and

(2.6) defines a tame function. Moreover we can readily calculate [by expressing (2.6) in terms of Hermite polynomials for instance] that

$$E(\|p_n(P\omega)\|^2) = \sum_{\nu=0}^{[n/2]} \left(\frac{n!}{(n-2\nu)!\, 2^\nu \nu!}\right)^2 \times \sum_{i_{2\nu+1}=1}^{m} \sum_{i_n=1}^{m} \left\| \sum_{i_1=1}^{m} \cdots \sum_{i_\nu=1}^{m} a_{i_1,i_1,\ldots,i_\nu,i_\nu,i_{2\nu+1},\ldots,i_n} \right\|^2 \tag{2.7}$$

But from Lemma 2.1, we have that

$$\|p_n(P\omega)\|^2 \leq [S_m\omega, \omega]^n \tag{2.8}$$

where $S_m = PSP$, and is of course trace-class and finite dimensional. Hence

$$E[\|p_n(P\omega)\|^2] \leq E([S_m\omega, \omega]^n) \tag{2.9}$$

Let ψ_k, $k = 1, \ldots, \nu$, be the orthonormalized eigenvectors of S_m with corresponding nonzero eigenvalues λ_k. Then

$$[S_m\omega, \omega] = \sum_1^\nu \lambda_i[\psi_i, \omega]^2$$

and we have

$$E([S_m\omega, \omega]^n) = f(\text{tr } S_m, \text{tr } S_m^2, \ldots, \text{tr } S_m^n)$$

where $f(\cdot)$ is a fixed continuous function. Of course tr S_m^j is monotone in m for each j and converges to tr S^j. Hence it follows that

$$E[\|p_n(P\omega)\|^2] < \infty$$

for *all* finite-dimensional projections.

To prove the converse, suppose (2.2) holds. Then (2.7) holds for every m, and taking $\nu = 0$ therein, we obtain that

$$\sum_{i_1}^{\infty} \cdots \sum_{i_r}^{\infty} \|k_n(\phi_{i_1}, \ldots, \phi_{i_n})\|^2 < \infty \tag{2.10}$$

for every orthonormal sequence $\{\phi_i\}$. Hence $p_n(\cdot)$ is Hilbert–Schmidt. Of course

$$\|p_n(\omega)\|^2 \leq M\|\omega\|^{2n} \tag{2.11}$$

Define S now by

$$[S\omega, \omega] = (\|p_n(\omega)\|^2)^{1/n}$$

Then S is Hilbert–Schmidt by (2.10). For any finite-dimensional projection P,

$$E[SP\omega, P\omega] = E[PSP\omega, \omega] = E((\|p_n(P\omega)\|^2)^{1/n})$$

and hence

$$\sup_{P} E[PSP\omega, \omega] < \infty$$

But taking the orthonormal basis of eigenvectors of S, it follows that S is trace-class.

It follows from Theorem 2.1 that if a homogeneous polynomial is uniformly continuous in the S-topology, the corresponding random variable has finite second moment.

For a homogeneous polynomial of degree 2 with range in R^1 ($H_r = R^1$ or R^m more generally) we can prove that continuity at the origin in the S-topology is sufficient to make it a random variable. For from (2.7) we have

$$E[\|p_2(P\omega)\|^2] = \sum_1^m \sum_1^m |k_2(\phi_i, \phi_j)|^2 + \left|\sum_1^m k_2(\phi_i, \phi_i)\right|^2 < \infty$$

and hence

$$\sum_1^\infty |k_2(\phi_i, \phi_i)| < \infty$$

for any orthonormal system. Hence it follows that

$$E[\|p_2(P_n\omega) - p_2(P_m\omega)\|^2]$$

is Cauchy. This suffices for our purposes here. See Balakrishnan [6] for further details, in particular the relation to multiple Ito integrals.

3. RADON–NIKODYM DERIVATIVES OF WEAK DISTRIBUTIONS

Let ω denote white noise samples as in Section 2 and let

$$y(\omega) = f(\omega) + \omega \tag{3.1}$$

where $f(\cdot)$ is a random variable mapping into W. Then $\{y(P_m\omega)\}$ is a Cauchy sequence in probability (being the sum of two such sequences) and the limit is independent of the particular sequence $\{P_m\}$ chosen. Hence $y(\omega)$ induces a weak distribution on W, which we shall call μ_y. As finitely additive measures, μ_y is said to be absolutely continuous with respect to μ_G if given any $\varepsilon > 0$, we can find $\delta > 0$ such that for any cylinder set C,

$$\mu_y(C) < \varepsilon$$

as soon as

$$\mu_G(C) < \delta$$

However, the definition of the derivative is more involved. For our purposes, we shall be concerned with the case in which the derivative is a random variable. That is, there exists a function $f(\omega)$ mapping W into R^1 such that $f(\omega)$ is a random variable and for any cylinder set C,

$$\mu_y(C) = \lim_m \int_C f(P_m \omega)\, d\mu_G$$

where $\{P_m\}$ is any monotone sequence of finite-dimensional projections converging strongly to the identity.

Let

$$W_s = L_2[(0, T); H_s]$$

where H_s is a separable Hilbert space. Let $\mu_G{}^s$ denote the Gauss measure thereon, and let ω_s denote points in W_s (the subscript s stands for signal). Let

$$W_2 = W_s \otimes W$$

the Cartesian product and induce the product Gauss measure μ_2 on W_2:

$$\mu_2(C_s \otimes C) = \mu_G{}^s(C_s)\mu_G(C)$$

where C_s is a cylinder set in W_s and C a cylinder set in W. Denote points in W_2 by ω_2:

$$\omega_2 = \begin{bmatrix} \omega_s \\ \omega \end{bmatrix}$$

Let

$$y(\omega_2) = f(\omega_s) + \omega \tag{3.2}$$

where $f(\cdot)$ is a random variable mapping W_s into W. Let μ_y denote again the (finitely additive) measure induced by $y(\cdot)$. We wish to prove the absolute continuity of the measure $\mu_y(\cdot)$ with respect to the measure $\mu_G(\cdot)$ and to find the corresponding derivative.

For the Wiener process version of (1.2), such a result appears to have been first developed by Duncan [7] for the case in which $f(\omega_s)$ is a diffusion process (see also Kallianpur and Striebel [8]). As may be expected, our result has a superficial similarity to the Stratanovich version [9, Eq. 12].

Let H be finite dimensional: $H = R^n$.

Theorem 3.1. Let $f(\omega_s)$ denote a random variable mapping W_s into W such that

$$E(\|f(\omega_s)\|^2) < \infty \tag{3.3}$$

Let $y(\omega_2)$ be as defined by (3.2).

Then μ_y is absolutely continuous with respect to μ_G and the derivative is a random variable (white noise integral) corresponding to the function $g(\omega)$ defined by

$$g(\omega) = \int_W [\exp(-\tfrac{1}{2}\{\|x\|^2 - 2[x, \omega]\})] \, d\mu_s \tag{3.4}$$

where x is a dummy variable denoting points in W, and $\mu_s(\cdot)$ is the countably additive measure induced by $f(\cdot)$ on the Borel sets of W. More precisely,

$$\lim_m E(\exp\{i[f(P_m \omega_s), h]\}) = C(h) = \int_W e^{i[\omega, h]} \, d\mu_s$$

where P_m is any monotone sequence of finite-dimensional projections converging strongly to the identity.

Proof. With μ_s denoting the (countably additive) measure induced by $f(\omega_s)$ on W, define for each ω:

$$g(\omega) = \int_W \exp(-\tfrac{1}{2}\{\|x\|^2 - 2[x, \omega]\}) \, d\mu_s$$

This is well defined since the integrand is continuous in x, nonnegative, and bounded by

$$\exp(\tfrac{1}{2}\|\omega\|^2)$$

Moreover $g(\omega)$ is actually a continuous functional on W. For, given $\varepsilon > 0$, we can find a closed bounded K_ε such that

$$\mu_s(K_\varepsilon) > 1 - \varepsilon$$

Then

$$\int_{K_\varepsilon} \exp(-\tfrac{1}{2}\{\|s\|^2 - 2[s, \omega]\}) \, d\mu_s$$

is continuous in ω and on the complement K_ε, and the integral is less than or equal to $(\exp\{\|\omega\|^2/2\})\varepsilon$.

Now let us show that $g(\omega)$ is a random variable. Let $\{P_m\}$ denote a monotone sequence of finite-dimensional projections on W strongly convergent to the identity. Let $\{\phi_i\}$ denote a corresponding orthonormal basis, with the range of P_m being the span of the first m members of the sequence. Let us note that we can write

$$g(P_m \omega) = \int_W [\exp(-\tfrac{1}{2}\|x - P_m x\|^2 - \tfrac{1}{2}\{\|P_m x\|^2 - 2[P_m x, \omega]\})] \, d\mu_s$$

and hence

$$g(P_m\omega) \leq \int_W \exp(-\tfrac{1}{2}\{\|P_m x\|^2 - 2[P_m x, \omega]\})\, d\mu_s$$

Let

$$g_m(\omega) = \int_W \exp(-\tfrac{1}{2}\{\|P_m x\|^2 - 2[P_m x, \omega]\})\, d\mu_s = g_m(P_m\omega)$$

Then

$$\int_W g_m(\omega)\, d\mu_G = \int_W \left(\int_W \exp(-\tfrac{1}{2}\{\|P_m x\|^2 - 2[P_m x, \omega]\})\, d\mu\right) d\mu_s = 1$$

Next

$$\begin{aligned} E(|g(P_m\omega) - g_m(\omega)|) &= \int_W \int_W (1 - \exp\{-\tfrac{1}{2}\|x - P_m x\|^2\}) \\ &\qquad \times \exp(-\tfrac{1}{2}\{\|P_m x\|^2 - 2[P_m x, \omega]\})\, d\mu_s\, d\mu_G \\ &= \int_W (1 - \exp\{-\tfrac{1}{2}\|x - P_m x\|^2\})\, d\mu_s \\ &< \varepsilon \qquad \text{for all} \quad m > m(\varepsilon) \end{aligned} \tag{3.5}$$

Hence the convergence properties of $\{g(P_m\omega)\}$ are the same as those of $\{g_m(\omega)\}$. The latter sequence is a martingale. At this point rather than repeat traditional arguments, we shall exploit them and thereby also show the connection to the Wiener process version. Thus let

$$[y(\omega), \phi_i] = y_i = x_i + \zeta_i$$

where

$$x_i = [x, \phi_i], \qquad \zeta_i = [\omega, \phi_i]$$

Here the ζ_i, $i = 1, \ldots, n$, for any finite n are independent zero mean, unit variance Gaussians. We can create a *probability space* with a countably additive measure on it such that for any finite number of coordinates we have the same distributions: namely R^∞ for the space, and the σ-algebra β generated by cylinder sets for the Borel sets. Equivalently, we could use $C[0, T]$, the Banach space of continuous functions with range in R^n (with the usual sup norm), as the space by defining the mapping W into $C[0, T]$ by

$$S(t) = \int_0^t x(\sigma)\, d\sigma, \qquad 0 \leq t \leq T$$

and $W(t)$ to be standard Wiener process on $C[0, T]$, and defining

$$Y(t) = \int_0^t x(\sigma)\,d\sigma + W(t) \tag{3.6}$$

with the Wiener measure and the measure induced by $S(\cdot)$ independent. In this way we get a "coordinate free" representation, and we note that the variables

$$\int_0^T [\phi_i(t), dY(t)] = y_i$$

have the same finite-dimensional distributions as before. Moreover, the variables $g_m(\omega)$ have a corresponding interpretation and have the same distribution for any finite m, and under the condition (3.3), we know that the measure induced by $Y(\cdot)$ is absolutely continuous with respect to Wiener measure, the martingale sequence converging to it in the mean of order 1. The derivative itself is given by (see Duncan [7])

$$E_x\left[\exp\left\{-\frac{1}{2}\left[\int_0^T x(t)^2\,dt - 2\int_0^T x(t)\,dW(t)\right]\right\}\right] \tag{3.7}$$

where $E_x[\;]$ denotes expectation with respect to the measure induced by the process $S(\cdot)$ on $C[0, T]$, the (Ito) integral (the processes being independent)

$$\int_0^T x(t)\,dW(t)$$

being the same as

$$\sum_1^\infty x_i\zeta_i \qquad \text{where} \quad x_i = \int_0^T [\phi_i(t), dS(t)], \qquad \zeta_i = \int_0^T [\phi_i(t), dW(t)]$$

We have thus proved that $g(P_m\omega)$ is Cauchy in the mean of order 1, and such sequences are equivalent as we change basis. Moreover, it readily follows that for cylinder sets C,

$$\mu_y(C) = \lim_m \int_C g_m(\omega)\,d\mu_G$$

$$= \lim_m \int_C g(P_m\omega)\,d\mu_G$$

This concludes the proof of the theorem.

Corollary. For any t, $0 \le t \le T$, let

$$\mathbf{W}(t) = L_2[[0, t]; R^n], \qquad \mathbf{W}_s(t) = L_2[[0, t]; H]$$

Let $\mu_G{}^t$ denote the Gauss measure on $\mathbf{W}(t)$ and $\mu_s{}^t$ similarly the projection of μ_s on the sub-σ-algebra of Borel sets in $\mathbf{W}(t)$. Then the statement of the theorem applied to measures on $\mathbf{W}(t)$ reads

$$g(t;\, P(t)\omega) = \int_W \exp(-\tfrac{1}{2}\{\|P(t)x\|^2 - 2[P(t)x,\, \omega]\})\, d\mu_s$$

where $P(t)$ denotes the projection of $\mathbf{W}$ on $\mathbf{W}(t)$.

Proof. The proof is immediate. We state the corollary to note that we cannot take derivatives (with respect to t) in this formula as we can in the Wiener process version.

Remark. The theorem holds for any countably additive measure μ_s on the Borel sets of W, not necessarily generated by a random variable $f(\omega_s)$.

Let us note that the main virtue of the theorem is not so much formula (3.4) but rather that the derivative is a random variable. The latter has been proved for a related but more general problem in Balakrishnan [10] under additional assumptions. We explore this in the following section.

3.1. The Linear Case

Mostly to illustrate the ideas involved, let us consider the special case in which $f(\omega_s)$ is linear. Thus let

$$y(\omega_2) = L\omega_s + \omega \tag{3.8}$$

where now we allow H in the definition of W to be infinite dimensional, and where L is a linear bounded transformation on W_s into W. Then we note that in order for $L\omega_s$ to be a random variable it is necessary and sufficient that L be Hilbert–Schmidt. Hence let L be Hilbert–Schmidt. Then $y(\omega_2)$ being Gaussian, it is completely characterized by the corresponding covariance operator:

$$I + LL^*$$

Since LL^* is certainly Hilbert–Schmidt (actually of course trace-class), we can apply the Krein factorization theorem to obtain the representation

$$(I + LL^*)^{-1} = (I - \mathscr{L}^*)(I - \mathscr{L})$$

where $\mathscr{L}$ is a Hilbert–Schmidt Volterra operator:

$$\mathscr{L}f = g; \qquad g(t) = \int_0^t k(t, s) f(s)\, ds, \qquad \text{a.e.} \quad 0 < t < T$$

mapping W into itself. In particular we note that

$$z(\omega_2) = y(\omega_2) - \mathscr{L}y(\cdot)$$

also defines white noise, and defining

$$(I + \mathcal{M}) = (I - \mathcal{L})^{-1}$$

where $\mathcal{M}$ must then also be Hilbert–Schmidt and Volterra, we note that we can represent $y(\omega_2)$ also as

$$y(\omega_2) = \mathcal{M}z(\omega_2) + z(\omega_2) \tag{3.9}$$

In this form we can seek the derivative of the weak distributions induced by $y(\cdot)$ to Gaussian measure induced by $z(\omega_2)$ but the processes are no longer independent. However, it is shown in Balakrishnan [10] that the derivative is a random variable if and only if

$$\mathcal{M} + \mathcal{M}^*$$

is trace-class. But in the present instance this readily follows from the fact that LL^* is trace-class, since

$$LL^* = \mathcal{M}\mathcal{M}^* + (\mathcal{M} + \mathcal{M}^*)$$

In other words, in the model (3.9) the condition that $\mathcal{M}$ be trace-class is always satisfied if it is deduced from the model (3.1). Incidentally, it is of interest to note that the derivative is given by

$$g(\omega) = \exp(-\tfrac{1}{2}[\|\mathcal{M}\omega\|^2 - 2[\mathcal{M}\omega, \omega] + \mathrm{tr}(\mathcal{M} + \mathcal{M}^*)]) \tag{3.10}$$

and can be deduced from (3.4). Also, it should be noted that

$$\mathrm{tr}(\mathcal{M} + \mathcal{M}^*) = \mathrm{tr}(\mathcal{L} + \mathcal{L}^*)$$

and also

$$\mathrm{tr}(\mathcal{M} + \mathcal{M}^*) = E[\|x - \mathcal{L}y\|^2]; \qquad x = L\omega_s \tag{3.11}$$

Equation (3.11) is particularly interesting since it has a variational interpretation. Since $L\omega_s$ is such that the covariance LL^* is trace-class, we can formulate the problem of minimizing

$$E[\|L\omega_s - Ky(\omega_2)\|^2] \tag{3.12}$$

over the class of all Hilbert–Schmidt *Volterra* operators K. But to show that a minimum exists and is given by the H–S Volterra operator K_0, it is enough to show that

$$\frac{d}{d\lambda} E(\|L\omega_s - (K_0 + \lambda K)y(\omega_2)\|^2]\Bigg|_{\lambda=0}^{=0}$$

$$= \frac{d}{d\lambda} \sum_1^\infty E[L\omega_s - (K_0 + \lambda K)y(\omega_2), \phi_i]^2\Bigg|_{\lambda=0} = 0$$

or

$$\frac{d}{d\lambda}\operatorname{tr}(K_0 + \lambda K)(I + LL^*)(K_0 + \lambda K)^* - 2LL^*(K_0 + \lambda K)^* = 0$$

which yields

$$\operatorname{tr}(K_0(I + LL^*) - LL^*)K^* = 0$$

for which it is necessary and sufficient that

$$K_0(I + LL^*) - LL^*$$

be the adjoint of a H–S Volterra operator. But substituting $\mathscr{L}$ for K_0, we see that

$$\begin{aligned} y(I + LL^*) - LL^* &= (\mathscr{L} - I)(I + LL^*) + I \\ &= -(I - \mathscr{L}^*)^{-1} + I = -(I + \mathscr{M}^*) + I = \mathscr{M}^* \end{aligned}$$

Hence $\mathscr{L}$ yields the optimal minimizing H–S Volterra operator. The main point to be noted here is that the optimal solution to the minimizing problem (3.12) is equivalent to that of the Krein factorization. Whether or not L is Volterra plays no role.

3.2. Conditional Expectation: Bayes' Formula

Let us now note one important by-product of Theorem 3.1. Let $\phi(\cdot)$ be any element of W. Then by

$$E[[f(\omega_s), \phi] \,|\, y(\omega_2)] \tag{3.13}$$

we shall mean the limit of the Cauchy sequence (in the mean of order 2):

$$E[[F(\omega_s), \phi] \,|\, P_n y(\omega_2)] \tag{3.14}$$

where P_n is a sequence of monotone increasing finite-dimensional projections converging strongly to the identity. It is implicit that this limit is independent of the particular sequence P_n chosen. We can then state (Bayes' formula)

Theorem 3.2

$$E[[f(\omega_s), \phi] \,|\, y(\omega_2)] = \frac{\int_W [S, \phi] \exp(-\frac{1}{2}\{\|S\|^2 - 2[S, y(\omega_2)]\})\, d\mu_s}{\int_W \exp(-\frac{1}{2}\{\|S\|^2 - 2[S, y(\omega_2)]\})\, d\mu_s} \tag{3.15}$$

Remark. Note that (3.15) is defined for every y in W.

Proof. Given the monotone sequence of finite-dimensional projections

$\{P_n\}$, we may consider an orthonormal basis $\{\phi_n\}$ for W such that P_n corresponds to the space spanned by the first n. Then we can calculate

$$E[[f(\omega_s), \phi_1] \mid P_n y(\omega_2)]$$

by the (finite-dimensional) Bayes' rule:

$$= \frac{\int_W [S, \phi_1] \exp(-\frac{1}{2}\{\|P_n S\|^2 - 2[P_n S, y]\}) \, d\mu_s}{\int_W \exp(-\frac{1}{2}\{\|P_n S\|^2 - 2[P_n S, y]\}) \, d\mu_s}$$

and obtain in the limit, the formula (3.15) with ϕ_1 for ϕ. The formula for arbitrary ϕ is then immediate.

Corollary. Let $P(t)$ denote the projections W onto $W(t)$. Then for any ϕ in W, and $0 \leq t \leq T$,

$$E[P(t)f(\omega_s), P(t)\phi] \mid P(t)y(\omega_2)]$$

$$= \frac{\int_W [P(t)S, P(t)\phi] \exp(-\frac{1}{2}\{\|P(t)S\|^2 - 2[P(t)S, P(t)y]\}) \, d\mu_s}{\int_W \exp(-\frac{1}{2}\{\|P(t)S\|^2 - 2[P(t)S, P(t)y]\}) \, d\mu_s} \tag{3.16}$$

Proof. The proof is immediate.

3.3. Likelihood Ratio: General Case

Let us now consider the general case in which the signal process is not necessarily Gaussian. Let

$$y(t) = S(t) + N(t), \qquad 0 \leq t \leq T < \infty$$

where $S(\cdot)$ and $N(\cdot)$ are independent processes. We shall assume that the signal $S(\cdot)$ has finite energy [corresponding to (3.3)]

$$\int_0^T E(\|S(t)\|^2) \, dt < \infty$$

For each t, $0 < t \leq T$, let

$$W(t) = L_2[R_n; (0, t)]$$

We shall shorten $W(T)$ to simply W. Under condition (3.3), the process $S(\cdot)$ induces a countably additive measure on W [and hence on $W(t)$ for each t]. (In other words, the cylinder measure on W can be extended to be countably additive; this is a consequence of the Sazonov theorem.) Thus $y(\cdot)$ defines a weak distribution on W defined by the characteristic function:

$$E[e^{i[y,h]}] = C_s(h) \exp\{-\tfrac{1}{2}\|h\|^2\} \tag{3.17}$$

where

$$C_s(h) = E[e^{i[S,h]}]$$

and we have used the inner product notation:

$$[S, h] = \int_0^T [S(t), h(t)]\, dt, \qquad h \in W$$

Then the cylinder measure μ_y induced by $y(\cdot)$ is absolutely continuous with respect to the Gaussian measure μ_G and the Radon–Nikodym derivative is defined by the function

$$f(\omega) = \int_W \exp(-\tfrac{1}{2}\{\|S\|^2 - 2[S, \omega)\}) \, d\mu_s \tag{3.18}$$

Thus for any cylinder set C,

$$\mu_y(C) = \lim_{n\to\infty} \int_C f(P_n\omega)\, d\mu_G$$

where P_n is any sequence of finite-dimensional projections strongly convergent to the identity.

Let $\{\phi_n\}$ be an orthonormal basis in W and let L denote the mapping of W into l_2:

$$Lx = a; \qquad a_n = \int_0^T [x(\sigma), \phi_n(\sigma)]\, d\sigma$$

Let $LS = \zeta$. Let μ_ζ denote the measure induced on l_2 by this mapping. Then we can rewrite (3.18) in the form

$$f(\omega) = \int_{l_2} \exp(-\tfrac{1}{2}\{[\zeta, \zeta] - 2[\zeta, L\omega]\})\, d\mu_\zeta \tag{3.19}$$

It must be emphasized that (2.6) is defined for every element ω in W. Note also that (3.19) can be defined with respect to any orthonormal system $\{\phi_n\}$.

Let us next consider the likelihood functional $f(y)$ where $y(\cdot)$ is the observation. For this purpose, let (3.19) be defined with respect to the orthonormal system $\{\phi_n\}$. For each t, $0 < t \leq T$, define the operators $\wedge(t)$, mapping into l_2 by

$$\wedge(t)x = a; \qquad a_n = \int_0^t [\phi_n(\sigma), x(\sigma)]\, d\sigma$$

Let

$$R(t) = \wedge(t) \wedge(t)^*$$

Then the Radon–Nikodym derivative of the measure induced by the process $y(\cdot)$ over $[0, t]$ with respect to the Gaussian measure on $W(t)$ is given by

$$f(t, \omega) = \int_{l_2} \exp(-\tfrac{1}{2}\{[R(t)\zeta, \zeta] - 2[\zeta, \wedge(t)\omega]\})\, d\mu_\zeta \tag{3.20}$$

Note that $\wedge(T) = L$. Let P_n denote the projection operator corresponding to the first n basis functions $\{\phi_i\}$, $i = 1, \ldots, n$. Then we define

$$\hat{\zeta}(t) = \lim_n E[\zeta \mid \wedge(t) P_n y]$$

As we have seen, we have (Bayes formula) that

$$\hat{\zeta}(t) = \frac{\int_{l_2} \zeta \exp(-\frac{1}{2}\{[R(t)\zeta, \zeta] - 2[\zeta, \wedge(t)y]\})\, d\mu_\zeta}{\int_{l_2} \exp(-\frac{1}{2}\{[R(t)\zeta, \zeta] - 2[\zeta, \wedge(t)y]\})\, d\mu_\zeta}$$

Note that, by the Schwarz inequality,

$$\|\hat{\zeta}(t)\|^2 \leq \frac{\int_{l_2} \|\zeta\|^2 \exp(-\frac{1}{2}\{[R(t)\zeta, \zeta] - 2[\zeta, \wedge(t)y]\})\, d\mu_\zeta}{\int_{l_2} \exp(-\frac{1}{2}\{[R(t)\zeta, \zeta] - 2[\zeta, \wedge(t)y]\})\, d\mu_\zeta}$$

$$= \frac{\int_{l_2} \|\zeta\|^2 \exp\{-\frac{1}{2}\|R(t)\zeta - \wedge(t)y\|^2\}\, d\mu_\zeta}{\int_{l_2} \exp\{-\frac{1}{2}\|R(t)\zeta - \wedge(t)y\|^2\}\, d\mu_\zeta}$$

$$\leq cE[\|\zeta\|^2] \exp\{+\tfrac{1}{2}(\|\wedge(t)y\| + k)^2\}, \quad 0 < c < \infty, \quad 0 < k < \infty \tag{3.21}$$

It should be noted that such an estimate is not available in the Wiener process version. Moreover we shall show that (3.20) is actually absolutely continuous in t with an L_2 derivative. Let $\phi(t)$ be infinitely differentiable with compact support in $(0, T)$. Then

$$\int_0^T [f(t, \omega)\phi'(t)]\, dt = \int_{l_2} \int_0^T [\exp(-\tfrac{1}{2}\{[R(t)\zeta, \zeta] - 2[\zeta, \wedge(t)\omega]\}\phi'(t)\, dt])\, d\mu_\zeta$$

$$= \int_{l_2} \left(\int_0^T -\tfrac{1}{2} \left\| \sum_1^\infty \phi_i(t)\zeta_i \right\|^2 + \left[\sum_1^\infty \phi_i(t)\zeta_i, \omega(t) \right] \right)$$
$$\times (\exp(-\tfrac{1}{2}\{[R(t)\zeta, \zeta] - 2[\zeta, \wedge(t)\omega]\})\phi(t)\, dt)\, d\mu_\zeta$$

where we note that both

$$\left\| \sum_1^\infty \phi_i(t)\zeta_i \right\|^2 \quad \text{and} \quad \left[\sum_1^\infty \phi_i(t)\zeta_i, \omega(t) \right]$$

are in $L_2[0, T]$ for each ζ in l_2. Hence the derivative is (defined a.e. $0 < t < T$)

$$\int_{l_2} \left(-\tfrac{1}{2}\left\|\sum_1^\infty \phi_i(t)\zeta_i\right\|^2 + \left[\sum_1^\infty \phi_i(t)\zeta_i,\ \omega(t)\right]\right)$$
$$\times \exp(-\tfrac{1}{2}\{[R(t)\zeta, \zeta] - 2[\zeta, \ \wedge(t)\omega]\})\, d\mu_\zeta$$

We shall next prove that

$$g_N(t) = \sum_1^N \phi_i(t)\hat{\zeta}_i(t), \qquad 0 \le t \le T$$

converges in the norm of W. But this is immediate from the fact that, analogous to (3.21),

$$\|g_N(t)\|^2 \le cE\left[\left\|\sum_1 \phi_i(t)\zeta_i\right\|^2\right] \exp\{+\tfrac{1}{2}\| \wedge(t)y\|^2\}, \qquad \text{a.e.} \quad 0 < t < T$$

Let

$$\hat{S}(t) = \sum_1^\infty \phi_i(t)\hat{\zeta}_i(t)$$

and

$$\widehat{\|S(t)\|^2} = \frac{\int_{l_2} \|\sum_1^\infty \phi_i(t)\zeta_i\|^2 \exp(-\frac{1}{2}\{[R(t)\zeta, \zeta] - 2[\zeta, \wedge(t)y]\})\, d\mu_\zeta}{\int_{l_2} \exp(-\frac{1}{2}\{[R(t)\zeta, \zeta] - 2[\zeta, \wedge(t)y]\})\, d\mu_\zeta}$$

Then from (2.13) we can write

$$\frac{d}{dt}\log f(t, y) = -\tfrac{1}{2}\{\|\hat{S}(t)\|^2 - [\hat{S}(t), y(t)2 + \widehat{\|S(t)\|^2} - \|\hat{S}(t)\|^2\}$$

and hence finally, for the log likelihood functional,

$$\log f(y) = -\tfrac{1}{2}\left\{\int_0^T \|\hat{S}(t)\|^2\, dt - 2\int_0^T [\hat{S}(t), y(t)]\, dt\right.$$
$$\left. + \int_0^T [\widehat{\|S(t)\|^2} - \|\hat{S}(t)\|^2]\, dt\right\} \tag{3.22}$$

We note that the third term can also be expressed as

$$\lim_{n\to\infty} E[\|S(t) - \hat{S}(t)\|^2 \mid \wedge(t)P_n y$$

The formula (3.22) differs from the Wiener process version in the appearance of the third term; in the case in which $S(t)$ is Gaussian, we know that this reduces to

$$E[\|S(t) - \hat{S}(t)\|^2]$$

which is then also independent of the observation $y(\cdot)$ as we have already seen.

4. DYNAMIC SYSTEMS

4.1. Finite-Dimensional Case

We wish now to specialize our results to the case in which $S(\theta, t)$ has a stochastic differential system representation:

$$S(\theta, t) = C(\theta)x(\theta, t)$$

$$\frac{dx(\theta, t)}{dt} = A(\theta)x(\theta; t) + F(\theta)\omega(t); \qquad x(\theta, 0) = 0 \tag{4.1}$$

and the observation process has the form

$$y(\theta, t) = S(\theta, t) + G\omega(t) \tag{4.2}$$

where we shall first consider the finite-dimensional case so that $C(\theta)$, $A(\theta)$, $F(\theta)$, G are all rectangular matrices with

$$A(\theta)\colon \quad m \times m, \qquad F(\theta)\colon \quad m \times n$$

$$F(\theta)G^* = 0$$

$$GG^* = \text{identity matrix}$$

We take $\omega(\cdot)$ as sample functions of white noise in

$$W = L_2[(0, T); R_n]$$

Now, Eq. (4.1) for each fixed θ has (see Balakrishnan [10]) the unique solution:

$$x(\theta, t) = \int_0^t e^{A(\theta)(t-s)}F(\theta)\omega(s)\, ds, \qquad 0 < t < T$$

and

$$x(\theta, t) = L\omega$$

defines a Hilbert–Schmidt operator on W into

$$W_S = L_2((0, T); R_m)$$

In that case

$$\hat{S}(\theta, t) = C(\theta)\hat{x}(\theta, t)$$

where

$$\dot{\hat{x}}(\theta, t) = A(\theta)\hat{x}(\theta, t) + P(\theta; t)C(\theta)^*[y(t) - \hat{x}(\theta, t)]\hat{x}(\theta, 0) = 0 \quad (4.3a)$$

and $P(\theta; t)$ satisfies the (Riccati) equation

$$\dot{P}(\theta, t) = A(\theta)P(\theta, t) + P(\theta, t)A^*(\theta) + F(\theta)F^*(\theta) - P(\theta, t)(\theta)^*C(\theta)P(\theta, t)$$

$$P(\theta, 0) = 0 \quad (4.3b)$$

And, finally, the likelihood functional becomes

$$\exp\left(-\tfrac{1}{2}\left\{\int_0^T \|\hat{S}(\theta, t)\|^2 \, dt - 2\int_0^t [\hat{S}(\theta, t)y(t)] \, dt + \int_0^T \operatorname{tr} C(\theta)P(\theta, t)C(\theta)^* \, dt\right\}\right) \quad (4.4)$$

This result was apparently first obtained by Schweppe [11] by proceeding formally from the time-discrete case.

In estimating parameters by maximizing (4.4) we proceed as follows. Let

$$q(\theta, T) = \int_0^T \|\hat{S}(\theta, t)\|^2 \, dt - 2\int_0^T [\hat{S}(\theta, t), y(t)] \, dt + \int_0^T \operatorname{tr} C(\theta)P(\theta, t)C(\theta)^* \, dt$$

We use the iteration (modified Newton–Raphson):

$$\theta_{n+1} = \theta_n - M(\theta_n)^{-1}\nabla_\theta q(\theta_n; T)$$

where $M(\theta; T)$ is the matrix with components

$$m_{ij}(\theta) = \int_0^T \left[\frac{\partial}{\partial \alpha_i}\hat{S}(\theta, t), \frac{\partial}{\partial \alpha_j}\hat{S}(\theta, t)\right] dt$$

where $\{\alpha_i\}$ denote the *components* of θ. We assume that $M(\theta; T)$ is positive definite on the set of admissible parameters θ.

4.2. Infinite-Dimensional Case

The extension of (4.1) to the infinite-dimensional case (corresponding to partial differential equations) can take many forms. One version is treated in Balakrishnan [10]. For each θ, $A(\theta)$ in (4.1) is now the infinitesimal generator of a strongly continuous semigroup over a separable Hilbert space H_S.

Equation (4.1) remains formally the same, with $F(\theta)$ being a linear bounded transformation for each θ, mapping H into H_S, $\omega(\cdot)$ denoting white noise in

$$W = L_2[(0, T]; H]$$

H being a separable Hilbert space. Similarly, $C(\theta)$ is assumed to be linear bounded and G linear bounded with

$$F(\theta)G^* = 0; \qquad GG^* = \text{identity}$$

In that case the finite-dimensional version (4.4) goes over without change provided we assume that

$$\int_0^T E(\|C(\theta)x(\theta; t)\|^2)\, dt < \infty$$

This in particular implies that

$$C(\theta)P(\theta, t)C(\theta)^*$$

is trace-class a.e. and that

$$\int_0^T \operatorname{tr} C(\theta)P(\theta, t)C(\theta)^*\, dt < \infty$$

However, from the practical point of view we need to consider the case in which $C(\theta)$ is allowed to be unbounded, and unclosable (corresponding to *boundary* or *pointwise* observations in distributed parameter systems).

Here we shall consider such an extension that takes care of the application to the case of turbulence with nonrational spectrum (see Section 5). Actually the model we shall study represents a wide variety of situations assuming only linearity. Thus we take

$$y(t) = S(t, \theta) + n_1(t), \qquad 0 < t < T \tag{4.5}$$

where $n_1(\cdot)$ is white noise in

$$W_0 = L_2((0, T); R_m)$$

and $S(t, \theta)$ has the form

$$S(t, \theta) = \int_0^t B(\theta; t - s)u(S)\, ds + \int_0^t F(\theta; t - s)n_2(s)\, ds \tag{4.6}$$

where $n_2(\cdot)$ is white noise [independent of $n_1(\cdot)$] in

$$W_S = L_2((0, T); R_n)$$

$u(\cdot)$ is a known (deterministic) function, and $\int_0^\infty \|u(t)\|^2 \, dt < \infty$, and for each θ,

$$\int_0^\infty \|B(\theta; \sigma)\|^2 \, d\sigma + \int_0^\infty \|F(\theta; \sigma)\|^2 \, d\sigma < \infty \tag{4.7}$$

Note that (4.1) and (4.2) form a special case of (4.5), (4.6), and (4.7), where the Laplace transforms of $B(\theta, \sigma)$ and $F(\theta, \sigma)$ are constrained to be rational functions. To handle the generalization when one (or both) is not necessarily rational, we proceed (see Balakrishnan [12]) as follows. We show that we can rewrite (4.7) in terms of a partial differential equation representation. Thus let

$$H = L_2[0, \infty; R_p]$$

where p is the dimension of the observation. Let A denote the generator of the shift semigroup over H:

$$\mathscr{D}(A) = [f \in H \mid f(\cdot) \text{ is absolutely continuous and the derivative } f'(\cdot) \in H]$$

and

$$Af = f'$$

Let $u(t)$ be an $m \times 1$ matrix function. Let $B(\theta)$ denote a linear bounded operator mapping R_m into H defined by

$$B(\theta)u = g; \qquad g(t) = B(\theta, t)u, \qquad 0 < t < \infty, \quad u \in R_m$$

Let

$$\omega(t) = \begin{pmatrix} n_1(t) \\ n_2(t) \end{pmatrix}$$

so that $\omega(\cdot)$ is white noise in

$$L_2((0, T); R_p \times R_n)$$

Let $\mathscr{F}(\theta)$ denote the linear bounded operator mapping $R_p \times R_n$ into H defined by

$$\mathscr{F}(\theta)v = g; \qquad g(t) = F(\theta; t)n_2, \qquad v = \begin{pmatrix} n_1 \\ n_2 \end{pmatrix}, \qquad n_1 \in R_p, \quad n_2 \in R_n$$

Finally let G denote the mapping of $R_p \times R_n$ into R_p defined by

$$Gv = w, \qquad w = n_1, \qquad v = \begin{pmatrix} n_1 \\ n_2 \end{pmatrix}, \qquad n_1 \in R_p, \quad n_2 \in R_n$$

Then we claim that (4.6) is representable as

$$\dot{x}(\theta, t) = Ax(t) + B(\theta)u(t) + \mathscr{F}(\theta)\omega(t); \qquad x(\theta, 0) = 0$$
$$y(t) = Cx(\theta, t) + G\omega(t) \tag{4.8}$$

where it should be noted that

$$\mathscr{F}(\theta)G^* = 0; \qquad GG^* = I$$

where C is the operator defined by

$$\text{domain of } C = [f \in H \mid f(\cdot) \text{ is continuous in } 0 \leq t < \infty]$$
$$Cf = f(0)$$

We assume that $B(\theta, t)$, $F(\theta, t)$ are locally continuous in $0 \leq t < \infty$. We can readily see that $x(\theta, t)$ is then in the domain of C for each t. That (4.8) is the same as (4.7) follows from the representation

$$x(\theta, t) = \int_0^t S(t - \sigma)B(\theta)u(\sigma)\, d\sigma + \int_0^t S(t - \sigma)\mathscr{F}(\theta)\omega(\sigma)\, d\sigma$$

where $S(t)$ is the semigroup (shift) generated by A. Even though C is not closable, $Cx(\theta, t)$ is defined and is locally continuous in $0 \leq t < \infty$, for each $u(\cdot)$ and $\omega(\cdot)$. We can then (see Balakrishnan [12]) deduce the analog of (4.3a), (4.3b) as

$$\hat{x}(\theta, t) = A\hat{x}(\theta, t) + (CP(\theta, t))^*[y(t) - C\hat{x}(\theta, t)]\hat{x}(\theta, 0) = 0 \tag{4.9}$$

where $P(\theta, t)$ satisfies

$$[\dot{P}(\theta, t)x, y] = [P(\theta, t)x, A^*y] + [P(\theta, t)y, A^*x]$$
$$+ [\mathscr{F}(\theta)^*x, \mathscr{F}(\theta)^*y] - [CP(\theta, t)x, CP(\theta, t)y]$$
$$P(\theta, 0) = 0, \qquad x, y \in \text{domain of } A^* \tag{4.10}$$

In particular $P(\theta, t)$ maps into the domain of C, and

$$CP(\theta, t)$$

is linear bounded (even though C is not closed; see Balakrishnan [12]) for each t. Moreover (cf. [12]): $(CP(\theta, t))^* \in$ domain of C and $C(CP(\theta, t))^*$ is bounded (and automatically trace-class, being finite dimensional). The Radon–Nikodym derivative formula (4.4) now becomes

$$\exp -\tfrac{1}{2}\left\{\int_0^T \|\hat{S}(\theta, t)\|^2\, dt - 2\int_0^T [\hat{S}(\theta, t), y(t)]\, dt\right.$$
$$\left. + \int_0^T \operatorname{tr} C(CP(\theta, t))^* dt\right\} \tag{4.11}$$

In this version it is important to note that the steady state solution of (4.10) exists:

$$P(\theta, \infty)x = \lim_{t \to \infty} P(\theta, \infty)x$$

$$0 = [P(\theta, \infty)x, A^*y] + [P(\theta, x)y, A^*x]$$
$$+ [\mathscr{F}(\theta)^*x, \mathscr{F}(\theta)^*y] - [CP(\theta, \infty)x, CP(\theta, \infty)y] \quad (4.12)$$

provided

$$\int_0^\infty \int_0^\infty \|F(\theta, \sigma + t)\|^2 \, d\sigma \, dt < \infty \qquad (4.13)$$

5. APPLICATION

We turn now to an application: estimation of stability and control derivatives from flight test data. The dynamic system considered arises from the longitudinal mode perturbation equations for an aircraft in windgust (turbulence) (Rediess Taylor, see Balakrishnan [13]). We use the Dryden version of the spectrum of turbulence, which is rational, so that the total system is finite dimensional. Leaving the many essential details to the comprehensive work of Iliff [14], the state space formulation of the problem is (see also Balakrishnan [12])

$$\dot{x}(t) = Ax(t) + Bu(t) + Fn_2(t)$$
$$V(t) = Cx(t) + Du(t) + Gn_1(t)$$

where $n_1(\cdot)$ and $n_2(\cdot)$ are independent white Gaussian, and the matrices in the equations have the form

$$A = \begin{bmatrix} Z_1 & 0 & 1 & Z_1 \\ 0 & 0 & 1 & 0 \\ M_1 & 0 & M_3 & M_1 \\ 0 & 0 & 0 & -\dfrac{\bar{v}}{1000} \end{bmatrix}, \qquad B = \begin{bmatrix} Z_4 & & 0 \\ 0 & & 0 \\ M_4 & F = \sigma & 0 \\ 0 & & \dfrac{1}{20\bar{v}} \end{bmatrix}$$

$$C = \begin{bmatrix} 0 & 0 & 1 & 0 \\ 0 & 1 & 0 & 0 \\ \dfrac{10M_1 - \bar{v}Z_1}{g} & 0 & 10M_3 & \dfrac{10M_1 - \bar{v}Z_1}{g} \\ k_1 & 0 & \dfrac{32K_1}{\bar{v}} & k_1 \end{bmatrix}, \qquad D = \begin{bmatrix} 0 \\ 0 \\ \dfrac{10M_4 - \bar{v}Z_4}{g} \\ 0 \end{bmatrix}$$

where g is the acceleration due to gravity and $G = \mathrm{diag}[0.0005, 0.0001, 0.01, 0.0001]$. The lettered entries (stability and control derivatives) are unknown, except for $\bar{v}$, which is 1670. Note that the turbulence power is an unknown parameter also.

The sampling interval was 0.02 sec and the data bandwidth about 5 Hz. Figure 1 shows the complete time history of the observation $v(t)$ (four components) subdivided into various regions for later identification, as well as the input time history. Estimates were computed over each of the various subregions by three methods:

Method I Neglecting the measurement noise on the angle of attack measurement (v_4) and following the corresponding maximal likelihood

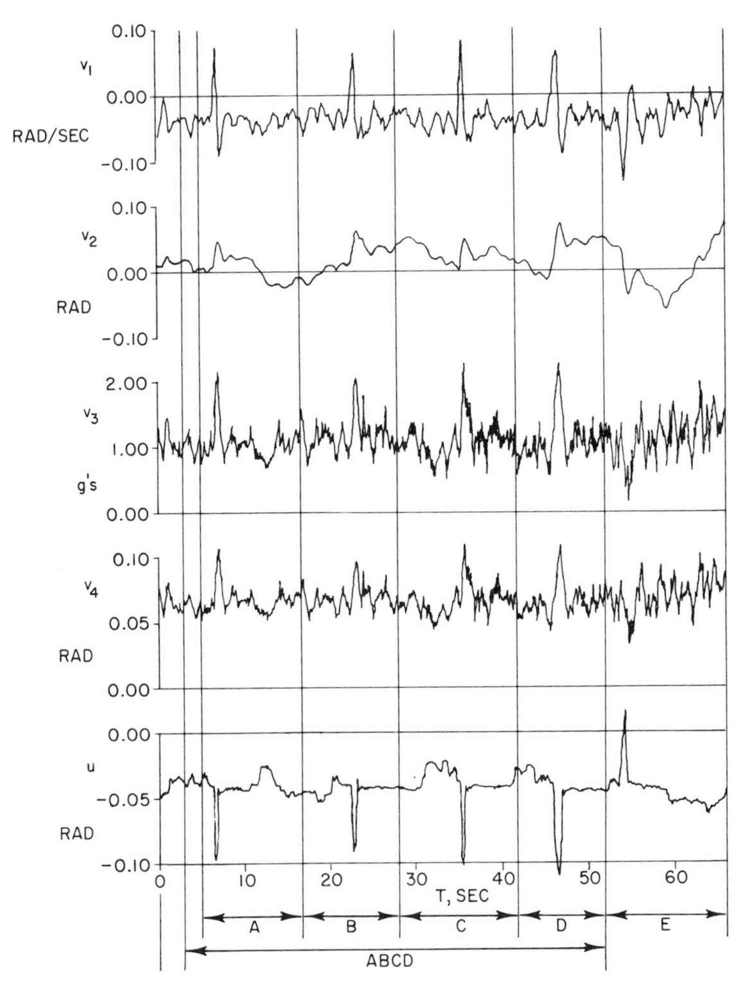

FIG. 1. Total Jetstar turbulence time history showing intervals of each maneuver.

technique developed in Balakrishnan [13]. This is reasonable for this particular example at high turbulence levels.

Method II This is the method developed herein.

Method III This was a *check* method, in which the turbulence was ignored completely in the model.

The results are summarized in Fig. 2. Sample means and variances of the estimates obtained over the different data regions are shown, along with the

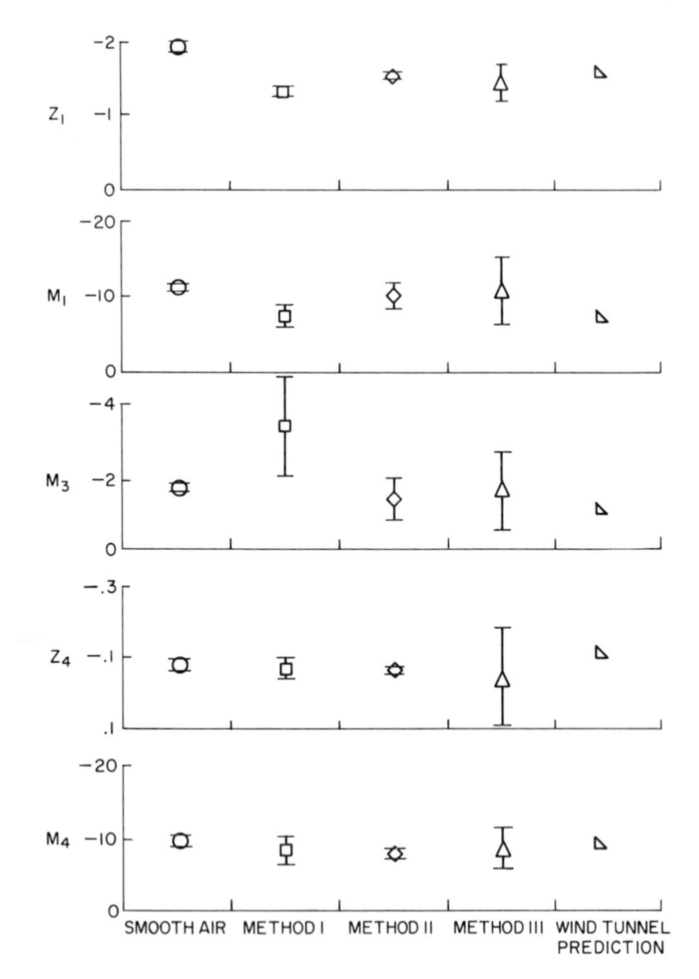

FIG. 2. Means and standard deviations for five methods of estimating coefficients. ○, Smooth air; □, method I; ◇, method II; △, method III; ◺, wind-tunnel prediction.

wind-tunnel values as well as estimates obtained on other turbulence-free (smooth air) flights. It can be seen that method II yields the most consistent estimates. It also turns out that method II requires the least computational time, the estimates converging in fewer iterations. It can also be seen that ignoring the turbulence leads to the worst results. For a further discussion see Iliff [14]. The remaining illustrations indicate the nature of the "fit" obtained using the estimated coefficients to the observed data. Figure 3 shows the close agreement provided by method II. Figures 4 and 5 indicate how much worse the agreement is on the same stretch of data if the turbulence is *not* accounted for.

If we use the nonrational (Kolmogorov) version of the spectrum of

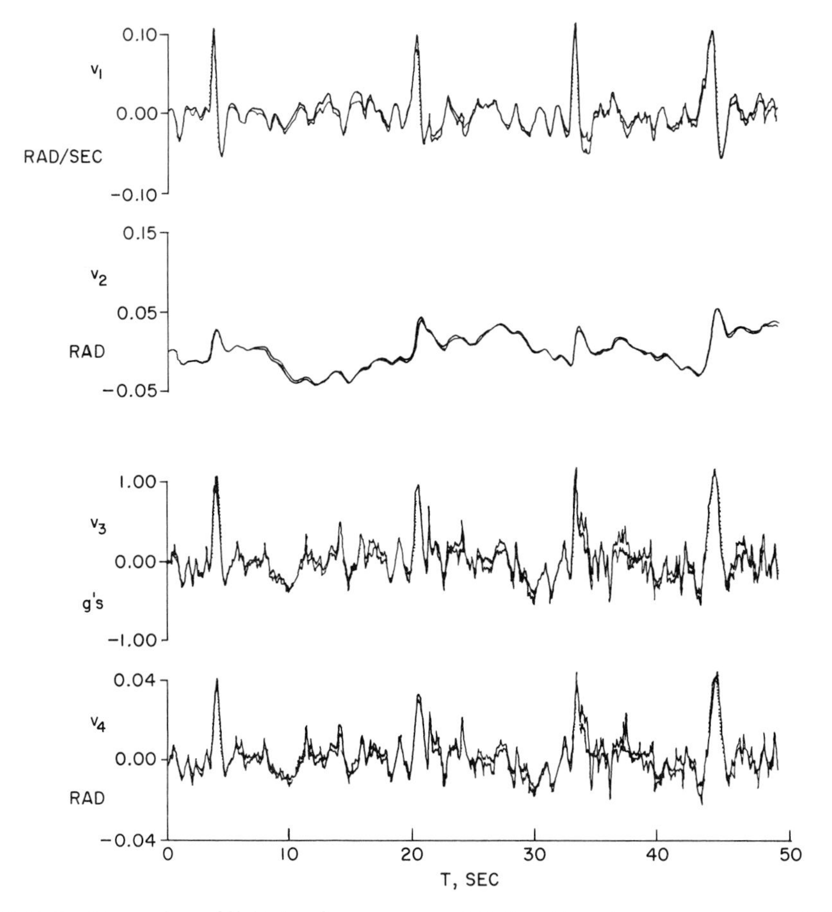

FIG. 3. Comparison of flight data from maneuver ABCD and the estimated data—method II.

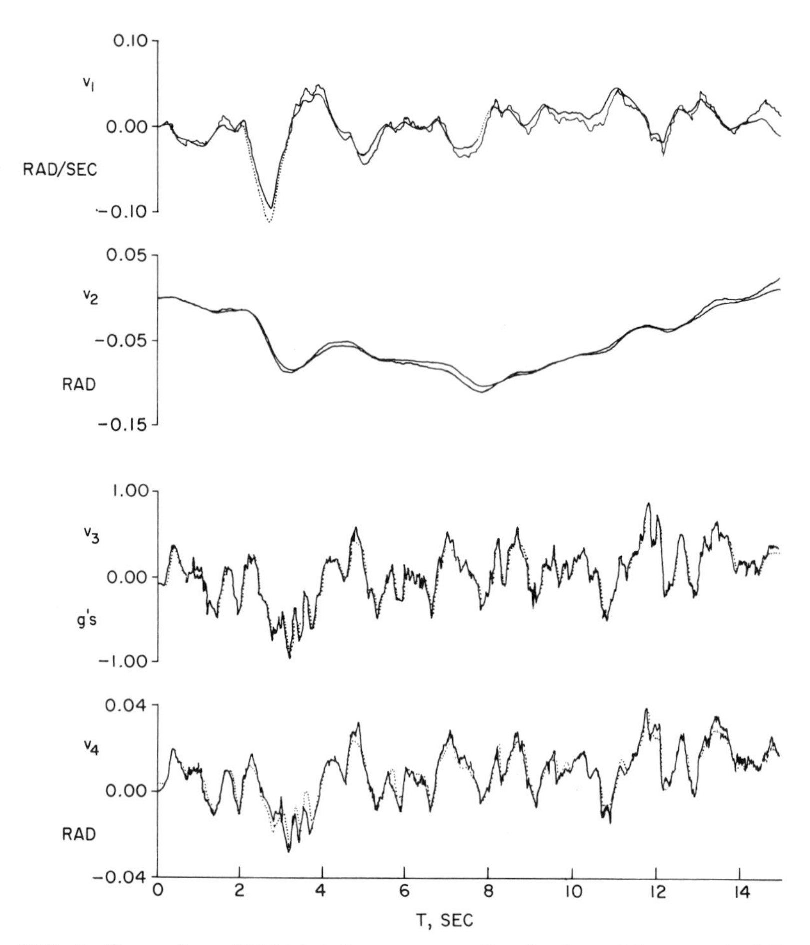

FIG. 4. Comparison of flight data from maneuver E and estimated data—method II.

turbulence, we have to use (4.11). In particular, in this case $F(\theta, t)$ has the form

$$F(\theta, t) = (a(\theta)t^{5/6} + b(\theta)t^{-1/6})e^{-kt}$$

corresponding to the spectral density of the form (cf. [15])

$$\frac{1 + cf^2}{(1 + df^2)^{11/6}}$$

The possibility of using flight-test data to distinguish between the two models of the spectral density is an intriguing one at the present time.

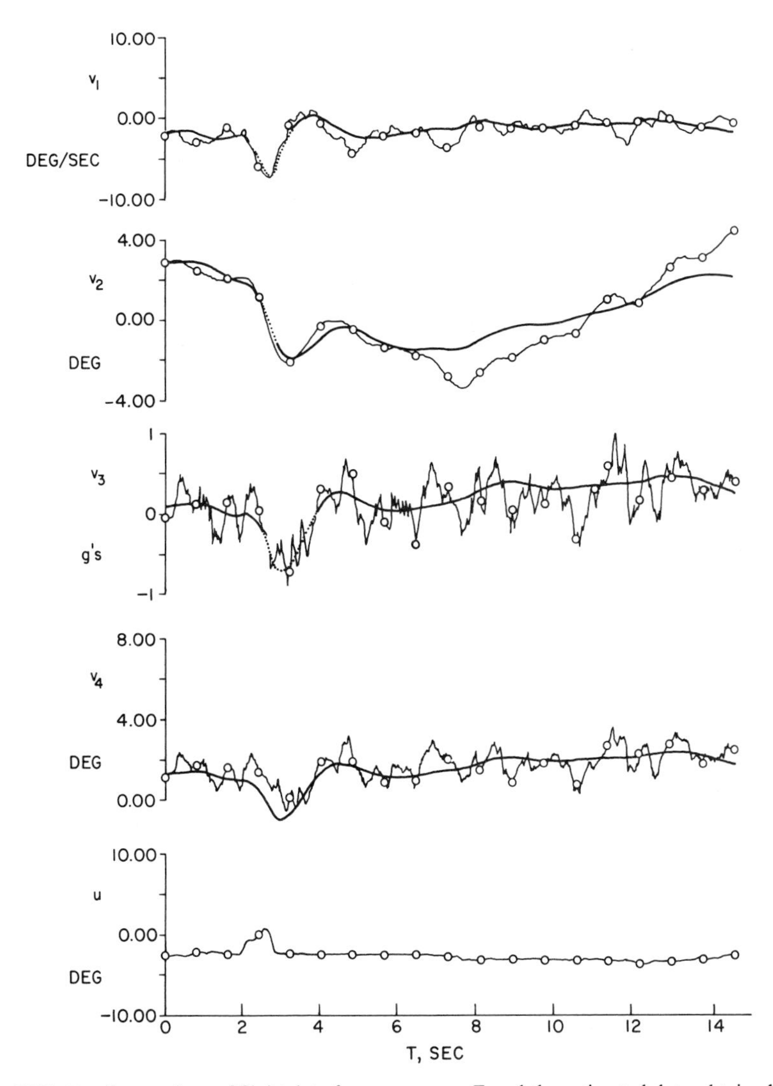

FIG. 5. Comparison of flight data from maneuver E and the estimated data obtained by method III. —○— Flight; ····, estimated.

REFERENCES

1. *Proc. IFAC Symp. Identificat. Syst. Parameter Estimat.*, 1976, 1973, 1970.
2. Liptser, Y., and Shiryayev, A., "Statistics of Random Processes." Nauka, Moscow, 1974.
3. Skorokhod, A. V., "Integration in Hilbert Space." Springer-Verlag, New York, 1975.
4. Gross, L., Harmonic analysis on Hilbert space, *Mem. Amer. Math. Soc.* No. 46 (1963).

5. Sazonov, V., Remark on characteristic functionals, *Theor. Probability Appl.* **3** (1958) 188–192.
6. Balakrishnan, A. V., A white noise version of the Girsanov formula, *Proc. Symp. Stochastic Differential Equations, Kyoto, Japan* (July 1976).
7. Duncan, T. E., Evaluation of likelihood functionals, *Information and Control* **13**, No. 1 (1968).
8. Kallianpur, G., and Striebel, C., Stochastic differential equations occurring in the estimation of continuous parameter stochastic processes, *Theory Probability and Appl.* **14** (1969) 567–594.
9. Sosulin, Yu, and Stratanovich, K. C., Optimum detection of a diffusion process in white noise, *Radiotech. i. Electron.* **10** (1965) 704–714.
10. Balakrishnan, A. V., "Applied Functional Analysis." Springer-Verlag, New York, 1976.
11. Schweppe, F. C., Evaluation of likelihood functions for Gaussian signals, *IEEE Trans. Information Theory* **11** (1965) 61–70.
12. Balakrishnan, A. V., Stochastic filtering and control of linear systems: A general theory, *In* "Control Theory of Systems Governed by Partial Differential Equations" (A. K. Aziz, J. W. Wingate, and M. J. Balas, eds.), Academic Press, New York, 1977.
13. Balakrishnan, A. V., Identification and adaptive control, an application to flight control systems, *J. Optimization Theory Appl.* (31) (1972) 187–213.
14. Iliff, K. W., Dissertation, School of Engineering, UCLA (1973) (UCLA-ENG-7340, May 1973).
15. Houbolt, J. C., *et al.*, Dynamic Response of Airplanes to Atmospheric Turbulence Including Flight Data on Input and Response, NASA TR R-199 (June 1964).

Comparative Aspects of the Study of Ordinary Time Series and of Point Processes†

DAVID R. BRILLINGER

DEPARTMENT OF STATISTICS
UNIVERSITY OF CALIFORNIA AT BERKELEY
BERKELEY, CALIFORNIA
AND
DEPARTMENT OF MATHEMATICS
UNIVERSITY OF AUCKLAND
AUCKLAND, NEW ZEALAND

† This research was partially supported by the J. S. Guggenheim Memorial Foundation and National Science Foundation Grant MCS76-06117.

ISBN 0-12-4266 01-0

1. INTRODUCTION

Both the literature concerning the subject of ordinary time series analysis and that of point process analysis are large and are rapidly growing. The following books may be mentioned: Bartlett (1966), Whittle (1963), Rozanov (1966), Box and Jenkins (1970), Hannan (1970), Anderson (1971), Koopmans (1974), and Brillinger (1975a) in the case of ordinary time series, and Harris (1963), Cox and Lewis (1966), Kerstan *et al.* (1974), Srinivasan (1974), Snyder (1975), and Murthy (1974) in the case of point processes. Generally speaking, the literatures of these two subjects have developed quite independently of each other, although Bartlett (1966) and Brillinger (1972) are exceptions. There are many similarities between the two subjects and it appears that each can benefit from a consideration of the methods of the other. The intention of this chapter is to indicate cases in which the concepts and procedures of ordinary time series (or point processes) have direct analogs in the study of point processes (or ordinary time series). Certain cases in which one subject has unique facets and there are no apparent immediate analogs will also be mentioned. There are gains to be had from adopting a unified approach. Indeed, nowadays data are being collected that are hybrid, part ordinary time series and part point process (e.g., see Bryant and Segundo, 1975), so some unified method of analysis is clearly called for. The structure of this chapter is one of parallel sections indicating corresponding results for ordinary time series and for point processes. The reader is generally referred to the original literature for detailed statements of theorems and most general results. About half of the material was presented in lectures to graduate students in mathematics at the University of Auckland during March to July 1976. Alan J. Lee made helpful comments concerning the manuscript.

1.1. Time Series and Some History of Their Study

The general phenomena studied and recorded by scientists depend on time. In many circumstances this dependence on time may be ignored. The intent of this work is to study phenomena that depend on time in an essential manner. Measurements corresponding to continuous real-valued functions of time are called *time series* and are denoted by $X(t)$, $0 \leq t < \infty$, assuming their domain of definition may be thought of as the interval $[0, \infty)$. Simple examples of time series include the current at a particular junction in an electric circuit as it varies in time, the displacement of the needle of a seismometer from its rest position as a function of time, the temperature at a given location on the earth's surface as a function of time, and finally the height of a sea's surface along a given parallel of latitude as a function of longitude. (In this last example the "time" parameter is really a distance parameter.) The books mentioned in Section 1 may be consulted for specific references to a wide variety of interesting time series.

The scientific analysis of time series has a very long history. It may be thought of as having commenced in 1664 when Issac Newton decomposed a light signal (or time series) into frequency components by passing the signal through a glass prism. A multicolored image was cast upon an opposite wall. He called this the spectrum. This analysis corresponds to the bandpass filtering operation discussed in Section 2.6. Newton did not carry out a quantitative analysis of his time series. However, in 1800 W. Herschel did, by using thermometers. He measured the average energy in various frequency bands of the sunlight's spectrum by placing thermometers along that spectrum. Mathematical foundations began to be laid in the mid-1800s for the analysis of time series when Gouy represented white light as a Fourier series. Later Rayleigh replaced the series by an integral. In 1881 S. P. Langley refined Herchel's experiment considerably by measuring the light energy with a spectral bolometer. (A device that he had invented, it makes use of electric current generated in a wire by incident radiation.)

In 1872 Lord Kelvin built a harmonic analyzer and a harmonic synthesizer for use in the analysis and prediction of the series $X(t)$ of the height of the tide at a particular location and time t. His devices were mechanical, based on pulleys. During the same time period a variety of workers (e.g., G. G. Stokes) were carrying out numerical Fourier analyses using computation schedules. In particular, in 1891 S. C. Chandler carried out an analysis of the variation of latitude with time. His analysis led him to suggest that the motion of the earth's pole of rotation was composite, containing components of period 12 and approximately 14 months.

A substantial advance in the analysis of time series corresponding to light signals occurred in 1891 when A. A. Michelson invented the interferometer. This device allowed the measurement of the average value

$$\left\{\int_0^T [X(t) + X(t+u)]^2 \, dt\right\} \Big/ T \tag{1.1.1}$$

for large values of T and nominated values of the lag u. In consequence, it allowed the estimation of the autocovariance function (see Section 2.4) of the incident signal. In 1898 Michelson and Stratton described a harmonic analyzer (based on springs) and used it to obtain the Fourier transform of the function (1.1.1). This Fourier transform provided an estimate of the power spectrum of the signal. Michelson envisaged the signal as being a sum of cosines. He saw the estimated spectra as descriptive statistics of the light emitting sources.

In 1894 M. I. Pupin invented the electric wave filter (also a sort of bandpass filter). This device considerably broadened the frequency domain over which time series could be analyzed. The power of an electric signal could now be measured in a range of frequency bands.

In a series of papers written during the years 1894–1898, A. Schuster proposed and discussed the periodogram statistic

$$\left| \sum_{t=1}^{T} \exp\{-i\lambda t\} X(t) \right|^2 \tag{1.1.2}$$

based on an observed stretch of time series $X(t)$, $t = 1, 2, \ldots, T$. His motivation was a search for "hidden periodicities." In the succeeding years periodograms and their equivalents were computed for a variety of phenomena by many workers.

In 1922 Crandall and MacKenzie at Bell Telephone Laboratories used resonance tubes to measure the energy distribution of speech as a function of frequency. Once again, the frequency domain over which phenomena could be studied was usefully broadened. Also during the 1920s time series of turbulent flow were investigated by G. I. Taylor (who may have been the first to define the autocovariance function of a time series). Certain of Taylor's methods were applied to meteorological series by G. Walker. During the 1920s and 1930s the field of quantum mechanics and its related form of spectral analysis underwent considerable development.

In the time period 1930–1950 substantial developments in the area of time series analysis were provided by N. Wiener, H. Cramér, A. N. Kolmogorov, M. S. Bartlett, and J. W. Tukey. Details of their contributions may be found in the books mentioned in Section 1.

The range of problems studied by time series analysts covers smoothing, forecasting, control, seasonal adjustment, detection, parameter estimation, discrimination of series from different populations, checking for association between series, and isolation of more elementary series. A variety of these topics are discussed in this chapter.

1.2. Point Processes and Some History of Their Study

The preceding section was concerned with phenomena evolving continuously in time. Point processes refer to isolated events occurring haphazardly in time. A stochastic point process is a random, nonnegative, integer-valued measure. If I is an interval of the real line and ω is a random element, then the values of this measure may be denoted by $N(\cdot, \omega)$ with $N(I, \omega)$ denoting the number of points in the interval I for the realization corresponding to ω. Here the atoms of the measure correspond to a particular set of points. Throughout this chapter, the notation $N(t) = N((0, t], \omega)$

will be employed. Then $dN(t)$ refers to the number of points in the small interval $(t, t + dt]$.

The variety of data that arise in practice and have point process character is staggering. Subject areas leading directly to the collection of point process data include traffic systems, queues, neuronal electrical activity, microscopic theory of gases, resistance noise, heartbeats, population growth, radioactivity, seismology, accident or failure processes, telephone systems, cosmic rays, and fluctuation of photoelectrons. A wide variety of examples are discussed by Snyder (1975) and in Lewis (1972). A recent paper discussing point processes in seismology is that of Udias and Rice (1975), and one in neurophysiology is by Brillinger *et al.* (1976).

Point process data are typically stored either in terms of actual times of events (locations of points) or in terms of lengths of successive intervals between events. In the case of points of several types, the data may be stored separately for each type or by the actual times of events with a "flag" to indicate the event type. Point process data are processed in both analog and digital manners.

The beginnings of the study of point processes may be found in the history of population mathematics. J. Graunt (1620–1674) constructed a life table. Such a table corresponds to the superposition of many independent point processes, each containing a single point at the time of death of an individual. Other workers in this area were C. Huygens (1629–1695) and E. Haley (1656–1742). Much of their concern was over the mean duration of life, the disadvantages of such a measure, and the calculation of the values of life annuities. Specific functional forms were proposed for the force of mortality by de Moivre, Lambert, Gompertz, and Makeham among others. This early history is described in some detail by Westergaard (1968).

The next area of point process research activity related to the Poisson distribution and process. The distribution is credited to de Moivre in 1718 and Poisson in 1837 (see Haight, 1967, p. 113). The name of von Bortkiewicz (1868–1931) is closely associated with the Poisson distribution, especially because of his use of it to model the frequency of death by horse-kick in the Prussian Army. The Poisson process was introduced over a long period. In 1868 Boltzmann determined the expression $\exp\{-\mu t\}$ for the probability of no points in an interval of length t (Haight, 1967, p. 114), and in 1910 Bateman determined the counting distributions by solving a set of differential equations (Haight, 1967, p. 120). In 1903 F. Lundberg investigated a process of which the Poisson is a particular case (see Cramér, 1976).

In 1909 Erlang applied the Poisson process to traffic studies, proposed the truncated Poisson, and considered the process with intervals made up of right displaced exponentials (Haight, 1967, p. 121–123). Erlang's interest was in building better telephone systems, for example, determining the optimum number of circuits. He may be said to have initiated the study of queueing systems—involving input and output point processes corresponding to times of arrival and departure of customers. Later workers who made substantial contributions to queueing theory include Molina, Fry, Khinchin, Palm, and Pollaczek (see Bhat, 1969).

Another class of point processes with a long history of study is that of the renewal processes. In these the successive intervals between points are independent nonnegative variates. Lotka (1957) ascribes the first serious investigation of these processes to Herbelot in 1909. Other historically important references may be found in Lotka's paper. A modern reference is Cox (1962).

The work of Bateman, mentioned earlier, was stimulated by problems of particle physics. The late 1930s saw the commencement of substantial developments in the modeling of point processes by physicists. Point processes occur in radioactive decay, in particle bombardment experiments, and in coincidence experiments among other areas of physics. In 1937 Bahba and Heitler and simultaneously Carlson and Oppenheimer applied stochastic methods to the cascade phenomena of cosmic ray showers. The Bhabha–Heitler approach led to the Poisson process. Furry criticized it and proposed a simple birth process instead. This approach was criticized in turn by Scott and Uhlenbeck who proposed a further model. Explicit details of this work may be found in the books by Bharucha-Reid (1960) and Srinivasan (1969).

A class of point correlation functions was introduced by Yvon (1935) to study the dependency properties of certain point processes. Later multidimensional product density functions were introduced by Ramakrishnan (1950) to study the higher order dependencies of point processes. These had appeared earlier in a specific situation in an article by Rice (1945).

The doubly stochastic Poisson process may be said to have been introduced by Quenouille (1949). The name of D. R. Cox is often associated with it also. The early history of branching processes, another form of point process, is described in some detail by Kendall (1975).

A point process may often be characterized by its conditional intensity function $\gamma(t, \omega)$. McFadden (1965) introduced this concept and used it to show that the rate of change of the entropy of a point process may be written

$$E\{\gamma(t, \omega)[1 - \log \gamma(t, \omega)]\}$$

Wang (1968) provides an early example of a $\gamma(t, \omega)$ of nontrivial form.

1.3. Some Notation

It is convenient to collect in this section some of the notation used throughout this chapter. When a concept is introduced its name will be displayed in italics: $\in$ means "is an *element of*." The *semiopen interval* of real numbers t satisfying $a < t \leq b$ is denoted $(a, b]$. If I is the set of real numbers $\{r : r \in I\}$, then $I + t$ represents the set of corresponding *translated numbers* $\{r + t : r \in I\}$. If I is an interval, $|I|$ is its *length*. The set of *all real numbers*, $(-\infty, \infty)$, is denoted by R. If s is a complex number, Re s denotes its *real part*; i is $\sqrt{-1}$. Given two sets, I and J, $I \times J$ denotes their *direct product* made up of pairs (r, s) with $r \in I$ and $s \in J$; log refers to the logarithm base e. $\prod_{j \in J} a_j$ represents the *product* of the a_j with $j \in J$. Given Q_{jk}, with $j = 1, \ldots, J$ and $k = 1, \ldots, K$, $[Q_{jk}]$ represents the *matrix* with the element Q_{jk} in row j and column k. The notation $N^{(K)}$ means $N(N-1) \cdots (N-K+1)$. When a domain of integration is not indicated, it is to be taken as the whole space. $\delta(t)$ denotes the *Dirac delta function* with the property

$$\int \delta(t) f(t)\, dt = f(0)$$

for functions $f(t)$ continuous at 0.

Turning to probabilistic considerations, Ω denotes a *sample space*. $B(\Omega)$ is the smallest *Borel field* generated by certain of the subsets of Ω (e.g., when Ω has a topology, by all open sets). P denotes a *probability measure* with values $P(A)$ for $A \in B(\Omega)$. *Integrals* with respect to P are given either by $\int f(\omega)\, dP(\omega)$ or $\int f(\omega) P(d\omega)$. Given a random variable $Y(\omega)$, its *expected value* is denoted by $EY(\omega)$ (or by EY when there is little chance for confusion). Its *variance* is var Y. The *covariance* of two random variables is written as cov$\{Y_1, Y_2\}$. The joint *cumulant* of K random variables is cum$\{Y_1, \ldots, Y_K\}$. A *stochastic time series* is a function-valued random variable $X(t, \omega)$, $t \in R$, $\omega \in \Omega$. It is sometimes denoted by $X(t)$ and sometimes by X, when confusion seems unlikely. A *stochastic point process* is a non-negative integral step-function-valued random variable $N(t, \omega)$, $t \in R$, $\omega \in \Omega$. On occasion it is written as $N(t)$ or N. Differential notation is often used with $dN(t) = N(t + dt) - N(t)$ for infinitesimal increments dt. When θ is a parameter, $\hat{\theta}$ stands for an *estimate* of θ. T means the length of the time period of observation of a time series or point process. In asymptotic studies, $T \to \infty$.

2. FOUNDATIONS

2.1. Formal Definitions of an Ordinary Time Series

Several approaches are available to the theory of *ordinary time series*. The intent of each approach is to provide a structure within which one can manipulate and deal with real-valued functions $X(t)$ for t in some index set. [In this work the index set is the interval $[0, \infty)$ and it is assumed that $X(t)$ is a continuous function of t.] The various constructions allow the definition of certain useful parameters.

One means of proceeding is to consider a function of two variables $X(t, \omega)$, with t in $[0, \infty)$ and ω in Ω, where $(\Omega, B(\Omega), P)$ is a probability space, $X(t, \omega)$ is a measurable function of ω for each t, and $X(t, \omega)$ is a continuous function of t for almost all ω. The section $X(\cdot, \omega)$ for fixed ω is called a *realization* or *trajectory* of the stochastic time series X. In this setup $X(\cdot, \omega)$ may be regarded as a random continuous function. An example of proceeding in this manner would be to define

$$X(t, \omega) = \rho \cos(\lambda t + \omega), \qquad 0 \le t < \infty$$

where ρ, λ are fixed and ω is a uniform variate on $(-\pi, \pi)$. A more complicated example would be a diffusion process defined through an integral equation involving Brownian motion.

A second means of proceeding is to imagine a collection of consistent *finite-dimensional distribution functions*

$$F_K(x_1, \ldots, x_K; t_1, \ldots, t_K) \tag{2.1.1}$$

$-\infty < x_1, \ldots, x_K < \infty; 0 \le t_1, \ldots, t_K < \infty$, with (2.1.1) thought of as being

$$\text{Prob}\{X(t_1) \le x_1, \ldots, X(t_K) \le x_K\} \tag{2.1.2}$$

The Kolmogorov extension theorem now indicates the existence of a probability space $(\Omega, B(\Omega), P)$ and $X(t, \omega)$, $0 \le t < \infty$, $\omega \in \Omega$, such that (2.1.2) equals (2.1.1) for all $t_1, \ldots, t_K$, $K = 1, 2, \ldots$, with $X(t) = X(t, \cdot)$. Further conditions must be set down in order to ensure that $X(\cdot, \omega)$ is almost surely a continuous function. An example of proceeding by this approach is when a Gaussian process is defined by stating that all the finite-dimensional distributions of the process are consistent multivariate normals.

A third means of proceeding is to imagine being given $C[0, \infty)$ as the space of continuous functions on $[0, \infty)$, with an appropriate Borel field of sets for the function space, and then to assume the existence of a probability measure on the space. In terms of the previous notation, $\Omega = C[0, \infty)$. Many useful theorems have been developed for this structure. This approach is considered by Billingsley (1968) and Nelson (1959) for example.

In some circumstances one proceeds through conditional distribution or transition probability functions, for example, by assuming the existence of the functions

$$\text{Prob}\{X(t) \in A \mid X(t_1), \ldots, X(t_K); \qquad 0 \le t_1 < t_2 < \cdots < t_K < t\} \tag{2.1.3}$$

for A in $(-\infty, \infty)$, $K = 1, 2, \ldots$, and of an initial probability function

$$\text{Prob}\{X(0) \in A\}$$

This is the usual means by which a Markov process on $[0, \infty)$ is defined. It may be used to define an autoregressive scheme of finite order also. Far-reaching generalizations of the notion of studying processes through conditional variates are given by Schwartz (1973) and Knight (1975).

Another means by which a stochastic time series may be characterized is through its *characteristic functional* (or, equivalently, the distribution of a broadly defined linear functional). The characteristic functional is discussed, e.g., by Bochner (1960) and Prohorov (1961). It may be thought of as being given by

$$C[\xi] = E \exp\left\{i \int \xi(t)X(t)\, dt\right\} \tag{2.1.4}$$

Provided a functional $C[\cdot]$ satisfies certain regularity conditions for $\xi(\cdot)$ in a sufficiently large function space, there is a corresponding time series. One thing needed is consistent characteristic functions for all finite-dimensional distributions. Stable processes are usually defined through the characteristic functional (see Bochner, 1960).

As a final means of introducing stochastic time series, consider the class of *harmonizable processes*. This is made up of series of the form

$$X(t, \omega) = \int_{-\infty}^{\infty} \exp\{i\lambda t\}\, dZ(\lambda, \omega) \tag{2.1.5}$$

for $Z(\lambda, \omega)$ a complex-valued random process, with the integral of (2.1.5) defined in a probabilistic manner (e.g., as a limit in the mean of approximating Stieltjes sums). Harmonizable processes are discussed by Loève (1955, p. 474). $X(t, \omega)$ is seen to be a linear function of $Z(\lambda, \omega)$ so its moments can be given directly in terms of those of $Z(\lambda, \omega)$. This manner of introducing a time series proves especially useful in the case of stationary series.

By a *stationary time series* is meant one with the property that the characteristics of finite collections of its values

$$\{X(t + u_1), \ldots, X(t + u_K)\} \tag{2.1.6}$$

do not depend on t for $0 \leq t + u_1, \ldots, t + u_K < \infty$ and $K = 1, 2, \ldots$. In particular, the finite-dimensional distribution of the variate (2.1.6) should not depend on t. In the stationary case there are alternative ways of beginning.

One is to suppose that a probability space $(\Omega, B(\Omega), P)$ is given as well as a semigroup of measure-preserving transformations of Ω. Then the series is constructed through setting

$$X(t, \omega) = Y(U_t \omega) \tag{2.1.7}$$

for some random variable $Y(\omega)$, with U_t, $t \in [0, \infty)$, the generator of the semigroup. This approach is discussed, e.g., by Doob (1953, p. 507). It is apparent that (2.1.5) implies, for example,

$$X(t+s, \omega) = Y(U_t U_s \omega) = X(t, U_s \omega), \qquad t, s \geq 0$$

Immediate ways of constructing stationary time series include beginning with a consistent family of finite-dimensional distributions that are invariant under shifts $(u_1, \ldots, u_K) \to (t + u_1, \ldots, t + u_K)$ or beginning with a characteristic functional $C[\cdot]$ unaffected by the transformation $\xi(t) \to \xi(t-u)$, $0 \leq t, t+u$.

An important means of introducing stationary time series is through the spectral representation (2.1.5). One can assume that the characteristic functional of the process $Z(\lambda, \omega)$, namely

$$E \exp\left\{i \int \xi(\lambda)\, dZ(\lambda, \omega)\right\}$$

is invariant under the transformations $\xi(\lambda) \to \xi(\lambda) \exp\{i\lambda u\}$ or that the process is determined by its moments and the moment measures

$$E\{dZ(\lambda_1) \cdots dZ(\lambda_K)\} \tag{2.1.8}$$

are concentrated in the hyperplanes

$$\lambda_1 + \cdots + \lambda_K = 0, \qquad K = 1, 2, \ldots$$

Each of the preceding approaches has been stochastic in character. An observed stretch of time series is to be thought of as a segment of some realization of a stochastic process. Wiener (1930) introduced a structure based on single nonstochastic functions. For a given function $X(t)$, $0 \leq t < \infty$, limits such as

$$\lim_{T\to\infty} \left\{\int_0^T X(t)\, dt\right\} \Big/ T, \qquad \lim_{T\to\infty} \left\{\int_0^T X(t+u) X(t)\, dt\right\} \Big/ T$$

are assumed to exist (and in the case of the latter to be continuous at 0). This approach goes under the name of *generalized harmonic analysis.*

Throughout this section, the basic time domain has been taken to be the semi-infinite interval $[0, \infty)$. In the stationary case, the series may always be extended to be stationary on the whole line $(-\infty, \infty)$. One simply takes for the finite-dimensional distribution at any collection of time points $-\infty < t_1 < t_2 < \cdots < t_K < \infty$, the distribution at $0, t_2 - t_1, \ldots, t_K - t_1$. The Kolmogorov extension theorem then indicates the existence of a process defined for $-\infty < t < \infty$. The extension of a discrete time series ($t = 0, 1, 2, \ldots$) is considered by Breiman (1968, p. 105). An alternative means of carrying out an extension in the case of continuous time is presented by Furstenberg (1960).

This section concludes by mentioning that certain function spaces provide important representations of time series. To begin, consider a time series $X(t, \omega)$ having finite second-order moments. Then $L_2(X; t)$ is defined to be the closed linear manifold spanned by the values $X(u, \omega)$ with $u \leq t$ and the norm of an element U taken to be $E\,|\,U\,|^2$. That is, it is the smallest Hilbert space which contains all random variables of the form

$$c_1 X(t_1, \omega) + \cdots + c_K X(t_K, \omega)$$

for real numbers $c_1, \ldots, c_K$, for $t_1, \ldots, t_K \leq t$, and $K = 1, 2, \ldots$. The spaces $L_2(X; t)$ are most useful in problems concerning linear prediction. They are all contained in $L_2(X) = L_2(X, \infty)$. Using these spaces Cramér has developed many important properties. Cramér's work may be found in the book by Ephremides and Thomas (1973). A Gaussian process is determined by its first- and second-order moments. Consequently, when studying such a process, it is often enough to consider these spaces $L_2(X; t)$.

In the general case, however, it is often necessary to consider $N_2(X; t)$, the smallest Hilbert space which contains random variables with finite second-order moment and of the form

$$g(X(t_1), \ldots, X(t_K))$$

for $t_1, \ldots, t_K \leq t$ and $K = 1, 2, \ldots$ (see Parzen, 1962). The space $N_2(X; t)$ is contained in $L_2(\Omega) = N_2(X; \infty)$ for each t. It is especially useful in problems concerning nonlinear prediction.

Before indicating the next class of function spaces, it is necessary to mention the existence of *reproducing kernel Hilbert spaces.* Let $R(\psi, \phi)$ be a real nonnegative definite function on the product space $\Phi \times \Phi$, for some index space Φ. Aronszan (1950) has shown the existence of a Hilbert space $H(R)$ of real-valued functions $f(\phi)$, ϕ in Φ, with the following properties: (i) $R(\cdot, \phi)$ belongs to $H(R)$ for each ϕ; (ii) if $\langle\ ,\ \rangle$ denotes the inner product of $H(R)$, then $\langle f(\cdot), R(\cdot, \phi)\rangle = f(\phi)$ for every f in $H(R)$; (iii) $H(R)$ is generated by linear combinations, and limits of these, of the functions $R(\cdot, \phi)$, ϕ in Φ. This space, $H(R)$, is called the reproducing kernel Hilbert space corresponding to the *kernel* R. Some of the reproducing kernel Hilbert spaces important in time series analysis include those corresponding to (a) $\Phi = [0, \infty)$ and $R(\psi, \phi) = E\{X(\psi)X(\phi)\}$ (see Parzen, 1959). The members of this $H(R)$ have the form $E\{UX(\phi)\}$ with U in $L_2(X)$. (b) $\Phi = \{\phi(t)$ for suitable $\phi(\cdot)\}$ and

$R(\psi, \phi) = E \exp\{i \int X(t)[\psi(t) - \phi(t)]\, dt\} = C[\psi - \phi]$ (see Hida, 1970). (c) $\Phi = \{(\phi_0, \phi_1(t_1), \phi_2(t_1, t_2), \phi_3(t_1, t_2, t_3), \ldots\}$ and

$$R(\psi, \phi) = \sum_{J,K} \int \cdots \int \psi_J(s_1, \ldots, s_J)\phi_K(t_1, \ldots, t_K)$$
$$\times E\{X(s_1) \cdots X(s_J)X(t_1) \cdots X(t_K)\}\, ds_1 \cdots ds_J\, dt_1 \cdots dt_K \quad (2.1.9)$$

In this case and when the series X is determined by its moments, $H(R)$ is congruent to $L_2(\Omega) = N_2(X, \infty)$.

Hida (1960) used the reproducing kernel space of (a) to show that purely nondeterministic series may be represented as

$$X(t) = \sum_{i=1}^{N} \int_{-\infty}^{t} g_i(t, u)\, dz_i(u) \quad (2.1.10)$$

where N $(\leq \infty)$ is the multiplicity of $X(t)$, the $dz_i(u)$ are mutually orthogonal random innovation processes, and the $g_i(t, u)$ are nonstochastic. The representation (2.1.10) is useful in that the best linear predictor of the value $X(t + h)$, $h > 0$, based on $\{X(u), u \leq t\}$ may be written

$$\sum_{i=1}^{N} \int_{0}^{t} g_i(t + h, u)\, dz_i(u)$$

2.2. Formal Definitions of Point Processes

There are a number of approaches to the foundations of point processes and certain of these correspond directly with the preceding constructions for ordinary time series. Once again assume that the time domain is $[0, \infty)$. Assume also that the processes have no multiple points.

Initially, consider a probability space $(\Omega, B(\Omega), P)$ and take a (stochastic) *point process* to be a function $N(t, \omega)$, $0 \leq t < \infty$, ω in Ω, with $N(t, \omega)$ a measurable function of ω for each t; $N(0, \omega) = 0$ or 1 (depending on whether or not there is an initial point at the origin); for each ω, $N(t, \omega)$ is finite, nondecreasing, purely discontinuous but right continuous, with unit jumps. This is one way of defining a point process.

Instead of dealing with functions in t, another means is to deal with atomic measures and to consider $N(I, \omega)$, $I \in B[0, \infty)$, $\omega \in \Omega$, with $N(I, \omega)$ a measurable function of ω for each I; and for each ω, $N(\cdot, \omega)$ is a measure with discrete support and mass 1 at each point of the support. The $N(t, \omega)$ of the preceding paragraph may be determined from this $N(\cdot, \omega)$ through the correspondence

$$N(t, \omega) = N([0, t], \omega) \quad (2.2.1)$$

The correspondence (2.2.1) may likewise be used to deduce $N(I, \omega)$ from $N(t, \omega)$.

A procedure analogous to the second one of the preceding section is to assume a consistent family of finite-dimensional probability functions

$$f(n_1, \ldots, n_K; I_1, \ldots, I_K) \tag{2.2.2}$$

for $n_1, \ldots, n_K = 0, 1, 2, \ldots$; bounded $I_1, \ldots, I_K$ in $B[0, \infty)$, $K = 1, 2, \ldots$, to be thought of as giving

$$\text{Prob}\{N(I_1) = n_1, \ldots, N(I_K) = n_K\} \tag{2.2.3}$$

Consistency now includes requirements necessary for the realizations of the process to be a measure such as, for disjoint I_1, I_2,

$$\begin{aligned} f(n_1, n_2, n_3; I_1, I_2, I_1 \cup I_2) &= 1 \quad \text{if} \quad n_1 + n_2 = n_3 \\ &= 0 \quad \text{otherwise} \end{aligned}$$

(see Jagers, 1974). The Poisson process is usually defined in this particular manner.

When they exist, the infinitesimal probability functions, or *product densities*,

$$p_K(t_1, \ldots, t_K)\, dt_1 \cdots dt_K = \text{Prob}\{dN(t_1) = 1, \ldots, dN(t_K) = 1\} \tag{2.2.4}$$

$t_1, \ldots, t_K$ distinct, corresponding to (2.2.2) and (2.2.3), often prove very useful and even characterize a point process under certain conditions. (However, there are examples of distinct point processes having the same product densities; see Ruelle, 1969, p. 106.) Product densities are discussed, e.g., by Macchi (1975) and Brillinger (1975a). The conditional Poisson of Macchi (1975) is defined via product densities.

A surprising result concerning the theory of point processes is that point processes $N(I, \omega)$, without multiple points, may be characterized by their *zero probability functions*

$$\phi(I) = \text{Prob}\{N(I, \omega) = 0\} \tag{2.2.5}$$

for bounded I in $B[0, \infty)$ (see Kurtz, 1974; Jagers, 1974). Given a function $\phi(I)$ on $B[0, \infty)$ satisfying certain conditions, there is a corresponding point process satisfying (2.2.5). Equivalently, one only needs to specify $f(0, \ldots, 0; I_1, \ldots, I_K)$ of (2.2.2) for finite intervals $I_1, \ldots, I_K$ in order to specify the process. For example, the *Poisson process* with parameter measure $\mu(I)$ may be characterized as the point process with

$$\phi(I) = \exp\{-\mu(I)\}$$

for bounded I in $B[0, \infty)$ (see Renyi, 1967).

The zero probability function and product densities are related through

$$\phi(I) = \sum_{k=0}^{\infty} \frac{(-1)^k}{k!} \int_I \cdots \int_I p_k(t_1, \ldots, t_k)\, dt_1 \cdots dt_k$$

$$p_1(t)\, dt = 1 - \phi(t, t + dt)$$

$$p_2(t_1, t_2)\, dt_1\, dt_2 = 1 - \phi(t_1, t_1 + dt_1) - \phi(t_2, t_2 + dt_2) + \phi((t_1, t_1 + dt_1) \text{ or } (t_2, t_2 + dt_2))$$

$$p_K(t_1, \ldots, t_K)\, dt_1 \cdots dt_K = (-1)^k \Delta_{dt_1} \cdots \Delta_{dt_K} \phi(\varnothing)$$

where $\Delta_{dt}\, \phi(I) = \phi(I \text{ or } (t, t + dt)) - \phi(I)$ and $\varnothing$ denotes the empty set (see Kurtz, 1974).

A point process may also be characterized by giving a *probability generating functional* $G[\xi]$, to be thought of as

$$G[\xi] = E \exp\left\{\int \log \xi(t)\, dN(t)\right\} \tag{2.2.6}$$

satisfying certain regularity conditions (see Daley and Vere-Jones, 1972, Theorem 3.10). This is the way the Gauss–Poisson process is defined, for example (see Milne and Westcott, 1972), as well as the infinitely divisible point process. The probability generating functional may be written out in terms of the product densities, specifically

$$G[\xi] = \sum_k \frac{1}{k!} \int \cdots \int (\xi(t_1) - 1) \cdots (\xi(t_k) - 1) p_k(t_1, \ldots, t_k)\, dt_1 \cdots dt_k \tag{2.2.7}$$

(e.g., see Vere-Jones, 1968). The finite-dimensional distributions of (2.2.3) may be determined from $G[\xi]$ by setting

$$\xi(t) = z_k \quad \text{for} \quad t \text{ in } I_k$$
$$= 1 \quad \text{otherwise}$$

and then determining the coefficient of $z_1^{n_1} \cdots z_K^{n_K}$ in the Taylor series expansion of the functional.

An approach to the foundations of point processes which is related to those just given, but yet has a somewhat different character, involves taking Ω to be the set of all locally finite subsets, ω, of $[0, \infty)$. A stochastic point process is then defined to be a probability measure on Ω or, equivalently, a measurable map from a probability space into $(\Omega, B(\Omega))$. The connection with the previous approach is indicated by setting

$$N(I, \omega) = \text{card}(I \cap \omega)$$

for I in $B[0, \infty)$ and ω in Ω. The individual elements of the subsets are called points of the process.

Because points on the real line are ordered, it might be desired to consider, instead of the space of the preceding paragraph, the space $\mathcal{T} = \{(\tau_0, \tau_1, \ldots, \tau_N)\colon 0 < \tau_0 < \tau_1 < \cdots < \tau_N < \infty;\ N = 1, 2, \ldots\}$, and to consider a stochastic point process to be a measurable map into $\mathcal{T}$ from a probability space $(\Omega, B(\Omega), P)$. A realization of the process can then be denoted $\tau_0(\omega), \tau_1(\omega), \ldots$, with the $\tau_j(\omega)$ the points of the realization. The connection with the previous approach is seen through setting

$$N(I, \omega) = \text{the number of } \tau_j(\omega) \text{ in } I$$

In the reverse direction one has the following connection: $\tau_j(\omega)$ is the t such that $N[0, t) = j < N[0, t]$. If $e\{t\}$ denotes the point measure of mass 1 at t, then an alternative way of writing the realizations of the process is

$$N(\cdot, \omega) = \sum_j e\{\tau_j(\omega)\}$$

Turning to specific descriptions of the probabilistic structure of the process within this last framework, the process may clearly be described by the finite-dimensional distributions of the discrete process $\{\tau_0(\omega), \tau_1(\omega), \tau_2(\omega), \ldots\}$ or by the conditional distributions

$$G_j(t; t_j, j < J) = \operatorname{Prob}\{\tau_J > t \mid \tau_j = t_j, j < J\} \tag{2.2.8}$$

Macchi (1971) gives the joint probability density function of the q successive points following t being at $t_1 < t_2 < \cdots < t_q$ as

$$\sum_{k=0}^{\infty} \frac{(-1)^k}{k!} \int \cdots \int_{(t,t_q)^k} p_{k+q}(t_1, \ldots, t_q, u_1, \ldots, u_k)\, du_1 \cdots du_k \tag{2.2.9}$$

This expression may be determined directly from the probability generating functional (2.2.7) by using it to evaluate

$$\operatorname{Prob}\{N(t, t_1) = 0, dN(t_1) = 1, N(t_1, t_2) = 0, dN(t_2) = 1, \ldots, dN(t_q) = 1\}$$

A further space to consider is $[0, \infty) \times Y$ where $Y = \{(y_1, \ldots, y_N)\colon 0 < y_1, \ldots, y_N;\ N = 1, 2, \ldots\}$. These y terms are to be thought of as the distances between successive points of a realization of a stochastic point process, $y_j = \tau_j - \tau_{j-1}$, $j = 1, 2, \ldots$. The individual points of the process may be reconstructed from τ_0 and the y_j through the relationship $\tau_j = \tau_0 + y_1 + \cdots + y_j$, $j = 1, 2, \ldots$. Defining a measurable map from a probability space $(\Omega, B(\Omega), P)$ into $[0, \infty) \times Y$ is now seen to be a further means of constructing a stochastic point process. A realization of the process is now denoted $\tau_0(\omega), y_1(\omega), y_2(\omega), \ldots$. This is the means by which a renewal process is generally constructed, or the Markov interval process studied by Vere-Jones (1975a), or the autoregressive model of Gaver and Lewis (1976).

Given a stochastic point process N, certain conditional probabilities, the Palm probabilities, are occasionally important. These are conditional probabilities given that a point of the process occurs at a particular location, and may be denoted by

$$P(A \mid t) = \text{Prob}\{A \mid t \text{ is a point of the realization}\}$$

for events A in $B(\Omega)$. $P(A \mid t)$ may be defined as the Radon–Nikodyn derivative of the relationship

$$\int_A N(I, \omega)\, dP(\omega) = \int_I P(A \mid t)\, d\mu(t)$$

for I in $B[0, \infty)$, A in $B(\Omega)$, and where μ is the intensity measure of the process (see Jagers, 1974, Section 6). These probabilities may be chosen such that for all t, one has a corresponding point process which will be denoted by $N(\ \mid t)$. In fact the original process N is determined by the specification of $P(A \mid t)$ and μ. For example,

$$\text{Prob}\{N(I) = k\} = \left\{\int_I \text{Prob}\{N(I \mid t) = k\}\, d\mu(t)\right\} \Big/ k \qquad (2.2.10)$$

for $k = 1, 2, \ldots$. It is clear that the zero probability function of N may be determined using expression (2.2.10).

The product densities of the process $N(\ \mid t)$ may be determined directly from those of the process N, namely

$$p_K(t_1, \ldots, t_K \mid t) = p_{K+1}(t, t_1, \ldots, t_K)/p_1(t) \qquad (2.2.11)$$

Its probability generating functional is therefore given by

$$G[\xi \mid t] = \sum (1/k!) \int \cdots \int (\xi(t_1) - 1) \cdots (\xi(t_k) - 1)$$

$$\times\, p_{k+1}(t, t_1, \ldots, t_k)\, dt_1 \cdots dt_k/p_1(t) \qquad (2.2.12)$$

It follows from (2.2.9) and (2.2.12) that the joint probability density function of q successive points being at $t_1 < t_2 < \cdots < t_q$ following an immediately preceding point at t is given by

$$\frac{1}{p_1(t)} \sum \frac{(-1)^k}{k!} \int_{(t,t_q)^k} \cdots \int p_{k+q+1}(t, t_1, \ldots, t_q, u_1, \ldots, u_k)\, du_1 \cdots du_k$$

(see Macchi, 1971).

By analogy with the transition probability functions of (2.1.3), one might be led to consider a *conditional intensity function* $\gamma(t, \omega)$ defined by

$$\text{Prob}\{dN(t) = 1 \mid \mathscr{F}_t\} = \gamma(t, \omega)\, dt \qquad (2.2.13)$$

where $\mathscr{F}_t$ denotes the Borel field generated by the variates $N(u)$, $u \le t$. Such a $\gamma(t, \omega)$ need not always exist (an example of nonexistence is given by Segall and Kailath, 1975); however, when it does exist it often provides a very useful variate. If A is an event of $\mathscr{F}_t$, then from (2.2.13)

$$\text{Prob}\{dN(t) = 1 \mid A\} = \int_A \gamma(t, \omega)\, dP(\omega)\, dt / \text{Prob}\{A\}$$

and so

$$\text{Prob}\{A \mid dN(t) = 1\} = \left[\int_A \gamma(t, \omega)\, dP(\omega)\right] \bigg/ p_1(t) \tag{2.2.14}$$

where $p_1(t)$ is the intensity of the process at time t. The left-hand side of (2.2.14) may be interpreted as the *Palm probability* $P(A \mid t)$. Expression (2.2.14) indicates that the probability measure $P(A \mid t)$ is absolutely continuous with respect to the probability measure P over $\mathscr{F}_t$ and the Radon–Nikodyn derivative is $\gamma(t, \omega)/p_1(t)$. (In the stationary case this result is developed by Papangelou, 1974.)

In the case in which the distributions of (2.2.8) are absolutely continuous

$$\begin{aligned}\gamma(t, \omega) &= -g_{N(t)}(t;\, \tau_j, j < N(t))/G_{N(t)}(t;\, \tau_j, j < N(t)) \\ &= -\frac{\partial}{\partial t} \log G_{N(t)}(t;\, \tau_j, j < N(t))\end{aligned} \tag{2.2.15}$$

for $\tau_{N(t)-1} \leq t < \tau_{N(t)}$ and

$$G_J(t;\, t_j, j < J) = \exp\left\{-\int_{t_{J-1}}^{t} \gamma(u, \omega)\, du\right\} \tag{2.2.16}$$

(e.g., Snyder, 1975, p. 245). An important use of the conditional intensity is in calculating the joint probability function for the number of points in an interval $[0, T]$ and their locations. From (2.2.16) the expression is

$$\exp\left\{-\int_0^T \gamma(t, \omega)\, dt + \int_0^T \log \gamma(t, \omega)\, dN(t, \omega)\right\} \tag{2.2.17}$$

(see Rubin, 1972; Vere-Jones, 1975b).

Defining

$$\Gamma(t, \omega) = E\{N(t) \mid \mathscr{F}_t\} = \int_0^t \gamma(u, \omega)\, du$$

the process $N(t, \omega) - \Gamma(t, \omega)$ may be seen to be a martingale. This occurrence leads to many useful results (see Chou and Meyer, 1975; Segall and Kailath, 1975). Among the results is the fact that the process N may be transformed into a Poisson process N^* by the random time change

$$\Gamma^{-1}(u) = \inf\{t\colon \Gamma(t) > u\}$$

Specifically, the process $N^*(u) = N(\Gamma^{-1}(u))$ is Poisson with rate 1 (see Bremaud, 1974; Meyer, 1971; Aalen, 1975; Papangelou, 1974). In fact, the result is strongly suggested by expression (2.2.17) and may be seen from (2.2.15).

The product densities $p_K(\cdot)$ of (2.2.4) may be determined from $\gamma(t, \omega)$ as

$$p_K(t_1, \ldots, t_K) = E\{\gamma(t_1, \omega)\} \prod_{1 < k \leq K} E\{\gamma(t_k, \omega) | \text{points at } t_1, \ldots, t_{k-1}\}$$

(see Snyder, 1975, p. 275). In particular the intensity or rate function is simply given by $E\gamma(t, \omega)$.

The conditional intensity function, as defined by (2.2.13), was based on $\mathscr{F}_t$, the Borel field generated by $\{N(u), u \leq t\}$. On occasion it is convenient to use a larger Borel field $\mathscr{G}_t \supset \mathscr{F}_t$ (see Segall and Kailath, 1975). This allows one to study point processes that depend, for example, on some signal of interest.

It is by now apparent that there are many intertwined yet different manners of introducing stochastic point processes. The particular one adopted in any given situation relates to the peculiarities of the situation.

In the case in which the process is *stationary*, certain simplifications occur. The joint probability functions of (2.2.2) are now invariant under shifts of time:

$$f(n_1, \ldots, n_K; I_1 + t, \ldots, I_K + t) = f(n_1, \ldots, n_K; I_1, \ldots, I_K)$$

for t such that the intervals $I + t$ contain no negative values. The product densities of (2.2.4) depend on one less argument

$$p_K(t_1, \ldots, t_K) = p_K(t_1 - t_K, \ldots, t_{K-1} - t_K)$$

The zero probability function is invariant under shifts

$$\phi(I + t) = \phi(I) \tag{2.2.18}$$

for t such that $I + t$ contains no negative values. The probability generating functional (2.2.6) now has the property

$$G[S^t\xi] = G[\xi] \tag{2.2.19}$$

where $S^t\xi(u) = \xi(t + u)$. Conversely, invariance as in (2.2.18) and (2.2.19) implies stationarity of the corresponding process.

The process has so far been considered to have domain $[0, \infty)$. As in the ordinary time series situation, in the stationary case the process may be extended to be stationary on the whole line $(-\infty, \infty)$. This may be demonstrated by the Kolmogorov extension theorem; alternatively, the zero probability function may be used to construct the extended process.

In the case in which the process is characterized by the location τ_0 of the initial point and the interpoint distances $y_1, y_2, \ldots,$ stationarity has a very important consequence. Namely, $y_1, y_2, \ldots$ is a stationary sequence conditional on the location of τ_0, i.e., under the Palm measure $P(\ |\tau_0 = t)$. This may be seen directly from (2.2.12). General proofs may be found in the works by Ryll-Nardzewski (1961) and Neveu (1968). In this stationary case, the Palm measure $P(\ |\tau_0 = t)$ will not depend on t, and on many occasions τ_0 is taken to be at 0.

A stationary N may also be thought of as a random counting measure with the property $N(I + t, \omega) = N(I, U^t\omega)$, for a semigroup of measure preserving U^t of Ω. Using these maps U^t, simple expressions may be set down relating the Palm and unconditional probability measures, namely

$$P(A\,|\,0) = \int \left\{ \int_0^1 I_A(U^t\omega)\, dN(t) \right\} dP(\omega)$$

$$P(A) = p_1 \int \left\{ \int_0^{y_1} I_A(U^{-t}\omega)\, dt \right\} dP(\omega\,|\,0) \tag{2.2.20}$$

where U^t is a typical map, p_1 the intensity of the process, y_1 the distance to the first positive point, A is in $B(\Omega)$, and I_A denotes the indicator function of the event A (see de Sam Lazaro and Meyer, 1975, Ryll-Nardzewski, 1961).

A deterministic approach to the foundations of point processes in a "stationary" case is also available. Consider an increasing sequence of points $0 \leq \tau_0 < \tau_1 < \tau_2 < \cdots$ along the half-line. Let $N(t)$ denote the number of τ_k in the interval $[0, t]$. One now proceeds by assuming that limits such as

$$\lim_{T\to\infty} N(T)/T, \qquad \lim_{T\to\infty} \left\{ \int_0^T [N(y + u) - N(t)]\, dN(t) \right\} \Big/ T \tag{2.2.21}$$

exist (see Brillinger, 1973). The limit in (2.2.21) may also be interpreted as that of

$$\lim_{T\to\infty} \{\text{number of } (j, k) \text{ with } \tau_j - \tau_k \leq u\}/T$$

Function spaces analogous to those introduced at the end of Section 2.1 are also of use in the point process case. The Hilbert space $L_2(N;\, t)$ spanned by linear combinations of the $N(u)$, $u \leq t$, is introduced for prediction purposes in the article by Vere-Jones (1974). The space $N_2(N;\, t)$ may be introduced in a directly analogous manner, and also the reproducing kernels.

The kernel (2.1.7) appears in Ruelle (1969) in a point process version.

Certain ordinary series may be associated with a given point process in a direct manner, for example,

$$X(t) = \int a(t, u)\, dN(u)$$

$$X(t) = N(t, t + 1]$$

$$X(t) = \int \exp\{-\mu(t - u)\}\, dN(u)$$

The last series is discussed in some detail for the stationary case in the article by Vere-Jones (1974).

2.3. Other Formal Definitions

The preceding two sections have provided separate discussions of the foundations of ordinary time series analysis and point process analysis. In fact, there are several methods by which one may simultaneously lay foundations for time series analysis and point process analysis.

Both types of processes may be discussed as particular cases of the *random additive set functions* of Bochner (1960). In Bochner's theory one deals with random interval functions $\alpha(I, \omega)$, I an interval of $[0, \infty)$, ω in Ω with $(\Omega, B(\Omega), P)$ a probability space, and

$$\alpha(I \cup J, \omega) = \alpha(I, \omega) + \alpha(J, \omega) \quad \text{with probability 1}$$

for I and J disjoint intervals. The interval representation $N(I, \omega)$ of a point process is immediately seen to be of this character. If the ordinary series $X(t, \omega)$ is such that

$$\int_I E|X(t)|\, dt < \infty \tag{2.3.1}$$

for bounded intervals I, then an additive set function may be associated with X by setting

$$\alpha(I, \omega) = \int_I X(t)\, dt \tag{2.3.2}$$

Bochner (1960) develops an extension theorem, defines a characteristic functional, discusses stationarity, and develops an harmonic analysis for such random interval functions. Brillinger (1972) develops further aspects and considers the particular case of point processes in some detail.

An essentially equivalent development would result from considering time series and point processes as particular kinds of *random signed measures*

on $[0, \infty)$. In fact, a point process corresponds directly to a random measure. A time series with nonnegative values corresponds to a random measure through the relationship (2.3.2).

The theories of time series and point processes may be subsumed under the theory of *random generalized functions* (random Schwartz distributions). Let $\mathscr{D}$ denote the space of infinitely differentiable functions $\phi(t)$ of compact support. A continuous linear functional $Y(\phi)$ on $\mathscr{D}$ is called a generalized function (Schwartz distribution). A measurable map from a probability space $(\Omega, B(\Omega), P)$ to the space of generalized functions is called a random generalized function (see Yaglom, 1962; Gelfand and Vilenkin, 1964; Brillinger, 1974). Its image may be denoted $Y(\phi, \omega)$. A time series $X(t)$, satisfying (2.3.1), gives rise to the random generalized function

$$X(\phi, \omega) = \int \phi(t)X(t, \omega)\, dt, \qquad \phi \text{ in } \mathscr{D} \tag{2.3.3}$$

A point process $N(t)$ gives rise to the random generalized function

$$N(\phi, \omega) = \int \phi(t)\, dN(t, \omega), \qquad \phi \text{ in } \mathscr{D} \tag{2.3.4}$$

(see Daley, 1969). The generalized function of (2.3.4) may alternatively be denoted by

$$\sum_j \delta(t - \tau_j) \tag{2.3.5}$$

where $\delta(\cdot)$ is the Dirac delta function (a generalized function!) and the τ_j are the positions of the points of the process. Further developments in the theory of generalized random functions include work by Rozanov (1969), who developed linear prediction theory and defined the space $L_2(Y; t)$ spanned by the $Y(\phi, \omega)$ with ϕ zero to the right of t; Kailath (1971), who considered the reproducing kernel Hilbert space associated with the kernel $R(\psi, \phi) = EY(\psi, \omega)Y(\phi, \omega)$; and Hida (1970), who considered the Hilbert space associated with the kernel $R(\psi, \phi) = E \exp\{iY(\psi - \phi, \omega)\}$.

The point processes considered so far are particular cases of *marked point processes*. These are point processes with auxiliary values (marks) associated with each point. If the marks are constant, a marked point process is equivalent to an ordinary point process. If the marks are real valued, the process may be thought of as being piecewise constant, but jumping (up or down) by the value of the mark at the points of the process. If the mark is M_j at the point τ_j, then the marked point process may be denoted

$$\sum_j M_j\, \delta(t - \tau_j) \tag{2.3.6}$$

a notation extending that of (2.3.5). If the mark M_j is k-dimensional, then the marked point process may be thought of as a standard point process with domain R^{k+1}, i.e., a point process with points at the (τ_j, M_j). A vector-valued point process may be thought of as a marked point process with the mark giving the type of point occurring at a given time. A point process with multiple points may be thought of as a marked point process with the mark giving the multiplicity of the point. A useful generalized process related to the one of (2.3.6) is provided by

$$\sum_j X(\tau_j)\, \delta(t - \tau_j) \tag{2.3.7}$$

where $X(t)$ is an ordinary time series. The process of (2.3.7) corresponds to sampling the series X at stochastic times.

Incidentally, if the moments $EX(t)^k$ of an ordinary time series or the moments $EN(t)^k$ of a point process grow more slowly than some power of t as $t \to \infty$, then Fourier transforms of the moments and indeed of the processes themselves are well defined using the theory of generalized functions. This fact leads to a direct definition of certain useful parameters of the processes, even in the nonstationary case.

The sample paths of both the time series and the point processes considered are members of $D[0, \infty)$, the *Skorokhod space* of real-valued functions on $[0, \infty)$ which are right continuous and have left limits. In consequence, general results, such as central limit theorems, developed for random variables with values in $D[0, \infty)$ will apply directly to the processes considered in this chapter. One general reference is the book by Billingsley (1968).

Further, the processes considered are particular cases of processes with a more general time parameter, for example, of a *spatial series*

$$X(t_1, \ldots, t_p), \qquad t_1, \ldots, t_p \text{ in } [0, \infty)$$

or a *spatial point process*

$$N(I), \qquad I \subset B[0, \infty)^p$$

with $N(I)$ giving the number of points lying in the p-dimensional interval I. A not so obvious correspondence is the one between random hyperplanes in R^p and random points in the hypercylinder $R \times [0, 2\pi)^{p-1}$ (see Krickeberg, 1974). The correspondence is through the characterization of a hyperplane by its angles of orientation and distance from the origin.

Processes of structure simpler than that considered so far in this chapter are often very useful and illustrative. In particular, consider the discrete time series

$$X(th), \qquad t = 0, 1, 2, \ldots$$

for some $h > 0$. As $h \downarrow 0$, the behavior of these series comes close to that of continuous time series. The particular case in which $X(th)$ takes on only the values 0 and 1 is a simple analog of a point process. This analogy is made use of by Ryll-Nardzewski (1961), Breiman (1968), and Lewis (1970) for example.

In the stationary case, both $N(t, \omega)$ and $\int_0^t X(u, \omega)\,du$ are examples of processes with *stationary increments* and *helices*. A variety of useful results have been developed for such processes (e.g., see Doob, 1953, p. 551; Yaglom, 1962; de Sam Lazaro and Meyer, 1975; Masani, 1972, including the definition of Palm measures in the case in which the process is non-decreasing (see Geman and Horowitz, 1973). Further, in the "stationary" case, it is possible to construct a generalized harmonic analysis for individual generalized functions if so desired (see Pfaffelhuber, 1975). Certain limits are assumed to exist in this development. The results may be specialized to obtain results either for individual time series or individual step functions.

2.4. Parameters for Ordinary Time Series

The preceding sections have been concerned with the complete characterization of time series and point processes. In fact, much information is often contained in certain simple parameters related to the processes. In this context parameters will be viewed as quantities providing descriptive features of a probability distribution of interest.

Basic parameters of a time series include the *moment functions*

$$m_K(t_1, \ldots, t_K) = E\{X(t_1) \cdots X(t_K)\} \tag{2.4.1}$$

$K = 1, 2, \ldots$, when these exist. In some circumstances the moment functions taken in totality characterize the distribution of a time series. In terms of the finite-dimensional distribution functions (2.1.1), the moment function (2.4.1) is given by

$$\int \cdots \int x_1 \cdots x_K F_K(dx_1, \ldots, dx_K; t_1, \ldots, t_K)$$

Under regularity conditions (e.g., $E|X(t)|^K < C^K$, for some finite C and K sufficiently large), the characteristic functional (2.1.4) may be expanded in terms of the moment functions:

$$C[\xi] = \sum_{K=0}^{\infty} \frac{i^K}{K!} \int \cdots \int \xi(t_1) \cdots \xi(t_K) m_K(t_1, \ldots, t_K)\,dt_1 \cdots dt_K \tag{2.4.2}$$

and the individual moment functions may be determined by evaluating the Gateaux derivatives of $C[\xi]$. The moment functions are seen to be symmetric in their arguments from the definition (2.4.1).

Of the moment functions, those of first and second order, $m_1(t) = EX(t)$ and $m_2(t_1, t_2) = EX(t_1)X(t_2)$, are used most often. $m_1(t)$ gives the level about which the series fluctuates at time t. Its units are those of X; $m_2(t, t) = E|X(t)|^2$ gives an indication of the expected magnitude of the series at time t. The function $m_2(t_1, t_2)$ is nonnegative definite, i.e.,

$$\sum_j \sum_k \alpha_j \alpha_k m_2(t_j, t_k) \geq 0 \tag{2.4.3}$$

for times t_j and scalars α_j. That (2.4.3) is true is immediate, since the expression on the left is

$$E\left|\sum_k \alpha_k X(t_k)\right|^2$$

The functions $m_1(t)$ and $m_2(t_1, t_2)$ are most useful in the linear analysis of time series and in the anlysis of Gaussian series.

In cases in which the values of a series are becoming less and less dependent (statistically) as they are becoming further separated in time, the moment functions possess certain factorization properties such as

$$\begin{aligned} \lim_{U \to \infty} & m_{J+K}(t_1, \ldots, t_J, t_{J+1} + U, \ldots, t_{J+K} + U) \\ &= m_J(t_1, \ldots, t_J) m_K(t_{J+1}, \ldots, t_{J+K}) \end{aligned} \tag{2.4.4}$$

Of course, for an m-dependent series (wherein collections of values more than m time units apart are statistically independent) expression (2.4.4) holds without the limit, provided $U > m$.

The degree of dependence of the values of a time series is more directly measured by the *cumulant functions*

$$c_K(t_1, \ldots, t_K) = \text{cum}\{X(t_1), \ldots, X(t_K)\} \tag{2.4.5}$$

when they exist. These are defined, for example, by the functional expansion

$$\log C[\xi] = \sum_{k=1}^{\infty} \frac{i^K}{K!} \int \cdots \int \xi(t_1) \cdots \xi(t_K) c_K(t_1, \ldots, t_K)\, dt_1 \cdots dt_K \tag{2.4.6}$$

and have the property that $c_K(t_1, \ldots, t_K) = 0$ whenever any subset of the variates $X(t_1), \ldots, X(t_K)$ is independent of the remaining variates (for a discussion of this and further properties of cumulant functions, see Brillinger, 1975a). The degree of dependence of a time series (or degree of mixing) may be described by specifying the rate at which its cumulant functions fall to 0 as the $|t_k|$ increase.

As in the case of the moment functions, the *mean function* $c_1(t) = EX(t) = m_1(t)$ and the *autocovariance function*

$$c_2(t_1, t_2) = \text{cov}\{X(t_1), X(t_2)\} = m_2(t_1, t_2) - m_1(t_1) m_1(t_2)$$

are the ones most often used in practice. The autocovariance function is also nonnegative definite as

$$\text{var}\left\{\sum_k \alpha_k X(t_k)\right\} = \sum_j \sum_k \alpha_j \alpha_k c_2(t_j, t_k) \geq 0$$

The *variance function*, var $X(t) = c_2(t, t)$, measures the extent of fluctuations of the series $X(t)$ about its mean level $c_1(t)$ at time t.

It will be seen in Section 2.6 that both the moment and cumulant functions transform in a direct manner when a series is subjected to a linear transformation. Expressions for the moments and cumulants of polynomials in the values of a time series may also be set down, however not so directly as in the linear case.

The moment and cumulant functions are based on expected values. There are other expected values, based on time series, of some interest. Suppose that the finite-dimensional distribution function (2.1.1) is absolutely continuous with density

$$f_K(x_1, \ldots, x_K; t_1, \ldots, t_K) \tag{2.4.7}$$

Then it may be written as

$$E\{\delta(X(t_1) - x_1) \cdots \delta(X(t_K) - x_K)\} \tag{2.4.8}$$

where $\delta(\cdot)$ is the Dirac delta function. Expression (2.4.8) suggests how the density function (2.4.7) might be estimated by using an approximate delta function. This is useful to do in the case in which it is possible that a time series has finite-dimensional distributions of a specific form and it is desired to check this possibility.

Another expected value which has a direct connection with the parameters of point processes is provided by

$$E\left\{\int_I \cdots \int_I \delta(X(t_1) - a) \cdots \delta(X(t_K) - a)|X'(t_1)| \cdots |X'(t_K)|\, dt_1 \cdots dt_K\right\} \tag{2.4.9}$$
$$\neq t_k$$

where $X'(t)$ denotes the derivative of $X(t)$. Expression (2.4.9) represents the Kth factorial moment of the number of crossings of the level a, by the series $X(t)$, in the time interval I. If $g_K(x_1, \ldots, x_K, y_1, \ldots, y_K; t_1, \ldots, t_K)$ denotes the probability density of the variate $\{X(t_1), \ldots, X(t_K), X'(t_1), \ldots, X'(t_K)\}$ then, under regularity conditions, expression (2.4.9) is given by

$$\int_I \cdots \int_I \int_0^\infty \cdots \int_0^\infty y_1 \cdots y_K g_K(a, \ldots, a, y_1, \ldots, y_K; t_1, \ldots, t_K)$$
$$\times\, dy_1 \cdots dy_K\, dt_1 \cdots dt_K \tag{2.4.10}$$

Matters of this sort are investigated by Leadbetter (1972).

Of course, conditional expected values such as

$$E\{X(t+h)\,|\,X(u),\ u \le t\}, \qquad h > 0 \tag{2.4.11}$$

are very important in prediction theory. Specifically, the random variable V of (2.4.11) provides the minimum of

$$E\,|\,X(t+h) - V\,|^2$$

for measurable functions V of $\{X(u),\ u \le t\}$. Alternatively, (2.4.11) is the variate, based on the past, having the highest correlation with $X(t+h)$. The value (2.4.11) may be viewed as a limit of expected values based on the conditional distributions (2.1.3). Were the process Markov, these would simplify. In addition in the Markov case, one would be interested in estimating the transition probability elements:

$$\text{Prob}\{x < X(t+h) < x + dx\,|\,X(t) = x_0\}$$

Other parameters of occasional use in dealing with time series include the fractiles (especially the median) and the fractile ranges.

In a situation wherein one has a number of realizations of a time series available for analysis, one can proceed to estimate the parameters just defined directly. More often, however, one has available only a segment of a single realization. Even in the case in which the series involved is approximately stationary, one can still set down reasonable estimates of parameters of interest.

In the stationary case, the moment functions of (2.4.1) and (2.4.5) depend on one fewer argument. Specifically,

$$m_K(t_1, \ldots, t_K) = m_K(t_1 - t_K, \ldots, t_{K-1} - t_K, 0) \tag{2.4.12}$$

$$c_K(t_1, \ldots, t_K) = c_K(t_1 - t_K, \ldots, t_{K-1} - t_K, 0) \tag{2.4.13}$$

This has the advantage that now a sensible estimate of (2.4.12) may be based on the expression

$$\int_I \cdots \int_I X(t_1 + u) \cdots X(t_K + u)\, du$$

where the domain of integration is over the region of observation. In the case of $K = 1$, the mean level of the series may be estimated by

$$\left[\int_0^T X(u)\, du\right] \Big/ T \tag{2.4.14}$$

given values of the process in the interval $(0, T)$.

Because of the simplification of expressions (2.4.12) and (2.4.13), the "0" appearing in them will be dropped and, in the stationary case, the definitions

$$m_K(u_1, \ldots, u_{K-1}) = E\{X(t+u_1) \cdots X(t+u_{K-1})X(t)\}$$

$$c_K(u_1, \ldots, u_{K-1}) = \text{cum}\{X(t+u_1), \ldots, X(t+u_{K-1}), X(t)\}$$

will be stated.

The mean function $m_1 = m_1(t) = EX(t) = c_1(t) = c_1$ is constant in the stationary case. The series is seen to be fluctuating about a constant level c_1. The autocovariance function $c_2(u) = \text{cov}\{X(t+u), X(t)\}$ is particularly useful in practice. Its shape provides many indications of the inherent character of the series X. The value $c_2(u)$ measures the degree of linear statistical dependence of values of the series lag u time units apart.

In the case in which $c_2(u)$ dies off sufficiently rapidly as $|u| \to \infty$, the *power spectrum* at frequency λ, $f_2(\lambda)$, may be defined by

$$f_2(\lambda) = (2\pi)^{-1} \int_{-\infty}^{\infty} \exp\{-i\lambda u\} c_2(u)\, du \tag{2.4.15}$$

for $-\infty < \lambda < \infty$, with the accompanying inverse relationship:

$$c_2(u) = \int_{-\infty}^{\infty} \exp\{iu\lambda\} f_2(\lambda)\, d\lambda \tag{2.4.16}$$

In the general case in which $c_2(u)$ is finite and continuous at 0, Khintchine (1934) has shown the existence of a unique finite spectral measure F on the real line satisfying

$$c_2(u) = \int_{-\infty}^{\infty} \exp\{iu\lambda\} F(d\lambda) \tag{2.4.17}$$

This last means, for example, that $f_2(\lambda)$ of (2.4.15) is nonnegative. Interpretations of $f_2(\lambda)$ and $F(d\lambda)$ will be provided in a later section after the notion of a bandpass filter has been introduced. The usefulness of the parameter $f_2(0)$ is suggested by the remark that the large sample variance of the statistic (2.4.14) is $2\pi f_2(0)/T$.

In the case in which

$$\int \cdots \int |c_K(u_1, \ldots, u_{K-1})|\, du_1 \cdots du_{K-1} < \infty \tag{2.4.18}$$

as, for example, when values at a distance of the series are only weakly dependent, the cumulant spectrum of order K of the series may be defined by

$$f_K(\lambda_1, \ldots, \lambda_{K-1}) = (2\pi)^{-K+1} \int_{-\infty}^{\infty} \cdots \int \exp\{-i(\lambda_1 u_1 + \cdots + \lambda_{K-1} u_{K-1})\}$$

$$\times\, c_K(u_1, \ldots, u_{K-1})\, du_1 \cdots du_{K-1} \tag{2.4.19}$$

with the inverse relationship

$$c_K(u_1, \ldots, u_{K-1}) = \int_{-\infty}^{\infty} \cdots \int \exp\{i(u_1 \lambda_1 + \cdots + u_{K-1}\lambda_{K-1})\}$$

$$\times f_K(\lambda_1, \ldots, \lambda_{K-1})\, d\lambda_1 \cdots d\lambda_{K-1} \qquad (2.4.20)$$

for $K = 2, 3, \ldots$. In general, there is no relationship analogous to (2.4.17) for $K > 2$. However, provided the functions $c_K(u_1, \ldots, u_{K-1})$ are of slow growth, the Fourier transform of (2.4.18) may be defined as a generalized function, with an inverse relationship analogous to (2.4.20).

In fact, a stationary series $X(t)$ has a direct Fourier representation. Under the conditions leading to the representation (2.4.17), Cramér (1942) develops the result

$$X(t) = \int_{-\infty}^{\infty} \exp\{it\lambda\}\, dZ(\lambda) \qquad (2.4.21)$$

with the "integral" defined as a limit in mean of Stieltjes sums and with $Z(\lambda)$ a stochastic function having the properties

$$E\, dZ(\lambda) = \delta(\lambda)c_1\, d\lambda$$

$$\operatorname{cov}\{dZ(\lambda_1), dZ(\lambda_2)\} = \delta(\lambda_1 - \lambda_2)F(d\lambda_1)\, d\lambda_2$$

where $F(d\lambda)$ appears in (2.4.17). In the case in which (2.4.18) is satisfied,

$$\operatorname{cum}\{dZ(\lambda_1), \ldots, dZ(\lambda_K)\} = \delta(\lambda_1 + \cdots + \lambda_K) f_K(\lambda_1, \ldots, \lambda_{K-1})\, d\lambda_1 \cdots d\lambda_K \qquad (2.4.22)$$

This last expression provides an interpretation for the cumulant spectrum of order K. It is proportional to the joint cumulant of order K of the increments $dZ(\lambda_1), \ldots, dZ(\lambda_K)$ with $\lambda_1 + \cdots + \lambda_K = 0$. In the case $K = 2$, $f_2(\lambda)$ is seen to be proportional to the variance of $dZ(\lambda)$.

Returning to the *Cramér representation* (2.4.21), the time series $\exp\{it\lambda\}\, dZ(\lambda)$ appearing therein is called the *component of frequency* λ in the series $X(t)$. If the series $X(t)$ is extended to be stationary on the whole real line, then the function $Z(\lambda)$ may be determined as the limit in mean of the variate

$$(2\pi)^{-1} \int_{-T}^{T} X(t) \frac{1 - \exp\{-i\lambda t\}}{-it}\, dt$$

as $T \to \infty$.

In the case in which the series is not stationary, but its moments are of slow growth, a representation of the form (2.4.21) may still be given, but now involving random generalized functions. Now, however, the reduction of the values to the hyperplane $\lambda_1 + \cdots + \lambda_K = 0$ occurring in (2.4.22) will no longer take place. In the case in which the second-order moment function of $Z(\lambda)$ is of bounded variation, the series $X(t)$ is called *harmonizable.*

A class of time series that are nonstationary, strictly speaking, but yet that possess many of the properties of stationary series is comprised of those of the form $X(t) = a(t) + \varepsilon(t)$ where $a(t)$ is a nonstochastic, nonconstant function and $\varepsilon(t)$ is a zero mean stationary time series. For this $X(t)$, the mean function is $a(t)$, and so nonconstant; however, all of its cumulant functions of order greater than 1 are those of a stationary series. In the case in which $a(t)$ is slowly varying, $a(\cdot)$ is called the *trend function* of the series. In engineering literature $a(t)$ is sometimes called the *signal* and $\varepsilon(t)$ the *noise.* Commonly used forms for $a(t)$ include $a(t) = \alpha + \beta t$ and $a(t) = \rho \cos(\omega t + \phi)$. The first is called a *linear trend model,* the second a *hidden periodicity model.*

The cumulant spectra of a stationary series may be obtained from the expansion of the logarithm of the characteristic functional, specifically,

$$\begin{aligned}\log C[\xi] &= \log E \exp\left\{i \int \xi(t)X(t)\,dt\right\}\\ &= \log E \exp\left\{i \int \hat{\xi}(\lambda)\,dZ(\lambda)\right\}\\ &= \sum_{k=1}^{\infty} \frac{i^k}{k!} \int \cdots \int \hat{\xi}(\lambda_1) \cdots \hat{\xi}(\lambda_{k-1})\overline{\hat{\xi}(\lambda_1 + \cdots + \lambda_{k-1})}\\ &\quad \times f_k(\lambda_1, \ldots, \lambda_{k-1})\,d\lambda_1 \cdots d\lambda_{k-1}\end{aligned}$$

where $\hat{\xi}(\lambda) = \int \xi(t) \exp\{i\lambda t\}\,dt$.

2.5. Parameters for Point Processes

Point process parameters directly analogous to the product moment functions (2.4.1) of ordinary time series analysis, are provided by the *product densities* $p_K(t_1, \ldots, t_K)$ of

$$p_K(t_1, \ldots, t_K)\,dt_1 \cdots dt_K = E\{dN(t_1) \cdots dN(t_K)\} \tag{2.5.1}$$

$K = 1, 2, \ldots,$ when they exist. In practice, because of the step function character of $N(t)$, it is only reasonable to expect the absolute continuity of (2.5.1) to occur when $t_1, \ldots, t_K$ are distinct. It has already been assumed that the points of the process are isolated. Suppose, in fact, that

$$\text{Prob}\{N(I) = n\} < L_\delta |I|^n \tag{2.5.2}$$

for some finite L_δ, whenever I is an interval of length $|I| \leq \delta$. Then, for example,

$$\begin{aligned} E\{dN(t)\, dN(t)\} &= E\, dN(t) + O(dt^2) \\ &= p_1(t)\, dt + O(dt^2) \end{aligned} \tag{2.5.3}$$

with similar reductions occurring in the case of general K whenever some of the t_k are equal (see Brillinger, 1972).

The parameter $p_1(t)$ is called the *mean rate* of the process at time t. A representation for it, as an alternative to (2.5.1), is provided by

$$\int_0^t p_1(u)\, du = EN(t) \tag{2.5.4}$$

It is a parameter with an intimate connection to the appearance of realizations of the process. When $p_1(t)$ is large, there tend to be many points in the neighborhood of t. When $p_1(t)$ is small, points tend to be rare near t. In the case $K = 2$, it follows from (2.5.1) that

$$\begin{aligned} \int_0^t \int_0^t p_2(u_1, u_2)\, du_1\, du_2 &= E\left\{ \iint_{u_1 \neq u_2} dN(u_1)\, dN(u_2) \right\} \\ &= E\left\{ \left(\int_0^t dN(u) \right)^2 - \int_0^t (dN(u))^2 \right\} \\ &= E\{N(t)(N(t) - 1)\} \end{aligned} \tag{2.5.5}$$

using (2.5.3). That is, the integral of $p_2(t_1, t_2)$ gives the second factorial moment of $N(t)$. In general,

$$\int_I \cdots \int_I p_K(t_1, \ldots, t_K)\, dt_1 \cdots dt_K = EN(I)^{(K)} \tag{2.5.6}$$

where $N^{(K)} = N(N-1) \cdots (N - K + 1)$, $K = 1, 2, \ldots$. If desired, the ordinary moment $EN(I)^K$ may be deduced from the factorial moments of order less than or equal to K. The *factorial moment generating functional* of the point process N is defined to be

$$E \exp\left\{ \int \log[1 + \xi(t)]\, dN(t) \right\} \tag{2.5.7}$$

If the points of the process are represented by $\tau_0, \tau_1, \ldots$, then (2.5.7) may be written

$$E\left\{ \prod_j [1 + \xi(\tau_j)] \right\} = E\left\{ \sum_k \frac{1}{k!} \prod \xi(\tau_{j_1}) \cdots \xi(\tau_{j_k}) \right\} \tag{2.5.8}$$

where $\tau_{j_1}, \ldots, \tau_{j_k}$ are distinct. It follows that the functional (2.5.7) has the representation

$$\sum_k \frac{1}{k!} \int \cdots \int \xi(t_1) \cdots \xi(t_k) p_k(t_1, \ldots, t_k)\, dt_1 \cdots dt_k \tag{2.5.9}$$

In cases in which increments of the process, well separated in time, are only weakly dependent it is often more convenient to consider functions q_k defined by the expansion of the logarithm of (2.5.7), namely by

$$\log E \exp\left\{\int \log[1 + \xi(t)]\, dN(t)\right\}$$
$$= \sum_k \frac{1}{k!} \int \cdots \int \xi(t_1) \cdots \xi(t_k) q_k(t_1, \ldots, t_k)\, dt_1 \cdots dt_k \tag{2.5.10}$$

These functions are called the *cumulant densities* and have the interpretations

$$q_K(t_1, \ldots, t_K)\, dt_1 \cdots dt_K = \operatorname{cum}\{dN(t_1), \ldots, dN(t_K)\} \tag{2.5.11}$$

when the t_k are distinct. They provide direct measures of the degree of statistical dependence of a process N. Those of order 1 and 2 are given by $q_1(t) = p_1(t)$, $q_2(t_1, t_2) = p_2(t_1, t_2) - p_1(t_1)p_2(t_2)$. The factorial cumulant of order K of $N(I)$ may be written

$$\int_I \cdots \int_I q_K(t_1, \ldots, t_K)\, dt_1 \cdots dt_K \tag{2.5.12}$$

Conditions under which expansions such as (2.5.9) and (2.5.10) may be manipulated may be found in an article by Wescott (1972).

Under a condition like (2.5.2), small increments $dN(t)$ generally contain no or one point. It follows then that the product density has the further important interpretation

$$p_K(t_1, \ldots, t_K)\, dt_1 \cdots dt_K = \operatorname{Prob}\{dN(t_1) = 1, \ldots, dN(t_K) = 1\} \tag{2.5.13}$$

for distinct t_k and $K = 1, 2, \ldots$. In particular,

$$\operatorname{Prob}\{dN(t) = 0\} = 1 - p_1(t)\, dt$$
$$\operatorname{Prob}\{dN(t) = 1\} = p_1(t)\, dt$$

The first product density, $p_1(t)$, is seen to give the mean intensity with which points occur at time t. Its units are those of points per unit time. It is an important parameter to estimate in practice.

The product density of order 2, $p_2(t_1, t_2)$, provides a measure of the intensity with which points simultaneously occur near t_1 and near t_2. A related useful function is provided by

$$p_2(t_1, t_2)/p_1(t_1) = \text{Prob}\{dN(t_2) = 1 \mid N\{t_1\} = 1\}/dt_2 \tag{2.5.14}$$

This gives the intensity with which points are occurring near t_2 given that there is a point at t_1. It may be thought of as the mean rate, $p_1(t_2 \mid t_1)$, of the Palm process $N(t \mid t_1)$. The measure (2.5.14) may also be written in terms of cumulant densities as

$$q_1(t_2) + q_2(t_1, t_2)/q_1(t_1) \tag{2.5.15}$$

In the case in which well-separated increments of the process are approximately independent, the measure (2.5.14) is seen to be approximately $p_1(t_2) = q_1(t_2)$ since the second term in (2.5.15) is negligible.

The first- and second-order densities are important in constructing a further useful point process parameter, the *index of dispersion*,

$$\begin{aligned} I(t) &= \text{var } N(t)\,/EN(t) \\ &= 1 + \int_0^t \int_0^t q_2(u_1, u_2)\, du_1\, du_2 \Big/ \int_0^t q_1(u)\, du \end{aligned} \tag{2.5.16}$$

This index measures, to some extent, the departure of the process N from being a Poisson process (for which the index is identically 1). Also useful is the *variance time curve*

$$V(t) = \int_0^t q_1(u)\, du + \int_0^t \int_0^t q_2(u_1, u_2)\, du_1\, du_2 \tag{2.5.17}$$

A further parameter, based on expression (2.5.14), is the *renewal function*

$$\begin{aligned} U(u; t) &= E\{N(t, t+u] \mid N\{t\} = 1\} \\ &= \left\{\int_t^{t+v} p_2(v, t)\, dv\right\} \Big/ p_1(t) \end{aligned} \tag{2.5.18}$$

giving the expected number of points (or renewals) within u time units of a point at t. The higher order factorial moments

$$E\{N(t, t+u]^{(K)} \mid N\{t\} = 1\}, \qquad K = 1, 2, \ldots$$

find some use in practice as well. They may be determined by integrating the conditional product densities (2.2.11).

Certain other parameters based on the Palm process $N(\cdot \mid t)$ may be defined. The *survivor function* (or distribution of lifetime) is given as

$$\begin{aligned} &\text{Prob}\{\text{next point from an event at } t \text{ occurs after } u \text{ time units}\} \\ &= \text{Prob}\{N(t, t+u] = 0 \mid N\{t\} = 1\} \\ &= 1 - G(u; t) \end{aligned} \tag{2.5.19}$$

where $G(u; t) = \text{Prob}\{N(t, t+u] \geq 1 \mid N\{t\} = 1\}$. The *hazard function* or *force of mortality* is given by

$$\begin{aligned}\mu(u; t) &= \text{Prob}\{dN(t+u) = 1 \mid N\{t\} = 1 \quad \text{and} \quad N(t, t+u] = 0\}/dt \\ &= g(u; t)/[1 - G(u; t)] \qquad (2.5.20)\end{aligned}$$

where g is the derivative of G. The inverse relation to this last expression is

$$1 - G(u; t) = \exp\left\{-\int_t^{t+u} \mu(v; t)\, dv\right\}$$

These parameters are often used in reliability theory and population mathematics. Many specific functional forms have been proposed for them. The probability (2.5.19) is an example of the general Palm probability

$$\text{Prob}\{N(t, t+u] = n \mid N\{t\} = 1\}$$

As suggested by the result (2.2.10), these probabilities are directly related to unconditional probabilities. Further details concerning this are given by Cramér *et al.* (1971) and Daley and Vere-Jones (1972).

The parameters introduced so far have been based directly on the step functions $N(t)$. Certain other parameters are more easily introduced through the sequence τ_j, $j = 0, 1, 2, \ldots$, of locations of points and the sequence $y_j = \tau_{j+1} - \tau_j$, $j = 1, 2, \ldots$, of interpoint distances. The *forward recurrence time* is the variate $\tau_{N(t)} - t$. Its distribution is given by

$$\begin{aligned}\text{Prob}\{\tau_{N(t)} - t \leq u\} &= 1 - \text{Prob}\{N(t, t+u] = 0\} \\ &= 1 - \phi((t, t+u]) \qquad (2.5.21)\end{aligned}$$

This distribution is useful in extrapolating the behavior of the process ahead of t. Expression (2.5.21) may be given in terms of the product densities as

$$1 - \sum_{k=0}^{\infty} \frac{(-1)^k}{k!} \int_t^{t+u} \cdots \int p_k(t_1, \ldots, t_k)\, dt_1 \cdots dt_k$$

In the reverse direction, one has the *backward recurrence time* $t - \tau_{N(t)-1}$. Its distribution is given by

$$\text{Prob}\{t - \tau_{N(t)-1} \leq v\} = 1 - \text{Prob}\{N[t - v, t) = 0\}$$

The joint distribution of the two variates $\tau_{N(t)} - t$ and $t - \tau_{N(t)-1}$ is sometimes of use, as is the distribution of their sum $\tau_{N(t)} - \tau_{N(t)-1}$. The distribution of this last variate should not be confused with that of any of the $\tau_{j+1} - \tau_j$ unconditionally or based on a Palm measure.

Using the individual times of events, an alternative expression may be given for (2.5.14). Let $\sigma_1, \sigma_2, \ldots$ denote the times of successive points following t_1 for the Palm process $N(t\,|\,t_1)$. Let $g_k(\sigma; t)$ denote the probability density of σ_k. Then it is clear from first principles that

$$\text{Prob}\{\text{point in } (t, t+dt)\,|\,\text{point at } t_1\} = \sum_k g_k(t; t_1)\,dt$$

and so

$$\frac{p_2(t_1, t)}{p_1(t_1)} = \sum_k g_k(t; t_1) \tag{2.5.22}$$

In the stationary case, the product densities and the cumulant densities depend on one less parameter than in the general case. Specifically,

$$E\{dN(t+u_1)\cdots dN(t+u_{K-1})\,dN(t)\}$$
$$= p_K(u_1, p_{pK}(u_1, \ldots, u_{K-1})\,du_1 \cdots du_{K-1}\,dt \tag{2.5.23}$$
$$\text{cum}\{dN(t+u_1), \ldots, dN(t+u_{K-1}), dN(t)\}$$
$$= q_K(u_1, \ldots, u_{K-1})\,du_1 \cdots du_{K-1}\,dt \tag{2.5.24}$$

for $u_1, \ldots, u_{K-1}$, 0 distinct and $K = 1, 2, \ldots$. This reduction has the important implication that plausible estimates of the parameters can now be based on single realizations of the process. For example, the mean rate may be estimated by $\hat{p}_1 = N(T)/T$ given the data stretch $N(t)$, $0 < t \leq T$. As in the case of ordinary time series, the first- and second-order parameters seem to be the most important ones in practice. It is convenient to describe these by

$$E\,dN(t) = p_1\,dt = q_1\,dt$$
$$E\{dN(t+u)\,dN(t)\} = \{\delta(u)p_1 + p_2(u)\}\,du\,dt$$
$$\text{cov}\{dN(t+u), dN(t)\} = \{\delta(u)q_1 + q_2(u)\}\,du\,dt \tag{2.5.25}$$

In the case in which

$$\int_{-\infty}^{\infty} |q_2(u)|\,du < \infty \tag{2.5.26}$$

the *power spectrum* of the point process N may be defined to be

$$f_2(\lambda) = (2\pi)^{-1}\left[\int_{-\infty}^{\infty} \exp\{-i\lambda u\}\,\text{cov}\{dN(t+u), dN(t)\}\right/dt$$
$$= (2\pi)^{-1}q_1 + (2\pi)^{-1}\int_{-\infty}^{\infty} \exp\{-i\lambda u\}q_2(u)\,du \tag{2.5.27}$$

for $-\infty < \lambda < \infty$. The parameter λ here is called the *frequency*. One important manner in which the power spectrum of a process satisfying (2.5.26) differs from that of an ordinary series satisfying (2.4.18) is that

$$\lim_{\lambda \to \infty} f_2(\lambda) = q_1/2\pi \neq 0 \tag{2.5.28}$$

The power spectrum of a point process is similar to that of an ordinary time series in that it is symmetric, $f_2(-\lambda) = f_2(\lambda)$, and nonnegative. Inverse relations to the definition (2.5.27) are provided by (2.5.27) and

$$q_2(u) = \int_{-\infty}^{\infty} \exp\{iu\lambda\}[f_2(\lambda) - q_1/2\pi]\, d\lambda$$

In the higher order case one sets down the assumption

$$\int \cdots \int |q_K(u_1, \ldots, u_{K-1})|\, du_1 \cdots du_{K-1} < \infty \tag{2.5.29}$$

and the definition of the *cumulant spectrum*

$$f_K(\lambda_1, \ldots, \lambda_{K-1}) = (2\pi)^{-K+1}\left[\int \cdots \int \exp\{-i(\lambda_1 u_1 + \cdots + \lambda_{K-1}u_{K-1})\} \times \operatorname{cum}\{dN(t+u_1), \ldots, dN(t+u_{K-1}), dN(t)\}\right] \Big/ dt \tag{2.5.30}$$

$-\infty < \lambda_K < \infty$, $K = 2, 3, \ldots$. The definition is completed by setting $f_1 = p_1$. From (2.5.10) it is clear that the cumulant spectra may also be defined as "coefficients" in the expansion

$$\log E \exp\left\{\int \xi(t)\, dN(t)\right\} = \sum_{k=0}^{\infty} \frac{1}{k+1!} \int \cdots \int \hat{\xi}(\lambda_1) \cdots \hat{\xi}(\lambda_k)\hat{\xi}(-\lambda_1 - \cdots - \lambda_k) \times f_{k+1}(\lambda_1, \ldots, \lambda_k)\, d\lambda_1 \cdots d\lambda_k$$

where $\hat{\xi}(\lambda) = \int \exp\{i\lambda t\}\xi(t)\, dt$.

In the general case in which $N(t)$ is continuous in mean square, Kolmogorov (1940) has developed the representations

$$\operatorname{cov}\{N(t+u), N(t)\} = \int_{-\infty}^{\infty} \frac{\exp\{iu\lambda\} - 1}{i\lambda} F(d\lambda) \tag{2.5.31}$$

and

$$N(t) = \int_{-\infty}^{\infty} \frac{\exp\{it\lambda\} - 1}{i\lambda} dZ(\lambda) \tag{2.5.32}$$

where $F(d\lambda)$ is a measure satisfying $\int (1 + \lambda^2)^{-1} F(d\lambda) < \infty$ and $Z(\lambda)$ is a random process with the properties

$$E\, dZ(\lambda) = \delta(\lambda) p_1\, d\lambda$$

$$\operatorname{cov}\{dZ(\lambda_1), dZ(\lambda_2)\} = \delta(\lambda_1 - \lambda_2)F(d\lambda_1)\, d\lambda_2$$

In the case in which (2.5.29) holds, $Z(\lambda)$ has the further properties

$$\operatorname{cum}\{dZ(\lambda_1), \ldots, dZ(\lambda_K)\} = \delta(\lambda_1 + \cdots + \lambda_K) f_K(\lambda_1, \ldots, \lambda_{K-1})\, d\lambda_1 \cdots d\lambda_K \tag{2.5.33}$$

$K = 1, 2, \ldots$. This relationship is identical to the corresponding result (2.4.22) for ordinary time series. Expression (2.5.33) provides one possible means of interpreting the cumulant spectra of a point process.

The spectral representation (2.5.32) gives a means of interpreting certain ordinary series associated with point processes. Namely, if

$$X(t) = \int a(t-u)\, dN(u)$$

then

$$X(t) = \int \exp\{it\lambda\} A(\lambda)\, dZ(\lambda)$$

where $A(\lambda) = \int \exp\{-i\lambda t\} a(t)\, dt$. Using (2.5.33) it is now clear that the Kth-order cumulant spectrum of the series X is given by

$$A(\lambda_1) \cdots A(\lambda_{K-1}) A(-\lambda_1 - \cdots - \lambda_{K-1}) f_K(\lambda_1, \ldots, \lambda_{K-1}) \qquad (2.5.34)$$

It is seen that if $A(\cdot)$ does not vanish, then estimates of the cumulant spectra of N may be determined directly from estimates of the cumulant spectra of X.

In the case of a stationary point process, a whole new family of cumulant functions and spectra may be associated with the process through the Palm measure. It was seen in Section 2.2 that the random sequence $y_j = \tau_{j+1} - \tau_j$ of interpoint distances is a stationary sequence for the Palm measure $P(\cdot\,|\,0)$. From (2.2.12) or (2.2.20) the mean level of this sequence may be seen to be $Ey_j = 1/p_1$. As would be expected a high rate for the point process corresponds to a small average between-point distance.

The second-order product density and the power spectrum of the process N may be determined from the characteristics of the time series y_j, $j = 1, 2, \ldots$, via the relationship (2.5.22). The autocovariance function and the power spectrum of the series y_j, $j = 1, 2, \ldots$, may be determined from the characteristics of the process N via the relationship (2.2.12). These matters are discussed by Cox and Lewis (1966, Section 4.6). The relationships are most useful in the case of particular processes, for example renewal processes. Data benefit from an analysis both in terms of the step function $N(t)$ and the interpoint distances y_j. It must be remembered, however, that the latter is stationary for the measure $P(\cdot\,|\,0)$, and in practice it is the process $N(t)$ which is observed, not $N(\cdot\,|\,0)$. However, the two processes lead to the same asymptotic results, so in situations in which the data set is large, the sequence y_j, $j = 1, 2, \ldots$, may typically be treated as stationary.

2.6. Operations on Time Series

A variety of physical operations are applied to time series before and after their collection. Sometimes these operations are applied deliberately, but sometimes not. In any case, it is important that the effect of the operations be understood to the greatest degree possible.

One very common operation is that of forming some linear combination

$$Y = \int \xi(t) X(t)\, dt$$

for a suitable function $\xi(\cdot)$. The variate Y has already appeared in the definition of the characteristic functional (2.1.4). It is apparent that Y is well defined when, for example, $\int |\xi(t)| E\{|X(t)|\}\, dt < \infty$. The moments and cumulants of Y are given by

$$EY^K = \int \cdots \int \xi(t_1) \cdots \xi(t_K) m_K(t_1, \ldots, t_K)\, dt_1 \cdots dt_K$$

$$\text{cum}_K\, Y = \int \cdots \int \xi(t_1) \cdots \xi(t_K) c_K(t_1, \ldots, t_K)\, dt_1 \cdots dt_K \qquad (2.6.1)$$

when they exist.

The preceding operation formed a real-valued variate from a given time series. By far the most important operations form time series from given time series. Consider, for example, the linear operation

$$Y(t) = \int a(t, u) X(u)\, du \qquad (2.6.2)$$

for a suitable function $a(t, u)$. The moment and cumulant functions of this new series are given by

$$\int \cdots \int a(t_1, u_1) \cdots a(t_K, u_K) m_K(u_1, \ldots, u_K)\, du_1 \cdots du_K$$

$$\int \cdots \int a(t_1, u_1) \cdots a(t_K, u_K) c_K(u_1, \ldots, u_K)\, du_1 \cdots du_K \qquad (2.6.3)$$

while the characteristic functional is given by

$$E \exp\left\{i \int Y(t)\, d\theta(t)\right\} = C\left[\int^t \left\{\int a(u, v)\, d\theta(u)\right\} dv\right] \qquad (2.6.4)$$

in terms of the characteristic functional of the series $X(t)$. The characteristics of the two series are seen to be directly connected in this case.

A related linear operation is described by requiring that the series $Y(t)$ satisfies

$$\int b(t, u) Y(u)\, du = X(t) \qquad (2.6.5)$$

for some function $b(t, u)$, given the series $X(t)$. For example, the series $Y(t)$ might be defined to be the solution of the differential equation

$$\sum_{k=0}^{K} \beta_k \frac{d^k Y(t)}{dt^k} = X(t)$$

in the case in which $Y(t)$ is K times differentiable. This is of the form (2.6.5) with

$$b(t, u) = \sum_{k=0}^{K} \beta_k \frac{d^k \delta(t-u)}{dt^k}$$

In the case in which there exists $a(v, t)$ such that

$$\int a(v, t) b(t, u)\, dt = \delta(v - u)$$

the solution of (2.6.5) may be written as (2.6.2) and the relations (2.6.3) and (2.6.4) are seen to apply to this series $Y(t)$ also.

On other occasions, the series $Y(t)$ may be determined from the series $X(t)$ through a polynomial relationship of the form

$$Y(t) = \sum_{j=0}^{J} \int \cdots \int a_j(t, u_1, \ldots, u_j) X(u_1) \cdots X(u_j)\, du_1 \cdots du_j \quad (2.6.6)$$

(see Wiener, 1958). Relations analogous to (2.6.3) may be written down for the series $Y(t)$ in the case in which the series $X(t)$ is Gaussian; however, matters are complicated in the non-Gaussian case (see Shiryaev, 1963).

The operations introduced so far have been smooth in character. On occasion one wishes to apply a strongly nonlinear operator, for example, $Y(t) = G[X(t)]$, with $G[\cdot]$ discontinuous. An important example is the *hard limiter*

$$\begin{aligned} Y(t) &= \ \ 1 \quad \text{if} \quad X(t) > 0 \\ &= -1 \quad \text{if} \quad X(t) < 0 \end{aligned}$$

or the *discretizer*

$$Y(t) = j \qquad \text{if} \quad jh \le X(t) < (j+1)h$$

given $h > 0$, for $j = 0, \pm 1, \ldots$. Such operations must be investigated separately of any general theory.

The functions $a(\cdot)$ of (2.6.2), $b(\cdot)$ of (2.6.5), and $a_j(\cdot)$ of (2.6.6) have been assumed nonstochastic. On occasion it is appropriate to consider them to be stochastic and to consider series $Y(t)$ defined by expressions of the form

$$Y(t) = \int a(t, u, \omega) X(u, \omega)\, du$$

Expressions for the characteristics of the series $Y(t)$ in terms of those of the series $X(t)$ now appear to exist only for very specific series and operations.

An operation of alternate character from those considered so far is that of time substitution, specifically

$$Y(t) = X(\Gamma(t))$$

where Γ is a nondecreasing function with domain $[0, \infty)$. In the case in which Γ is linear, the series $Y(t)$ behaves essentially like the series $X(t)$. Nonlinear Γ leads to quite drastically altered characteristics on many occasions, especially when Γ is a random nonlinear function.

So far the discussion has been concerned with operations on a single real-valued series. In practice one often has to consider operations on vector-valued series. As in the case of a single series, linear operations may be described quite succinctly. For example, suppose the characteristic functional is given for the r vector-valued series $\{X_1(t), \ldots, X_r(t)\}$, namely

$$C[\theta_1, \ldots, \theta_r] = E \exp\left\{i \int X_1(t)\, d\theta_1(t) + \cdots + i \int X_r(t)\, d\theta_r(t)\right\} \tag{2.6.7}$$

Then the generalization of the relationship (2.6.4) to a series such as

$$Y(t) = \int a_1(t, u)X_1(t)\, dt + \cdots + \int a_r(t, u)X_r(t)\, dt$$

is immediately apparent. In particular, the characteristic functional of the superposed series

$$Y(t) = X_1(t) + \cdots + X_r(t)$$

is $C[\theta, \ldots, \theta]$. In the case in which the series are independent and identically distributed it is $C[\theta]^r$. In many circumstances, this latter result may be used to prove that the superposed series is asymptotically Gaussian as $r \to \infty$.

An important class of operations is now introduced. Consider an operation $\mathscr{A}$ carrying real-valued functions $X(t)$, $-\infty < t < \infty$, over into real-valued functions $Y(t)$, $-\infty < t < \infty$, with the properties of (i) linearity

$$\mathscr{A}[\alpha_1 X_1 + \alpha_2 X_2](t) = \alpha_1 \mathscr{A}[X_1](t) + \alpha_2 \mathscr{A}[X_2](t)$$

and (ii) time invariance

$$\mathscr{A}[S^u X](t) = \mathscr{A}[X](t + u)$$

where S^u denotes the shift operator $[S^u X(t) = X(t + u)]$. Then $\mathscr{A}$ has the property of carrying complex exponentials, $\exp\{i\lambda t\}$, over into complex exponentials, specifically $\mathscr{A}[e](t) = A(\lambda)e(t)$, $-\infty < t < \infty$, for $e(t) = \exp\{i\lambda t\}$ (e.g., see Brillinger, 1975a, Lemma 2.7.1). An operation with properties (i) and (ii) is called a (linear) *filter*. The function $A(\lambda)$, $-\infty < \lambda < \infty$, is called the *transfer function* of the filter, and it is generally complex-valued. Its amplitude, $|A(\lambda)|$, is called its *gain*, and its argument, $\arg A(\lambda)$, is called its *phase*. Many of the operations applied to a time series, by an analyst or nature, seem to be well approximated by filters.

An important class of filters of ordinary time series takes the form

$$Y(t) = \int X(t-u)\, da(u) \tag{2.6.8}$$

for some definition of the integral. The transfer function of this filter is

$$A(\lambda) = \int \exp\{-i\lambda u\}\, da(u) \tag{2.6.9}$$

If $a(u)$ is differentiable, its derivative is called the *impulse response* of the filter, since it is the output series when $X(t) = \delta(t)$ is the input. Important filters include the *bandpass filter* at frequency ν with bandwidth Δ and gain A where

$$\begin{aligned} A(\lambda) &= A \qquad \text{for} \quad |\lambda \pm \nu| < \Delta/2 \\ &= 0 \qquad \text{otherwise} \end{aligned} \tag{2.6.10}$$

and the *Hilbert transform*, with

$$\begin{aligned} A(\lambda) &= -i, \qquad \lambda > 0 \\ &= 0, \qquad \lambda = 0 \\ &= i, \qquad \lambda < 0 \end{aligned}$$

If the series $X(t)$ has a spectral representation

$$X(t) = \int \exp\{it\lambda\}\, dZ(\lambda) \tag{2.6.11}$$

for some process $Z(\lambda)$, then the output of the filter with transfer function $A(\lambda)$ has representation

$$Y(t) = \int \exp\{it\lambda\} A(\lambda)\, dZ(\lambda)$$

The output of the bandpass filter (2.6.10) with small Δ is seen to be approximately $A[\exp\{it\nu\}\, \Delta Z(\nu) + \exp\{-it\nu\}\, \Delta Z(-\nu)]$.

In the case of a stationary series $X(t)$, it is apparent that the series $Y(t)$ is also stationary with mean $A(0)EX(t)$ and cumulant spectra

$$A(\lambda_1) \cdots A(\lambda_{K-1}) A(-\lambda_1 - \cdots - \lambda_{K-1}) f_K(\lambda_1, \ldots, \lambda_{K-1}) \tag{2.6.12}$$

In particular, the power spectrum of the series $Y(t)$ is $|A(\lambda)|^2 f_2(\lambda)$. If

$$Y_k(t) = \int \exp\{it\lambda\} A_k(\lambda)\, dZ(\lambda)$$

$k = 1, \ldots, K$, then

$$\text{cum}\{Y_1(t), \ldots, Y_K(t)\} = \int \cdots \int A_1(\alpha_1) \cdots A_{K-1}(\alpha_{K-1}) A_K(-\alpha_1 - \cdots - \alpha_{K-1}) \times f_K(\alpha_1, \ldots, \alpha_{K-1})\, d\alpha_1 \cdots d\alpha_{K-1} \tag{2.6.13}$$

This last result provides an interpretation of the Kth-order cumulant spectrum f_K. Let $A_k(\cdot)$ be the transfer function of a bandpass filter at frequency λ_k, with small bandwidth Δ, $k = 1, \ldots, K$, and suppose $\lambda_1 + \cdots + \lambda_K = 0$. Then the cumulant of (2.6.13) is seen to be proportional to $\text{Re}\{f_K(\lambda_1, \ldots, \lambda_{K-1})\}$. In words, the real part of the Kth-order cumulant spectrum at $(\lambda_1, \ldots, \lambda_{K-1})$ may be interpreted as the scaled Kth-order joint cumulant of the output of a bank of narrow bandpass filters at frequencies $\lambda_1, \ldots, \lambda_K$ where $\lambda_1 + \cdots + \lambda_K = 0$. In particular, the power spectrum at frequency λ may be interpreted as the scaled variance of the output of a narrow bandpass filter at frequency λ. The imaginary part of f_K may be obtained by taking the transfer function of one of the filters to be the product of the Hilbert transform and the previous narrow bandpass filter and once again considering the joint cumulant (2.6.13).

An operation somewhat out of the ordinary may be defined by choosing $Y(t)$ to be the value that minimizes

$$\int_{t-U}^{t+U} |X(v) - Y(t)|\, dv$$

for some U. The series $Y(t)$ could be called the *running median* of the series $X(t)$.

2.7. Operations on Point Processes

In general, it appears that the class of interesting operations is much larger in the case of point processes than in the case of time series. Another distinction is that the typical operation appears to have stochastic character. The description of a specific operation is sometimes best given in terms of the step functions $N(t)$, sometimes best in terms of the sequence $\tau_0, \tau_1, \ldots$ of locations of points, sometimes in terms of the series $y_1, y_2, \ldots$ of intervals between events, and sometimes in terms of $\gamma(t, \omega)$, the conditional intensity function. The concern in this section is with operations carrying point processes over into point processes.

The analog of the linear operation (2.6.2) seems to be one carrying a process with points at $\tau_0, \tau_1, \ldots$ over into one with points at $\tau_j + u_i(\tau_j)$, $i = 1, 2, \ldots, I, j = 0, 1, \ldots$, for given functions $u_i(t)$. A representation for the new process is

$$\sum_j \sum_i \delta(t - \tau_j - u_i(\tau_j)) \tag{2.7.1}$$

the mean rate for which is

$$\sum_i p_1(t - u_i(t))$$

and the higher order product densities are given by

$$\sum_{i_1} \cdots \sum_{i_K} p_K(t_1 - u_{i_1}(t_1), \ldots, t_K - u_{i_K}(t_K))$$

In the case in which $I = 1$, this operation is called a *displacement*. In the case in which the $u_i(t)$ do not depend on t, the image series will be stationary when the domain series is.

This particular operation does not seem to have a great deal of use in practice; however, its stochastic analog, with the $u_i(t)$ random, is very important. For example, consider the case in which the point τ_j is replaced by the points $\tau_j + u_{ji}$, $i = 1, \ldots, I_j$, $j = 0, 1, \ldots$, with the u_{ji} and the I_j random. In the case in which $I_j = 1$, the operation might be thought of as corresponding to the action of a service system, with no waiting time, in which the τ_j are the arrival times of customers, u_j the service time of the jth customer, and $\tau_j + u_j$ his departure time. The process of points $\{\tau_j + u_{ji}, i = 1, \ldots, I_j, j = 0, 1, \ldots\}$ is called a *cluster process*. If $N^j(A)$ denotes the number of u_{ji} in the set A for $i = 1, \ldots, I_j$, then $N^j(\cdot)$ is seen to be a point process and the cluster process is seen to have the representation

$$\sum_j N^j(A - \tau_j) \tag{2.7.2}$$

Supposing that the process $N^j(\cdot)$ has intensity function $p_1{}^*(t)$ for each j and is independent of the process N, the process (2.7.2) is seen to have intensity function

$$\int p_1(u) p_1{}^*(t - u)\, du$$

This process is well defined if this integral is finite. In the case in which the process $N^j(\cdot)$ has probability generating functional $G^*[\xi]$ for each j, and where $N(\cdot)$, $N^0(\cdot)$, $N^1(\cdot), \ldots$ is a sequence of independent processes, the process (2.7.2) is seen to have probability generating functional

$$G[G^*[S^t\xi]] \tag{2.7.3}$$

where S^t is the shift operator. Expression (2.7.3) shows directly that the process (2.7.2) is stationary when the process N is.

A useful operation in the study of point processes is that of time substitution involving the replacement of the process $N(t)$ by the process $N(\Gamma(t))$, for $\Gamma(t)$ a nondecreasing function, which may either be fixed or stochastic. If the process $N(t)$ jumps at the points τ_j, then the image process jumps at the points

$$\sigma_j = \inf\{\sigma: \Gamma(\sigma) = \tau_j\} = \Gamma^{-1}(\tau_j)$$

In Section 2.2, it was seen that a random time substitution could, in certain circumstances, transform a given process into a Poisson process of rate 1. The intervals between the points of the two processes are related by

$$\tau_{j+1} - \tau_j = \Gamma(\sigma_{j+1}) - \Gamma(\sigma_j)$$

In the case in which the rate of the transformed process is high, this last relationship may be written approximately

$$\tau_{j+1} - \tau_j = \gamma(\sigma_j)(\sigma_{j+1} - \sigma_j) \tag{2.7.4}$$

In the case in which $\Gamma(\cdot)$ is deterministic, the product densities of the image process are given by

$$p_K(\Gamma(t_1), \ldots, \Gamma(t_K))\gamma(t_1) \cdots \gamma(t_K) \tag{2.7.5}$$

where $\gamma(t)$ is the derivative of $\Gamma(t)$. The probability generating functional of the image process is given by $G[\xi \circ \Gamma^{-1}]$. It is clear that, in this case, unless $\Gamma(\cdot)$ is linear, a stationary process will not be transformed into a stationary process.

A very useful means of generating further point processes from given point processes is to employ a stochastic $\Gamma(\cdot)$, independent of $N(\cdot)$. This particular procedure leads to the doubly stochastic Poisson when N is Poisson. From expression (2.7.5), the product densities of the transformed process are seen to be given by

$$E\{p_K(\Gamma(t_1), \ldots, \Gamma(t_K))\gamma(t_1) \cdots \gamma(t_K)\} \tag{2.7.6}$$

The time substitution of Section 2.2 depended on the process N in a very direct manner, however, and an expression such as (2.7.6) does not apply. An important use of time substitutions is in the transformation of a given nonstationary point process into one that is (approximately) stationary.

Moving on to a different class of transformations, suppose that a process $N(\cdot)$ has the representation

$$\sum_j \delta(t - \tau_j) \tag{2.7.7}$$

Consider the process with representation

$$\sum_j I(\tau_j)\,\delta(t - \tau_j) \tag{2.7.8}$$

where $I(\tau_j)$ is a random indicator variable taking on the value 0 or 1. The operation of forming the process (2.7.8) from the process (2.7.7) is called *thinning, skipping*, or *random deletion*. If $E\{I(\tau_j)\,|\,\tau_j\} = \pi(\tau_j)$, then the intensity function of the process (2.7.8) is given by

$$p_1(t)\pi(t) \tag{2.7.9}$$

If, in addition, $I(\tau_0)$, $I(\tau_1)$, ... given N is a sequence of independent random variables, then the probability generating functional of the process (2.7.8) is given by

$$\begin{aligned} E \prod \{\xi(\tau_j)^{I(\tau_j)}\} &= E \prod_N \{\xi(\tau_j)\pi(\tau_j) + 1 - \pi(\tau_j)\} \\ &= G[\xi\pi + 1 - \pi] \end{aligned} \tag{2.7.10}$$

Defining the factorial cumulant generating functional of the process by

$$H[\xi] = \log G[1 + \xi] = \sum_{k>0} \frac{1}{k!} \int \cdots \int \xi(t_1) \cdots \xi(t_k) q_k(t_1, \ldots, t_k)\, dt_1 \cdots dt_k$$

the effect of the operation may be described more directly by

$$H[\xi] \to H[\pi\xi] \tag{2.7.11}$$

The cumulant densities are given, in terms of those of the process N, by

$$\pi(t_1) \cdots \pi(t_K) q_K(t_1, \ldots, t_K) \tag{2.7.12}$$

A related operation is that of *censoring* in which, for example, only every other point is retained. In terms of the function $I(\cdot)$, this operation might correspond to requiring $I(\tau_{j+1}) = 1 - I(\tau_j)$.

An important transformation of a point process occurs when one is fed into a physical counter meant to record the times of its events. Many counters have the unpleasant but not unexpected property of becoming inoperative for a brief period (called the *dead time*) following each registration of an event. This property is analogous to the one for recording apparatus for ordinary time series that do not respond if the series is changing too rapidly. Two important classes of counters are usually distinguished. *Type I* has the property of no output events occurring during the dead time whatever. *Type II* is such that events occurring during the dead time make the dead time begin again (the counter is paralyzable). Applying the operation of the type

II counter is one means used to decluster point process realizations (for a seismological example, see Udias and Rice, 1975). In analytic terms, a type I counter may be described by

$$\begin{aligned} \text{Prob}\{dN(t) = 1 \mid N(u), M(u); u \leq t\} &= \gamma(t, \omega)\, dt, \qquad \tau_{N(t)-1} + \Delta \leq t \\ &= 0 \qquad \text{otherwise} \end{aligned} \tag{2.7.13}$$

where M denotes the input series, N the recorded series, Δ the dead time, and $\gamma(t, \omega)$ the conditional intensity of M. The operation corresponds to modulation by a 0–1 function. Clearly $p_2(u) = 0$ for $|u| \leq \Delta$ here. A type II counter may be described by

$$\begin{aligned} \text{Prob}\{dN(t) = 1 \mid N(u), M(u); u \leq t\} &= \gamma(t, \omega), \qquad \sigma_{M(t)-1} + \Delta \leq t \\ &= 0 \qquad \text{otherwise} \end{aligned}$$

with $\sigma_0, \sigma_1, \ldots$ denoting the times of input events. It is clear that the rate of the output of a type II counter may be determined from

$$\text{Prob}\{dN(t) = 1\} = \text{Prob}\{M(t - \Delta, t] = 0 \text{ and } dM(t) = 1\}$$

as

$$p_1{}^*(t) = \sum_{k=0}^{\infty} \frac{(-1)^k}{k!} \int_{t-\Delta}^{t} \cdots \int p_{k+1}(u_1, \ldots, u_k, t)\, du_1 \cdots du_k$$

using the expansion (2.2.7). The higher order product densities may be determined in a similar manner.

An interesting aspect of the operations just given is that the sample path of the transformed process is in each case absolutely continuous with respect to the path of the original process.

A further form of transformation occurs when a point process is passed through a service system, i.e., the points of the original process are envisaged as the times at which customers arrive at a queueing system with, say, k servers. If all the servers are busy, a customer must wait for a free server. When a server becomes available, the customer first in line experiences a random service time. The image process of the transformation is taken to be the point process corresponding to the exit times of the customers. In the case in which there are an infinite number of servers, the output process is the cluster process mentioned earlier, with one member per cluster. There is an extensive literature concerning queueing theory of particular situations, e.g., Cox and Smith (1961), Benes (1963), and Prahbu (1965).

If τ_j denotes the arrival time of the jth customer, ω_j his waiting time, and σ_j his service time, then the transformed process may be represented by

$$\sum_j \delta(t - \tau_j - \omega_j - \sigma_j)$$

Generally the ω_j have quite complicated stochastic structure and this is what leads to difficulties in the analysis of queues. If $M(t)$ denotes the arrival process and $N(t)$ the departure, then the number of customers in the system at time t is $M(t) - N(t)$. The number being served at time t is $\min\{M(t) - N(t), k\}$. The times at which the number of customers in the system falls to 0 are important in the analysis. They are called *points of regeneration*, since the system starts anew at these times (in a certain sense).

Suppose next that a point process M has conditional intensity function defined by

$$\begin{aligned}\gamma(t, \omega)\, dt &= \operatorname{Prob}\{dM(t) = 1 \mid M(u), u \le t\} \\ &= \gamma(t; \sigma_0, \ldots, \sigma_{M(t)-1})\, dt\end{aligned}$$

where $\sigma_0, \sigma_1, \ldots$ are the locations of its points. Suppose $\beta(t, \omega) = \beta(t; \sigma_0, \ldots, \sigma_{M(t)-1})$ is a nonnegative function of the past of the process. The operation of forming a point process N with conditional intensity function

$$\beta(t, \omega)\gamma(t, \omega) \tag{2.7.14}$$

provided $E\{\beta(t, \omega)\gamma(t, \omega)\} < \infty$, is called *modulation*, discussed, e.g., by Cox (1972) and Varaiya (1975). From (2.2.16) the conditional distribution functions of the points $\tau_0, \tau_1, \ldots$ of the process N are given by

$$\begin{aligned}&\operatorname{Prob}\{\tau_J > t \mid \tau_j = t_j, j < J\} \\ &\quad = \exp\left\{-\int_{t_{J-1}}^{t} \beta(u; t_j, j < J)\gamma(u; t_j, j < J)\, du\right\}\end{aligned} \tag{2.7.15}$$

Equation (2.7.15) indicates a means by which a modulation process might be simulated given $\beta(t, \omega)$ and $\gamma(t, \omega)$. The standard Poisson process has $\gamma(t, \omega) \equiv 1$. Expression (2.7.12) shows that a point process with general conditional intensity $\beta(t, \omega)$, say, is a modulated version of the standard Poisson process. Cox (1972) considers the particular case of nonstochastic $\beta(t)$ in some detail. The joint probability distribution for the number of points of the process N in the interval $[0, T]$ and their locations may be determined from (2.7.15) and (2.2.17). Varaiya (1975) shows that, if the probability measures corresponding to the processes M and N are mutually absolutely continuous, then there exists a $\beta(t, \omega)$ of this character, in many circumstances.

He also shows that the likelihood ratio of the process N with respect to the process M, on the interval $[0, T]$, may be written

$$\exp\left\{\int_0^T \log \beta(t, \omega)\, dN(t, \omega) - \int_0^T [\beta(t, \omega) - 1]\gamma(t, \omega)\, dt\right\} \tag{2.7.16}$$

The operations considered so far have been defined on univariate point processes. In the case of vector-valued processes, $N_1(t), \ldots, N_r(t)$, referring to points of r different types, an important operation is that of *pooling* or *superposition*. Here one forms the process

$$N_1(t) + \cdots + N_r(t)$$

If the probability generating functional of the original process is

$$G[\xi_1, \ldots, \xi_r] = E \exp\left\{\int \log \xi_1(t)\, dN_1(t) + \cdots + \int \log \xi_r(t)\, dN_r(t)\right\} \tag{2.7.17}$$

then that of the superposed process is clearly $G[\xi, \ldots, \xi]$. In the case in which the component series are independent realizations of a process with probability generating functional $G[\xi]$, the superposed process has probability generating functional $G[\xi]^r$. In a variety of circumstances, this last result may be used to demonstrate that the limit, as $r \to \infty$, of superposed and re-time-scaled point processes is a Poisson process (e.g., see Vere-Jones, 1968). The conditional intensity is here the sum of the individual intensities.

Finally, it is clear that many of the operations of discrete time series analysis may be applied to the interval series $\{y_1, y_2, \ldots\}$ in order to obtain a further interval series and in consequence a new point process.

2.8. The Identification of Time Series Systems

By a *time series system* is meant the collection of a space of input series, a space of output series, and an operation carrying an input series over into an output series. A common form of a time series system is provided by the specification

$$Y(t) = \mathscr{A}[X](t) + \varepsilon(t) \tag{2.8.1}$$

where $\mathscr{A}[\cdot]$ is a deterministic operator of the kind discussed in Section 2.6, X denotes an input series, Y the output series, and ε an unobservable stochastic error series. The problem of system identification is that of determining $\mathscr{A}$, or essential properties of $\mathscr{A}$, from a stretch of input and corresponding output data $\{X(t), Y(t)\}$, $0 < t < T$. Part of the motivation for studying this problem is the desire to be able to indicate properties of output series corresponding to specific input series. A system is called *time invariant* if the bivariate series $\{X(t), Y(t)\}$ is stationary when the series $X(t)$ is stationary. A system is called *causal* when $\mathscr{A}[X](t)$ depends only on $X(u)$, $u \leq t$, and the future error $\varepsilon(v)$, $v > t$, is independent of the past $\{X(u), Y(u)\}$, $u \leq t$.

In a few circumstances the identification problem has a fairly direct solution. For example, consider the (regression) system

$$Y(t) = \mu + \int X(t-u)\, da(u) + \varepsilon(t) \tag{2.8.2}$$

where $\varepsilon(t)$ is a zero mean stationary series independent of the stationary series $X(t)$. Then (2.8.2) leads to the relationship

$$f_{11}(\lambda) = A(-\lambda) f_{20}(\lambda) \tag{2.8.3}$$

where $A(\lambda) = \int \exp\{-i\lambda u\}\, da(u)$, $f_{20}(\lambda)$ is the power spectrum of the series X, and

$$f_{11}(\lambda) = (2\pi)^{-1} \int \exp\{-i\lambda u\} \operatorname{cov}\{X(t+u), Y(t)\}\, du \tag{2.8.4}$$

is the *cross spectrum* of the series X with the series Y. Expression (2.8.3) indicates that the transfer function $A(\lambda)$ of the linear filter of (2.8.2) may be estimated once estimates of the second-order spectra $f_{20}(\lambda)$ and $f_{11}(\lambda)$ are available. (The construction of such estimates is described by Brillinger, 1975.) This method of system identification was proposed by Wiener (1949).

The relationship (2.8.3) has a direct extension to a nonlinear system in one case. Suppose the system is determined by

$$Y(t) = \mu + \int X(t-u)\, da(u) + \iint X(t-u)X(t-v)\, db(u, v) + \varepsilon(t) \tag{2.8.5}$$

where $\varepsilon(t)$ is a zero mean stationary series independent of the stationary Gaussian series $X(t)$. The relationship (2.8.3) still holds for this system. In addition, one has

$$f_{21}(\lambda, \nu) = 2B(-\lambda, -\nu) f_{20}(\lambda) f_{20}(\nu) \tag{2.8.6}$$

where $B(\lambda, \nu) = \iint \exp\{-i(\lambda u + \nu v)\}\, db(u, v)$ and

$$f_{21}(\lambda, \nu) = (2\pi)^{-1} \iint \exp\{-i(\lambda u + \nu v)\} \operatorname{cum}\{X(t+u), X(t+v), Y(t)\}\, du\, dv \tag{2.8.7}$$

is a third-order joint cumulant spectrum of the bivariate series $\{X(t), Y(t)\}$. The functions $A(\lambda)$, $B(\lambda, \nu)$ may therefore be estimated once estimates of the second- and third-order spectra of the series $\{X(t), Y(t)\}$ are available.

Consider, next, the system defined by the equations

$$U(t) = \int X(t-u)\, da(u)$$

$$V(t) = G[U(t)]$$

$$Y(t) = \mu + \int V(t-u)\, db(u) + \varepsilon(t)$$

where $\varepsilon(t)$ is again a zero mean stationary series independent of the stationary Gaussian series $X(t)$, and where $G[\cdot]$ is a function from reals to reals. The system (2.8.2) corresponds to $G[\cdot]$ and $b(\cdot)$ identities. When $G[u]$ is a quadratic in u, this system is a particular case of (2.8.5). Now, for jointly normal variates U, V, W, and $G[\cdot]$ such that

$$E|G[U][1+|U|+|V|+|W|+|VW|+|U|^2]| < \infty,$$
$$\operatorname{cov}\{V, G[U]\} = \operatorname{cov}\{V, U\} \operatorname{cov}\{U, G[U]\}/\operatorname{var} U$$
$$\operatorname{cum}\{W, V, G[U]\} = [\operatorname{cov}\{V, U\} \operatorname{cov}\{W, U\} \operatorname{cum}\{U, U, G[U]\}]/\operatorname{var}^2 U$$

Hence for the previous system, assuming the series $X(t)$ Gaussian

$$f_{11}(\lambda) = L_1 A(-\lambda) B(-\lambda) f_{20}(\lambda)$$
$$f_{21}(\lambda, \nu) = L_2 A(-\lambda) A(-\nu) B(-\lambda-\nu) f_{20}(\lambda) f_{02}(\nu)$$

where $L_1 = \operatorname{cov}\{U(0), V(0)\}/\operatorname{var} U(0)$ and

$$L_2 = \operatorname{cum}\{U(0), U(0), V(0)\}/\operatorname{var}^2 U(0).$$

When either of the filters $a(\cdot)$, $b(\cdot)$ is the identity, the transfer function of the other may therefore be determined, up to a constant of proportionality, by $f_{11}(-\lambda) f_{20}(\lambda)^{-1}$. In the general case, the previous relations lead to

$$|A(\lambda)| \propto \{|f_{XXY}(\lambda, -\lambda)|\}^{1/2}/f_{XX}(\lambda)$$

on setting $\nu = -\lambda$. If $\phi(\lambda) = \arg A(\lambda)$, $\psi(\lambda, \nu) = \arg\{f_{XXY}(\lambda, \nu)/f_{XY}(\lambda+\nu)\}$, then

$$\phi(\lambda) = \left\{2 \int_0^\lambda \phi(\alpha)\, d\alpha + \int_0^\lambda \psi(\alpha, \lambda-\alpha)\, d\alpha \right\} \Big/ \lambda$$

provides a recursive procedure for obtaining $\phi(\lambda)$ given $\phi(\alpha)$, $0 \le \alpha < \lambda$. Korenberg (1973) considered this system and developed related expressions for the case of $X(\cdot)$ Gaussian white noise and $G[u]$ a polynomial in u.

A classical procedure that has been used to estimate the transfer function $A(\lambda)$ of the system (2.8.2) is to take for input series $X(t) = \exp\{i\lambda t\}$ (i.e., separately input the series cos λt and sin λt). Then the output series is

$$Y(t) \doteq \mu + A(\lambda) \exp\{i\lambda t\}$$

and $A(\lambda)$, $\lambda \neq 0$, may be estimated by

$$\left[\int_0^T [Y(t) - \hat{m}] \exp\{-i\lambda t\}\, dt\right] \Big/ T$$

for example, where $\hat{m}$ is the mean of the Y terms. This procedure has the disadvantage of requiring the use of a whole family of input series, $\exp\{i\lambda t\}$, covering the λ domain.

A variant of this procedure exists for the identification of a system such as (2.8.5) in which one takes the series $\exp\{i\lambda t\} \pm \exp\{ivt\}$ as input (see Brillinger, 1970, and the references cited therein).

It is worth noting that the filter with transfer function $A(\lambda)$ determined by (2.8.3) has an alternative interpretation. Given input and output series $X(t)$, $Y(t)$ of a system, consider the problem of determining the best linear approximant of the system, i.e., finding the $a(u)$ that minimizes

$$E\left[Y(t) - \mu - \int X(t-u)\, da(u)\right]^2 \tag{2.8.8}$$

In the case in which $\{X(t), Y(t)\}$ is a stationary series, it may be seen that one wants to choose μ and $A(\lambda)$ to satisfy (2.8.3) and $EY(t) = \mu + A(0)EX(t)$. When dealing with nonlinear filters, it is often useful to consider their approximate linear effect.

A system is called a *linear dynamic system* if it has the representation

$$dS(t) = F(t)S(t)\, dt + G(t)X(t)\, dt + K(t)\, dW_1(t)$$
$$dY(t) = H(t)S(t)\, dt + J(t)X(t)\, dt + L(t)\, dW_2(t) \tag{2.8.9}$$

$0 \leq t < \infty$, where $F(t), \ldots, L(t)$ are fixed (matrix-valued) functions, $W_1(t)$, $W_2(t)$ independent noise series, $S(t)$ an unobservable (vector-valued) state series, $X(t)$ the input series, and $Y(t)$ the corresponding output series. A considerable literature exists concerning the theory of such systems (e.g., see the December, 1974, number of the *IEEE Transactions on Automatic Control*, Vol. 19). With the series and functions appearing vector-valued, the model is exceedingly general. The functions $F(t), \ldots, L(t)$ are parameters of the system. In some cases they are fixed functions of a common set of parameters. The problem of identification then is concerned with the estimation of the parameters of some canonical form of the system, given stretches of input and corresponding output. The description (2.8.9) is often called the *state space* or *state variable* model. The state process $S(t)$ is meant to represent the totality of information from the past and present input to be transmitted to the future output. In practice, one seeks minimal realizations of the system, wherein the dimension of $S(t)$ is as small as possible. An appealing aspect of the description (2.8.9) is its clear indication of the dynamic development of the system.

Much of the literature of state space models is concerned with obtaining expressions for the series

$$\hat{S}(t) = E\{S(t) \mid X(u), Y(u); u \leq t\}$$

In the case in which (i) $W_1(t)$, $W_2(t)$ are independent Wiener processes, (ii) $S(0)$ is a normal variate independent of the process $\{W_1, W_2\}$, (iii) the series $X(t)$ is fixed, the results of Kailath (1970) and Balakrishnan (1973) show that the likelihood ratio of the series $Y(t)$ relative to the process $L(t)W_2(t)$ is given by

$$\exp\left\{\int_0^T L(t)^{-2}(H(t)\hat{S}(t) + J(t)X(t))\,dt \right.$$
$$\left. - \tfrac{1}{2}\int_0^T (H(t)\hat{S}(t) + J(t)X(t))^2 L(t)^{-2}\,dt\right\} \qquad (2.8.10)$$

This variate is useful in problems of estimation and detection. An important role is played in the analysis of the system (2.8.9) by the *innovations process* defined as

$$d\nu(t) = dY(t) - H(t)\hat{S}(t)\,dt - J(t)X(t)\,dt$$

The identification of systems like (2.8.9) has generally proceeded by parametrizing them in some canonical manner, and then maximizing the likelihood or some other criterion function involving the parameters.

The system (2.8.9) is linear. In the case in which the $F(t), \ldots, L(t)$ are constant and the noise series W has stationary increments, the system is time invariant. When the noise processes are absent, the system corresponds to a linear filter with matrix-valued transfer function $J + H(i\lambda I - F)^{-1}G$. The various entries of this matrix are rational functions of λ.

There is one class of nonlinear systems that may be identified quite readily, when the experimenter has the freedom to use any input series. It is represented by

$$Y(t) = \sum_{K=0}^{\infty} \int \cdots \int_{u_k \text{distinct}} a_K(u_1, \ldots, u_K) X(t - u_1) \cdots X(t - u_K)\,du_1 \cdots du_K + \varepsilon(t) \qquad (2.8.11)$$

with $a_K(u_1, \ldots, u_K)$ symmetric. Suppose that the experimenter takes as input series the (generalized) Gaussian series with mean 0 and $\text{cov}\{X(t+u), X(t)\} = \delta(u)$. Then as Wiener (1958) shows

$$a_K(u_1, \ldots, u_K) = E\{X(t-u_1) \cdots X(t-u_K)Y(t)\}/K!$$

for the u_k distinct. A number of the papers in the proceedings edited by McCann and Marmarelis (1975) are concerned with the details of (approximately) carrying out this identification procedure in practice.

Consider next a situation involving a trivariate time series $\{X_1(t), X_2(t), X_3(t)\}$. Suppose that there is an apparent association between the series X_1 and X_2, but it is felt that this association may simply be due to their individual associations with the series X_3. A system that allows an investigation of such a possibility is provided by

$$X_1(t) = \mu_1 + \int a_1(t-u)X_3(u)\,du + \varepsilon_1(t)$$
$$X_2(t) = \mu_2 + \int a_2(t-u)X_3(u)\,du + \varepsilon_2(t) \qquad (2.8.12)$$

where $\{\varepsilon_1, \varepsilon_2, X_3\}$ is a stationary series with the series X_3 independent of the series ε_1 and ε_2. In the case in which the series ε_1 and ε_2 were independent of each other, one could say that the association of the series X_1 and X_2 is simply due to the common influence of the series X_3.

Now, one measure of the degree of association of two stationary time series is provided by the *coherence* function defined as

$$|g_{12}(\lambda)|^2/[g_{20}(\lambda)g_{02}(\lambda)] \qquad (2.8.13)$$

where g_{20} and g_{02} are the power spectra of the series and g_{11} their cross spectrum. The values of the coherence function (2.8.13) may be shown to lie in the interval [0, 1], with 0 occurring if the series are independent and 1 occurring if the series are connected in a linear time invariant manner (see Brillinger, 1975a). For the system (2.8.12) it may be shown that the second-order spectra of the series $\varepsilon_1, \varepsilon_2$ are given by

$$g_{jk}(\lambda) = \{f_{jk0}(\lambda) - [f_{j01}(\lambda)f_{0k1}(\lambda)]\}/f_{002}(\lambda) \qquad (2.8.14)$$

$j + k = 2$, in terms of the second-order spectra of the process $\{X_1, X_2, X_3\}$. With the substitution (2.8.14), the expression (2.8.13) is called the *partial coherence* at frequency λ of the series X_1 and X_2 with the linear effects of the series X_3 removed. This function may be estimated once estimates of the second-order spectra are available.

2.9. The Identification of Point Process Systems

By a *point process system* is meant a collection of a space of input step functions $M(t)$, a space of output step functions $N(t)$, and an operation carrying input functions over into output functions. A common form of the point process system is given by

$$N(I) = \mathscr{A}[M](I) + E(I) \tag{2.9.1}$$

I in $B[0, \infty)$, where $\mathscr{A}$ is a point process operation of the kind discussed in Section 2.7 and $E(t)$ is a further point process. In contrast to the situation for ordinary time series systems, the operation $\mathscr{A}$ is generally stochastic [and it is sometimes convenient to consider the process $E(t)$ as part of it]. The problem of point process system identification is that of determining essential properties of the operator $\mathscr{A}$ from a stretch of input and output data $\{M(t), N(t)\}$, $0 < t \leq T$. A system is called *time invariant* if the bivariate process $\{M(t), N(t)\}$, $0 \leq t < \infty$, is stationary whenever the input process $M(t)$, $0 \leq t < \infty$, is. An operator $\mathscr{A}$ is called *causal* when $\mathscr{A}[M_j](t)$ is the same for two realizations $M_j(\cdot)$ satisfying $M_1(v, \omega) = M_2(v, \omega)$ for $v > t$. The system (2.9.1) is called causal when the operator $\mathscr{A}$ is causal and the future increments $E(v) - E(t)$, $v > t$, are independent of the past $\{M(u), N(u)\}$, $u \leq t$. A point process system will be said to have a *refractory period* if there exists a time interval immediately following an output event, during which there can be no further output events.

As an example of a point process system, consider the noisy random displacement. Suppose that input points occur at σ_j, $j = 0, 1, \ldots$. Suppose that σ_j is displaced, randomly, to $\sigma_j + u_j$, $j = 0, 1, \ldots$. Suppose that γ_j, $j = 0, 1, \ldots$, denote the points of a further point process $E(t)$. Let the points of the output process be the union of the $\sigma_j + u_j$ and the γ_j. The output process may be represented by

$$\sum_j \delta(t - \sigma_j - u_j) + \sum_j \delta(t - \gamma_j) \tag{2.9.2}$$

In the case in which the process $E(t)$ is stationary with mean rate μ and where u_j is independent of σ_j with density function $a(u)$, $j = 0, 1, \ldots$, it follows from (2.9.2) that

$$E\{dN(t) \mid M\} = \left[\mu + \int a(t - u)\, dM(u)\right] dt \tag{2.9.3}$$

This expression is seen to be analogous to the regression model (2.8.2) of ordinary time series analysis.

In the case in which the process M is stationary with power spectrum $f_{20}(\lambda)$ and where

$$f_{11}(\lambda) = (2\pi)^{-1} \left[\int_{-\infty}^{\infty} \exp\{-i\lambda u\} \operatorname{cov}\{dM(t + u), dN(t)\}\right] \Big/ dt \tag{2.9.4}$$

is the *cross spectrum* of the process M with the process N, it follows from (2.9.3) that

$$f_{11}(\lambda) = A(-\lambda)f_{20}(\lambda) \tag{2.9.5}$$

Hence the system with output (2.9.2) may be identified, in the sense that $A(\lambda)$, and hence $a(u)$, may be estimated once estimates of $f_{11}(\lambda), f_{20}(\lambda)$ are available. Incidentally, in terms of the random displacement model,

$$A(-\lambda) = \int \exp\{i\lambda u\}a(u)\,du = E\exp\{i\lambda u\}$$

is the characteristic function of the distribution of translations.

In the case in which there are a number of displacements u_{jk}, $k = 1, \ldots, K_j$, applied to each input point, the relationship (2.9.5) continues to hold with

$$A(\lambda) = E\sum_k \exp\{-i\lambda u_{jk}\}$$

provided this last function does not depend on j.

It should be remarked that the specification (2.9.3) does not characterize the distribution of the process completely, in the manner that (2.8.2) characterized that of Y. Were it given that

$$\text{Prob}\{dN(t) = 1 \mid N(u),\, u \le t,\, M\} = \left[\mu + \int a(t-u)\,dM(u)\right]dt \tag{2.9.6}$$

for the system, then the distribution of N could be characterized as time inhomogeneous Poisson given M.

A point process system analogous to those of Hawkes (1972) and suggested by (2.9.6) is the one defined by

$$\begin{aligned}\text{Prob}&\{dN(t) = 1 \mid N(u),\, u \le t,\, M\}\\ &= \left[\mu + \int a(t-u)\,dM(u) + \int^t b(t-u)\,dN(u)\right]dt \end{aligned}\tag{2.9.7}$$

For this system, the cross spectrum between the input and output processes is given by

$$f_{11}(-\lambda) = (1 - B(\lambda))^{-1}A(\lambda)f_{20}(\lambda)$$

where $B(\lambda) = \int_0^\infty \exp\{-i\lambda u\}b(u)\,du$. This expression generalizes (2.9.5). Rice (1975) discusses some aspects of the problem of identification of the system (2.9.7).

The modulation model of Section 2.7 is sometimes useful in describing certain point process systems. Suppose that $E(t)$ is a point process with conditional intensity function $\gamma(t, \omega)$. Suppose that $M(t)$ is a deterministic step function. Suppose that the output process $N(t)$ is to have conditional intensity function given by

$$\text{Prob}\{dN(t) = 1 \mid N(u), u \leq t, M\} = \beta(t, M, \omega)\gamma(t, \omega)\, dt \tag{2.9.8}$$

The likelihood ratio of the process N relative to the process E is here

$$\exp\left\{\int_0^T \log \beta(t, M, \omega)\, dN(t) - \int_0^T [\beta(t, M, \omega) - 1]\gamma(t, \omega)\, dt\right\} \tag{2.9.9}$$

In this situation, $\gamma(t, \omega)$ may be thought of as specifying the output process when there is no input M, and $\beta(t, M, \omega)$ as indicating the effect that input has upon the output. In the case of the model (2.9.6), $\gamma(t, \omega) = 1$ corresponding to a unit Poisson process, and

$$\beta(t, M, \omega) = \beta(t, M) = \mu + \int a(t - u)\, dM(u) \tag{2.9.10}$$

In the case of the model (2.9.7), $\gamma(t, \omega) = 1$ again, and

$$\beta(t, M, \omega) = \mu + \int a(t - u)\, dM(u) + \int^t b(t - u)\, dN(u) \tag{2.9.11}$$

In a situation in which the functions $a(\cdot)$ and $b(\cdot)$ of these last two cases depend on a parameter θ, the likelihood ratio (2.9.9) with the substitution (2.9.11) may be used to estimate θ and hence identify the system, in certain circumstances.

Consider next a system with the property that for each pair (σ_j, σ_k) of input points σ_j, σ_k, an output point appears at $\sigma_j + u_{jk}$ where u_{jk} is a random variable with density function $h(u, \sigma_k - \sigma_j)$. Then the output process may be represented by

$$\sum_{j<k} \delta(t - \sigma_j - u_{jk})$$

and it is the case that

$$\frac{1}{dt} E\{dN(t) \mid M\} = \sum_{j<k} h(t - \sigma_j, \sigma_k - \sigma_j) = \sum_{j<k} b(t - \sigma_j, t - \sigma_k)$$

$$= \iint_{u<v} b(t - u, t - v)\, dM(u)\, dM(v)$$

where $b(u, v) = h(u, u - v)$. The system in this case is seen to be "quadratic." Continuing in this manner a system may be constructed such that

$$\frac{1}{dt} E\{dN(t) \mid M\} = \sum_{K=0} \int \cdots \int_{u_k \text{distinct}} a_K(u_1, \ldots, u_K)\, dM(t - u_1) \cdots dM(t - u_K) \tag{2.9.12}$$

to parallel the ordinary time series system (2.8.11). Suppose that one is able to employ unit Poisson noise as input process M. Then from (2.9.12)

$$E\{dM'(t-u_1)\cdots dM'(t-u_K)\,dN(t)\} = a_K(u_1,\ldots,u_K)\,du_1\cdots du_K\,dt\,K! \tag{2.9.13}$$

in the case in which a_K is symmetric and where $M'(t) = M(t) - p_1 t$. Hence the system of (2.9.12) may be identified once one can estimate joint product densities of order K of bivariate point processes. This estimation problem is considered by Brillinger (1975b).

One general remark that may be made concerning point process systems of the character just considered is that a useful parameter for use in the modeling of systems is $\mu_M(t)$ defined by

$$E\{dN(t)\,|\,M\} = \mu_M(t)\,dt$$

The identification problem then comes down to, in part, the estimation of $\mu_M(t)$ given a stretch of input and output data of the system.

The linear dynamic system (2.8.9) may be paralleled to some extent through the point process system specified by

$$dS(t) = F(t)S(t)\,dt + G(t)\,dM(t) + K(t)\,dW(t)$$

$$\mathrm{Prob}\{dN(t) = 1\,|\,N(u),\,u \le t,\,S,\,M,\,W\} = [\mu + H(t)S(t)]\,dt \ge 0 \tag{2.9.14}$$

where $M(t)$ is the input process, $S(t)$ a (vector-valued) state process, and $W(t)$ a noise process. In this system, following expression (2.2.17), the likelihood function is given by

$$\exp\left\{\int_0^T \log[\mu + H(t)\hat{S}(t)]\,dN(t) - \int_0^T [\mu + H(t)\hat{S}(t)]\,dt\right\} \tag{2.9.15}$$

assuming the input process fixed. Here the variate $\hat{S}(t)$ is defined by

$$\hat{S}(t) = E\{S(t)\,|\,N(u),\,u \le t\}$$

and may be determined by Kalman–Bucy type of recursive equations. Some details concerning this sort of model are given by Snyder (1975). In the stationary case with $F(t), \ldots, J(t)$ constant and the process W uncorrelated with the process M, the second-order spectra of the input and output are related by

$$f_{11}(-\lambda) = H(i\lambda I - F)^{-1}Gf_{20}(\lambda) \tag{2.9.16}$$

In the case in which the $F, \ldots, J$ have been parametrized to make the system identifiable, the relationship (2.9.16) or the criterion (2.9.15) may be used on occasion to estimate the parameters.

Consider next the system

$$N_1(t) = M_1(t) + M_3(t), \qquad N_2(t) = M_2(t) + M_4(t) \tag{2.9.17}$$

where the processes M_1 and M_2 depend on a process N_3, but the bivariate process $\{M_3, M_4\}$ is independent of the process $\{M_1, M_2, N_3\}$. In particular, suppose that

$$\text{Prob}\{dM_j(t) = 1 \mid N_3\} = \left[\mu_j + \int a_j(t-u)\, dN_3(u)\right] dt \tag{2.9.18}$$

for $j = 1, 2$. One may be interested in the question of whether an association observed between the processes N_1 and N_2 indicates a proper connection, or whether it is simply due to their common association with the process N_3. The model (2.9.17), (2.9.18) is one means of examining this question. In the case in which the processes M_3 and M_4 are independent, the association would be apparent, not real. Now the degree of dependence of the processes M_3 and M_4 may be measured by their coherence function

$$|g_{11}(\lambda)|^2/[g_{20}(\lambda)g_{02}(\lambda)] \tag{2.9.19}$$

with g_{jk} given in terms of the second-order spectra of the process N_1, N_2, N_3 as

$$g_{jk}(\lambda) = \{f_{jk0}(\lambda) - [f_{j0k}(\lambda)f_{0k1}(\lambda)]\}/f_{002}(\lambda)$$

$j + k = 2$. The function (2.9.19) is here called the partial coherence at frequency λ of the processes N_1 and N_2 with the linear effects of the process N_3 removed. For networks of three nerve cells some estimates of partial coherences are presented by Brillinger *et al.* (1976). The parameter proved useful in investigating the connections between the cells, specifically whether each pair of cells had a direct link.

Numerous models of point process systems are provided by the various models that have been proposed for the operation of a nerve cell, driven by a spike train to emit a further spike train. The effect of the input train may be excitatory or inhibitory, corresponding to an increase or decrease of the instantaneous output rate. As an example, consider the following model of an excitatory system, of the kind discussed by Moore *et al.* (1966). Let the input spike train be denoted by $M(t)$. Let the output spike train be denoted by $N(t)$ and suppose its spikes occur at the times $\tau_0, \tau_1, \ldots$. Define the time series

$$X(t) = \alpha \int_{\tau_{N(t)-1}}^{t} \exp\{-\alpha(t-u)\}\, dM(u) \tag{2.9.20}$$

for $\tau_{N(t)-1} < t \le \tau_{N(t)}$ and $\alpha > 0$. Then $\tau_{N(t)}$ is defined to be

$$\tau_{N(t)} = \inf\{t\colon t > \tau_{N(t)-1},\ X(t) \ge \theta\}$$

The value θ is assumed to be positive and is called the *threshold* value. $X(t)$ is called the *postsynaptic potential* of the cell at time t. To be realistic a nerve cell system model must also have a refractory period. Other models of nerve cell activity are described by Feinberg (1974).

2.10. Some Particular Time Series

Many specific properties are known concerning certain types of ordinary time series. The class that has been subjected to the highest level of development is undoubtedly that of *Gaussian time series*. A series $X(t)$ is called Gaussian when all of its finite-dimensional distributions are multivariate normal. In the Gaussian case, the transition probability functions (2.1.2) will also be normal. The characteristic functional is given by

$$C[\xi] = \exp\left\{ i \int \xi(t) c_1(t)\, dt - \tfrac{1}{2} \iint \xi(t_1)\xi(t_2) c_2(t_1, t_2)\, dt_1\, dt_2 \right\} \qquad (2.10.1)$$

with $c_1(t) = EX(t)$ and $c_2(t_1, t_2) = \operatorname{cov}\{X(t_1), X(t_2)\}$. It follows from expression (2.10.1) that a Gaussian series has moments and cumulants of all orders and that the cumulant functions of order greater than 2 are identically 0. A Gaussian series is determined by its first- and second-order moment functions, and given any function $c_1(t)$ and continuous nonnegative definite $c_2(t_1, t_2)$, there exists a Gaussian series with these parameters. Linear operations on Gaussian series produce Gaussian series, and conditional distributions based on linear combinations of its values remain Gaussian. The Gaussian series appears as a limit when independent series satisfying finite second-order moment conditions are added. The importance of the use of Gaussian series in the identification of nonlinear systems was indicated in Section 2.7.

One important Gaussian series is the *Wiener process* $W(t)$ satisfying $W(0) = 0$, $c_1(t) = 0$, $c_2(t_1, t_2) = \min\{t_1, t_2\}$. The increments of the Wiener process are stationary and independent. The generalized process $dW(t)/dt$ is called *Gaussian white noise*. Its covariance function may be represented by $c_2(t_1, t_2) = \delta(t_1 - t_2)$. It provides a continuous time analog of a sequence of independent standardized normal variates.

A variety of conditions have been set down to ensure that the sample paths of a Gaussian series are continuous. A useful sufficient condition that ensures this (for any series) is

$$E|X(t_2) - X(t_1)|^a \leq b|t_2 - t_1|^{1+c} \qquad (2.10.2)$$

for some $a, b, c > 0$ and any t_1, t_2 (e.g., see Cramér and Leadbetter, 1967). This criterion shows that the Wiener process has continuous paths.

Another important class of time series is made up of the *diffusion processes.* These are Markov processes with continuous sample paths. In particular,

$$\text{Prob}\{X(v) \leq y \,|\, X(u), u \leq t\} = \text{Prob}\{X(v) \leq y \,|\, X(t)\}$$

for all $v > t$. Suppose the transition density function is given by

$$p(t, x; v, y) = \frac{d}{dy}\text{Prob}\{X(v) \leq y \,|\, X(t) = x\} \tag{2.10.3}$$

Then it satisfies

$$p(t, x; v, y) = \int p(t, x; u, z)p(u, z; v, y)\, dz$$

for $t < u < v$ and also $p(t, x; t, y) = \delta(x - y)$. Suppose

$$E\{dX(v) \,|\, X(t) = x\} = \mu(t, x)\, dv \tag{2.10.4}$$

and

$$E\{(dX(v))^2 \,|\, X(t) = x\} = \sigma^2(t, x)\, dv \tag{2.10.5}$$

The parameter $\mu(t, x)$ is called the *local mean* or *drift*. The parameter $\sigma^2(t, x)$ is called the *local variance*, and its reciprocal is called the *speed*. The transition density function satisfies certain differential equations:

$$-\frac{\partial f(t, x; v, y)}{\partial t} = \tfrac{1}{2}\sigma^2(t, x)\frac{\partial^2 f(t, x; v, y)}{\partial x^2} + \mu(t, x)\frac{\partial f(t, x; v, y)}{\partial x}$$

called the *backward Kolmogorov equation*, and

$$\frac{\partial f(t, x; v, y)}{\partial t} = \frac{1}{2}\frac{\partial^2[\sigma^2(v, y)f(t, x; v, y)]}{\partial y^2} - \frac{\partial[\mu(v, y)f(t, x; v, y)]}{\partial y}$$

the *forward Kolmogorov equation.* This diffusion process may be approximated by the following discrete model: Suppose $X(t) = x$. Then $X(t + dt) = x + dx$ with probability

$$\frac{1}{2} + \frac{\mu(t, x)}{2\sigma(t, x)}dt \tag{2.10.6}$$

and $X(t + dt) = x - dx$ with probability $1 - (2.10.6)$, $dx = \sigma(t, x)\sqrt{dt}$ independently of its behavior before t (see Prohorov and Rozanov, 1969, p. 263).

Particular diffusion processes include the Wiener process with $\mu(t, x) = 0$ and $\sigma(t, x) = 1$ and the *Ornstein–Uhlenbeck process* with $\mu(t, x) = -1$ and $\sigma^2(t, x) = 1$. The latter process is the most general Gaussian Markov stationary process.

The Wiener process may be used to derive a large class of diffusion processes. Ito (1951) suggests solving the equation

$$X(t) - X(0) = \int_0^t \mu(s, X(s))\, ds + \int_0^t \sigma(s, X(s))\, dW(s) \qquad (2.10.7)$$

where $W(t)$ is the Wiener process. Under appropriate conditions, the equation has a solution that is a diffusion process. Equation (2.10.7) is called a *stochastic Ito differential equation.* A considerable literature exists concerning first passage time distributions for diffusion processes. Differential equations may be set down for certain characteristic functions and expected values based on a diffusion process.

Certain transformations carry diffusion processes over into diffusion processes. Let g_x and g_t denote thé partial derivatives of $g(x, t)$. Suppose $g_x(x, t) > 0$. Let $X(t)$ be the diffusion process just discussed and $Y(t) = g(X(t), t)$. Then $Y(t)$ is a diffusion process with parameters

$$\mu_Y(t, x) = \mu(t, x)g_x(x, t) + \tfrac{1}{2}\sigma^2(t, x)g_{xx}(x, t) + g_t(x, t)$$

$$\sigma_Y(t, x) = \sigma(x, t)g_x(x, t)$$

(see Gihman and Skorohod, 1972). Alternatively, consider a random time transformation defined by

$$\frac{d\tau(t, \omega)}{dt} = \frac{1}{V(t, X(\tau))} > 0$$

and the process $Y(t) = X(\tau(t, \omega))$. Then $Y(t)$ is also a diffusion with the parameters

$$\mu_Y(t, x) = \mu(t, x)/V(t, x), \qquad \sigma_Y(t, x) = \sigma(t, x)/\{V(t, x)\}^{1/2}$$

The transformations described here may be used to carry a given process over into one with a simpler description.

Likelihood ratios may be determined for diffusion processes in certain cases. For example, suppose that $\sigma(t, x) = 1$; then under conditions including $X(0) = 0$, the likelihood ratio of the process of (2.10.7) relative to the process W is given by

$$\exp\left\{\int_0^T \mu(t, X(t))\, dX(t) - \tfrac{1}{2}\int_0^T \mu^2(t, X(t))\, dt\right\} \qquad (2.10.8)$$

(see Gihman and Skorohod, 1972, p. 90). This ratio is used to determine maximum likelihood estimates of the parameters $\theta_1, \ldots, \theta_K$ in the case in which

$$\mu(t, x) = \sum_{k=1}^{K} \theta_k \phi_k(t, x)$$

with the ϕ_k known (see Taraskin, 1974).

A generalization of the Gaussian Markov process is provided by the *N-tuple Markov Gaussian* of Hida (1960). It is defined as a Gaussian process with the property that the variates

$$E\{X(t_k) \mid X(u), u \leq t_0\}, \qquad k = 1, \ldots, K$$

$t_0 \leq t_1 < \cdots < t_K$ are linearly independent for $K = N$, but linearly dependent for $K = N + 1$. The process may be shown to have a representation

$$X(t) = \int_0^t \sum_{n=1}^{N} f_n(t) g_n(u)\, dW(u) \tag{2.10.9}$$

where $W(t)$ is a Wiener process. In the stationary case the kernel of (2.10.9) may be shown to be a linear combination of the functions $t^k u^{n-k} \exp\{-(\rho + i\mu)(t - u)\}$ (see Hida, 1960). The power spectrum is a rational function of λ.

As a final general class of time series consider the *linear processes* defined by $X(t) = \int a(t, u)\, dV(u)$, where $V(t)$ is a process with independent increments. Examples of independent increment processes include the Wiener and Poisson processes. The general increment process is completely characterized by the first-order distribution of $V(t)$ and the increment distribution of $V(t) - V(u)$. The process $V(t)$ may be shown to have the representation

$$V(t) = V_c(t) + \int_{|x| \leq 1} x[N([0, t] \times dx) - \mu([0, t] \times dx)]$$

$$+ \int_{|x| > 1} xN([0, t]\, dx) \tag{2.10.10}$$

where $V_c(t)$ is a continuous Gaussian process with independent increments and $N(t, x)$ a Poisson process in the plane with parameter $\mu(t, x)$ independent of the process V_c (see Prohorov and Rozanov, 1969). Supposing $E\, dV_c(t) = \alpha(t)\, dt$, $\operatorname{cov}\{dV_c(t_1), dV_c(t_2)\} = \beta(t_1)\, \delta(t_1 - t_2)\, dt_1\, dt_2$,

$$\mu((t, t + dt] \times dx) = m(t, x)\, dt\, dx,$$

the cumulant functions of the process V are given by

$$E\{dV(t)\} = \left[\alpha(t) + \int_{|x| > 1} xm(t, x)\, dx\right] dt$$

$$\operatorname{cov}\{dV(t_1), dV(t_2)\} = \left[\beta(t_1) + \int_{|x| > 1} x^2 m(t_1, x)\, dx\right] \times \delta(t_1 - t_2)\, dt_1\, dt_2$$

$$\operatorname{cum}\{dV(t_1), \ldots, dV(t_K)\} = \int x^K m(t_1, x)\, dx\, \delta(t_1 - t_2) \cdots \times \delta(t_1 - t_K)\, dt_1 \cdots dt_K$$

$K > 2$. The cumulant functions of the linear process may be derived directly from these expressions. In the case in which the process $V(t)$ has stationary increments and $X(t) = \int a(t-u)\, dV(u)$, $A(\lambda) = \int \exp\{-i\lambda u\}a(u)\, du$, the cumulant spectrum of order K of $X(t)$ will be proportional to

$$A(\lambda_1) \cdots A(\lambda_{K-1})A(-\lambda_1 - \cdots - \lambda_{K-1})$$

Certain additional aspects concerning linear processes may be found in the article by Westcott (1970).

2.11. Some Particular Point Processes

Foremost among the point processes is the Poisson process. The *Poisson process* with intensity function $p(t)$ is defined by the requirement that for $I_1, \ldots, I_K$ disjoint intervals, the variates $N(I_1), \ldots, N(I_K)$ are independent Poisson random variables with means $P(I_1), \ldots, P(I_K)$ where

$$P(I) = \int_I p(t)\, dt \tag{2.11.1}$$

The zero probability function also characterizes the Poisson. It is given by

$$\phi(I) = \exp\{-P(I)\} \tag{2.11.2}$$

for I in $B[0, \infty)$. The probability generating functional is given by

$$G[\xi] = \exp\left\{\int [\xi(t) - 1]p(t)\, dt\right\} \tag{2.11.3}$$

It follows from (2.11.3), or directly from the definition of the process, that the product densities of the Poisson are given by

$$p_K(t_1, \ldots, t_K) = p(t_1) \cdots p(t_K) \tag{2.11.4}$$

and the cumulant densities by $q_1(t) = p(t)$ and $q_K(t_1, \ldots, t_K) = 0$, $K = 2, 3, \ldots$. The conditional intensity function is constant in ω and given by

$$\gamma(t, \omega) = p(t) \tag{2.11.5}$$

If the points observed in the interval $[0, T]$ are $0 \le \tau_0 < \tau_1 < \cdots$, then the corresponding likelihood function is

$$\exp\left\{\int_0^T \log p(t)\, dN(t) - \int_0^T p(t)\, dt\right\} \tag{2.11.6}$$

and the likelihood ratio relative to the unit Poisson process is

$$\exp\left\{\int_0^T \log p(t)\, dN(t) - \int_0^T (p(t) - 1)\, dt\right\} \tag{2.11.7}$$

From expression (2.2.9), the joint density of the q successive intervals after τ_0 is given by

$$\sum_{k=0}^{\infty} \frac{(-1)^k}{k!} \int_0^{\infty} \int \cdots \int_{(0,t+y_1+\cdots+y_q)^k} p_{k+q+1}(t, t+y_1, \ldots, t+y_1+\cdots + y_q, u_1, \ldots, u_k)\, du_1 \cdots du_k\, dt \qquad (2.11.8)$$

In the Poisson case this reduces to

$$\int_0^{\infty} p(t)p(t+y_1)\cdots p(t+y_1+\cdots+y_q) \exp\left\{-\int_0^{t+y_1+\cdots+y_q} p(u)\,du\right\} dt \qquad (2.11.9)$$

In the stationary case, $p(t) = p$, and (2.11.9) becomes

$$p^q \exp\{-p(y_1+\cdots+y_q)\} \qquad (2.11.10)$$

and the successive intervals are seen to be independent exponentials with mean $1/p$. Expression (2.2.9) may also be used to show that τ_0 is independent of $\{y_1, \ldots, y_q\}$ and exponential with mean $1/p$.

From (2.2.12), the probability generating functional of the Palm process is given by

$$G[\xi \mid t] = \exp \int [\xi(t) - 1]p(t)\,dt$$

showing that in the Poisson case the distribution of the Palm process is the same as that of the original process. Among other things, this implies that the survivor function is given by

$$\text{Prob}\{N(t, t+u] = 0 \mid N\{t\} = 1\} = \exp\left\{-\int_t^{t+u} p(v)\,dv\right\}$$

and that the hazard function is given by

$$\text{Prob}\{dN(t+u) = 1 \mid N\{t\} = 1, N(t, t+u] = 0\}/du = p(t+u)$$

The forward recurrence time distribution is given by

$$\text{Prob}\{\tau_{N(t)} - t \le u\} = 1 - \exp\left\{-\int_t^{t+u} p(v)\,dv\right\}$$

The *unit Poisson* process has intensity $p(t)$ identically 1. If N is a Poisson process with intensity $p(t)$ and one sets

$$P(t) = \int_0^t p(u)\,du$$

then $N(P^{-1}(t))$ is a unit Poisson process. Conversely, if $M(t)$ is a unit Poisson, then $M(P(t))$ is Poisson with mean rate $p(t)$.

The general Poisson process is clearly a Markov process with state space 0, 1, 2, ... and transition distributions given by

$$\text{Prob}\{N(t + dt) = j + 1 \mid N(t) = j\} = p(t)\, dt$$

$$\text{Prob}\{N(t + dt) = j \mid N(t) = j\} = 1 - p(t)\, dt$$

$j = 0, 1, \ldots$. It is the simplest discontinuous Markov process.

When the Poisson process is thinned, as indicated in the discussion following (2.7.8), its probability generating functional becomes

$$G[\xi\pi + 1 - \pi] = \exp\left\{\int [\xi(t) - 1]\pi(t)p(t)\, dt\right\}$$

That is, it becomes a Poisson process with intensity $\pi(t)p(t)$.

When r independent Poissons, with intensities $p_1(t), \ldots, p_r(t)$, are superposed, as indicated by the discussion following (2.7.17), the probability generating functional of the superposed process is

$$\exp\left\{\int [\xi(t) - 1]p_1(t)\, dt + \cdots + \int [\xi(t) - 1]p_r(t)\, dt\right\}$$

i.e., the superposed process is Poisson with intensity $p_1(t) + \cdots + p_r(t)$.

If $\tau_0, \tau_1, \ldots$ are the points of a Poisson process with rate $p(t)$ and if independent random displacements U_j, $j = 0, 1, \ldots$, with cumulative distribution function (c.d.f.) $F(u)$ are applied to these points, then the process with points $\tau_j + U_j$, $j = 0, 1, \ldots$, has, from expression (2.7.3), probability generating functional (p.g.f.) given by

$$G\left[\int \xi(\cdot + u)\, dF(u)\right] = \exp \int \left[\int \xi(u + t)\, dF(u) - 1\right] p(t)\, dt$$

$$= \exp\left\{\int [\xi(v) - 1]\left(\int p(v - u)\, dF(u)\right) dv\right\}$$

i.e., the displaced process is also Poisson. It has mean rate $\int p(v - u)\, dF(u)$. This result is due to Mirasol (1963).

Next, consider a stationary Poisson process N with intensity p. Because of the independence properties of the increments of N one has

$$\text{cum}\{dN(t_1), \ldots, dN(t_K)\} = p\, \delta(t_1 - t_K) \cdots \delta(t_{K-1} - t_K)\, dt_1 \cdots dt_K$$

$K = 2, 3, \ldots$, and so from (2.5.30), the cumulant spectra of N are given by

$$f_K(\lambda_1, \ldots, \lambda_{K-1}) = (2\pi)^{-K+1} p \tag{2.11.11}$$

for $K = 2, 3, \ldots$. Being stationary, the process has a spectral representation

$$N(t) = \int \frac{\exp\{it\lambda\} - 1}{i\lambda} \, dZ(\lambda)$$

In view of (2.11.11) and (2.5.33), the process $Z(\lambda)$ has the interesting property

$$\text{cum}\{dZ(\lambda_1), \ldots, dZ(\lambda_k)\} = (2\pi)^{-K+1} p \, \delta(\lambda_1 + \cdots + \lambda_k) \quad (2.11.12)$$

When a type I counter with dead time Δ is applied to a stationary Poisson process, the interpoint distribution simply becomes the original exponential translated by Δ. To see this, note from (2.7.11) that the conditional intensity function of the output process is given by

$$\begin{aligned} \gamma(t, \omega) &= p \qquad \text{for} \quad \tau_{N(t)-1} + \Delta \leq t \\ &= 0 \qquad \text{otherwise} \end{aligned}$$

Therefore, from (2.2.16)

$$\begin{aligned} \text{Prob}\{\tau_J > t \,|\, \tau_j = t_j, j < J\} &= 1 \qquad \text{for} \quad t < t_{J-1} + \Delta \\ &= \exp\left\{-\int_{t_{J-1}+\Delta}^{t} p \, du\right\} \qquad \text{for} \quad t \geq t_{J-1} + \Delta \end{aligned}$$

giving the result.

Consider now a Poisson process $N_\theta(t)$ with intensity function $\theta p(t)$. Consider the characteristic functional of the process

$$X_\theta(t) = (N_\theta(t) - \theta p(t))/\sigma$$

where $\sigma^2 = \theta p(t)$. From (2.11.3) the functional is

$$\begin{aligned} E \exp\left\{i \int X_\theta(t)\xi(t) \, dt\right\} &= \exp\left\{-i \int \sigma\xi(t) \, dt\right\} \\ &\quad \times \exp\left\{\int (\exp\{i\xi(t)/\sigma\} - 1) \, \theta p(t) \, dt\right\} \\ &\to \exp\left\{-\int \xi(t)^2 \, dt\right\} \end{aligned}$$

as $\theta \to \infty$. This is the characteristic functional of the Wiener process. This result suggests that a Poisson process with high rate may be approximated by a simple function of a Wiener process.

Suppose now that $H[\xi]$ is the factorial cumulant generating functional of a stationary point process with finite intensity p. Let

$$N_n(t) = M_1(t/n) + \cdots + M_n(t/n)$$

where the M_j are independent realizations of M. Then the factorial cumulant generating functional of N_n is

$$nH[\xi(nt)] = n\left[\int \xi(nt)p\,dt + o\left(\frac{1}{n}\right)\right] \to p\int \xi(t)\,dt$$

as $n \to \infty$. This is the factorial cumulant generating functional of a Poisson process with intensity p. Hence the limit in distribution of rescaled superpositions of independent replicas of a point process is Poisson. This form of argument appears in the article by Vere-Jones (1968). More general results on the Poisson limit occurring when point processes are superposed appear in the chapter by Cinlar (1972).

Consider a process M with intensity p and factorial cumulant generating functional $H[\xi]$. Consider the effect of applying the thinning operation represented by (2.7.8) with $EI(t) = \pi$ and letting $\pi \to 0$. (This would be the case if the same thinning operation with $\pi < 1$ were applied repeatedly.) Let N_π denote the thinned process. Consider the rescaled process $N_\pi(t/\pi)$. From (2.7.10) its factorial cumulant generating functional (f.c.g.fl.) is given by

$$\begin{aligned} H[\pi\xi(t\pi)] &= \log E \exp\left\{\int \log[1 + \pi\xi(\pi t)]\,dM(t)\right\} \\ &\sim \log E \exp\left\{\int \pi\xi(\pi t)\,dM(t)\right\} \\ &\to \log E \exp\left\{(\lim_{\pi\to 0} \pi M(1/\pi)) \int \xi(u)\,du\right\} \end{aligned} \qquad (2.11.13)$$

using a Wiener type of formula. When the process M is ergodic

$$\lim_{\pi\to 0} \pi M(1/\pi) = p$$

almost surely and (2.11.13) becomes $p \int \xi(u)\,du$, the factorial cumulant generating functional of the Poisson process with intensity p. This result is suggestive of why the Poisson process is sometimes said to correspond to the law of rare events. It may also be developed via the zero probability function. More formal developments of thinning results may be found in the article by Kallenberg (1975).

As an alternative procedure leading to a Poisson limit consider the following result of Volkonskii and Rozanov (1959). Let $\mathscr{F}_{a,b}$ denote the Borel field generated by events of the form

$$\{N(v_1) - N(u_1) \le n_1, \ldots, N(v_K) - N(u_K) \le n_K\}$$

$a < u_k < v_k \le b$, n_k a nonnegative integer, $K = 1, 2, \ldots$. The process N is called *strong mixing* with mixing coefficient $\alpha(\tau)$ when

$$\alpha(\tau) = \sup\{|P(AB) - P(A)P(B)| : A \text{ in } \mathscr{F}_{-\infty,t},\ B \text{ in } \mathscr{F}_{t+\tau,\infty}\} \to 0$$

as $\tau \to \infty$. Volkonskii and Rozanov consider a sequence of stationary processes N^T, $T = 1, 2, \ldots$, with N^T (i) having intensity of the form $p\varepsilon_T$, $\varepsilon_T \to 0$, as $T \to \infty$; (ii) having mixing coefficient $\alpha^T(\tau) \to 0$ uniformly as $\tau \to \infty$, and such that $\alpha^T(\tau) \approx \alpha(\tau)$; and (iii) $E\{N^T(t/\varepsilon_T)[N^T(t/\varepsilon_T) - 1]\} = o(t)$ as $t \to 0$, $T \to \infty$. They show that under these conditions $N^T(t/\varepsilon_T)$ tends in distribution to a Poisson variate with mean pt.

The point process obtained by retaining every kth point of a stationary Poisson process is called an *Erlang process*. Its interpoint distributions will be independent gammas, in view of the exponential distributions of the Poisson. An Erlang process may be viewed as the output of a counter that remains paralyzed for $k - 1$ points. It is a renewal process. Useful properties may be determined as particular cases of the renewal process properties to be set down shortly.

Consider an ordinary time series with nonnegative sample paths $p(t, \omega')$, ω' in Ω', $0 \le t < \infty$. Having obtained a particular realization of this series, generate a Poisson process with intensity $p(t, \omega')$. The point process N obtained in this manner is called a *doubly stochastic Poisson*. The product densities of N are given by

$$p_K(t_1, \ldots, t_K) = \underset{\omega'}{E}\,\{p(t_1, \omega') \cdots p(t_K, \omega')\}$$
$$= m_K(t_1, \ldots, t_K) \tag{2.11.14}$$

the moments of the original series. In the case in which $p(t, \omega') = p(\omega')$ and the latter variate has c.d.f. $F(p)$, the product densities are constant in time and given by

$$p_K(t_1, \ldots, t_K) = \int p^K \, dF(p) \tag{2.11.15}$$

The cumulant densities of the process are given by

$$q_K(t_1, \ldots, t_K) = \text{cum}\{p(t_1, \omega'), \ldots, p(t_K, \omega')\}$$
$$= c_K(t_1, \ldots, t_K) \tag{2.11.16}$$

i.e., they are those of the original process. The zero probability function is

$$\phi(I) = \underset{\omega'}{E} \exp\left\{-\int_I p(t, \omega') \, dt\right\} \tag{2.11.17}$$

The probability generating functional is

$$G[\xi] = \underset{\omega'}{E} \exp\left\{\int [\xi(t) - 1] p(t, \omega') \, dt\right\} = C[-i[\xi(t) - 1]] \tag{2.11.18}$$

in terms of the characteristic functional of the original process.

The conditional intensity function is

$$\gamma(t, \omega) = \underset{\omega'}{E}\{p(t, \omega') \mid N(u), u \leq t\} \tag{2.11.19}$$

It is seen to be the minimum mean squared error estimate of the level of the intensity process at time t, given the history of the point process. One may write

$$\begin{aligned} G_J(t; t_j, j < J) &= \operatorname{Prob}\{\tau_J > t \mid \tau_j = t_j, j < J\} \\ &= \frac{E_{\omega'}\{p(t_0, \omega') \cdots p(t_{J-1}, \omega') \exp\{-\int_0^t p(u, \omega')\, du\}\}}{E_{\omega'}\{p(t_0, \omega') \cdots p(t_{J-1}, \omega') \exp\{-\int_0^{t_{J-1}} p(u, \omega')\, du\}\}} \end{aligned}$$

Using this representation and (2.2.15)

$$\gamma(t, \omega) = \frac{E_{\omega'}\{p(t, \omega') \exp\{\int_0^t \log p(u, \omega')\, dN(u) - \int_0^t p(u, \omega')\, du\}\}}{E_{\omega'}\{\exp\{\int_0^t \log p(u, \omega')\, dN(u) - \int_0^t p(u, \omega')\, du\}\}} \tag{2.11.20}$$

The likelihood function may be written in two distinct ways: as

$$\underset{\omega'}{E} \exp\left\{\int_0^T \log p(t, \omega')\, dN(t) - \int_0^T p(t, \omega')\, dt\right\} \tag{2.11.21}$$

or, from (2.2.17), as

$$\exp\left\{\int_0^T \log \gamma(t, \omega)\, dN(t) - \int_0^T \gamma(t, \omega)\, dt\right\}$$

with $\gamma(t, \omega)$ given by (2.11.19).

When the doubly stochastic Poisson is thinned, as at (2.7.8), its probability generating functional becomes

$$G[\xi\pi + 1 - \pi] = \underset{\omega'}{E} \exp\{[\xi(t) - 1]\pi(t)p(t, \omega')\, dt\}$$

and so the thinned process is doubly stochastic Poisson as well.

If $\{p_1(t, \omega'), \ldots, p_r(t, \omega')\}$ is an r vector-valued ordinary time series with nonnegative components, then suppose that independent Poissons are generated with intensities $p_1(t, \omega'), \ldots, p_r(t, \omega')$, respectively. Suppose that these Poissons are superposed. Then the probability generating functional of the superposed process is

$$\underset{\omega'}{E} \exp\left\{\int [\xi(t) - 1][p_1(t, \omega') + \cdots + p_r(t, \omega')]\, dt\right\}$$

and the superposed process is seen to be doubly stochastic Poisson.

Suppose next that $\tau_0, \tau_1, \ldots$ are the points of a realization of a doubly stochastic Poisson. Suppose the points are subjected to independent random displacements $U_0, U_1, \ldots$, respectively, with c.d.f. $F(u)$. Then from (2.7.3) and (2.11.18) the probability generating functional of the displaced process is

$$G\left[\int (\cdot + u)\, dF(u)\right] = \underset{\omega'}{E} \exp\left\{\int [\xi(v) - 1]\left(\int p(v - u, \omega')\, dF(u)\right) dv\right\}$$

and the displaced process is seen to be doubly stochastic Poisson.

Let us return to the discussion of thinning a general point process given preceding expression (2.11.13). Suppose now that

$$\lim_{\varepsilon \to 0} \varepsilon M(1/\varepsilon) = p(\omega)$$

with the limit depending on the particular realization. Then the limit of the factorial cumulant generating functional is

$$\log E \exp\left\{p(\omega) \int \xi(u)\, du\right\}$$

and the limit is seen to be a doubly stochastic point process in this case.

If the process $p(t, \omega')$ is stationary, then the form of the probability generating functional of N shows that N is stationary as well. Its intensity is m_1, the mean level of the process $p(t, \omega')$. From (2.11.16) its autointensity function is

$$p_2(u) = c_2(u) \tag{2.11.22}$$

and hence from (2.5.27), its power spectrum is given by

$$f_2(\lambda) = (2\pi)^{-1} m_1 + f_2'(\lambda) \tag{2.11.23}$$

where $f_2'(\lambda)$ is the power spectrum of the process $p(t, \omega')$. Expression (2.11.23) indicates a defect that doubly stochastic processes have regarding their use in the general modeling of point processes. Their power spectra are necessarily bounded below by $(2\pi)^{-1}$ times their intensity. This is a reflection of the fact that the doubly stochastic Poisson is more disorderly than a Poisson process with the same intensity. Expression (2.11.22) indicates that the autointensity function may be any function that can be the autocovariance function of a time series with nonnegative sample paths. If the higher order cumulant spectra of the series $p(t, \omega')$ are denoted by $f_K'(\lambda_1, \ldots, \lambda_{K-1})$, then using (2.11.16) the Kth-order cumulant spectrum of the process N is given by

$$f_K(\lambda_1, \ldots, \lambda_{K-1}) = \sum_{k=1}^{K} (2\pi)^{k-K} f_k'\left(\sum_{j \in \nu_1} \lambda_j, \ldots, \sum_{j \in \nu_{k-1}} \lambda_j\right) \tag{2.11.24}$$

where the summation extends over all partitions $(\nu_1, \ldots, \nu_k)$ of the set $(1, \ldots, K)$, $\lambda_K = -\lambda_1 - \cdots - \lambda_{K-1}$ and $f_1' = m_1$ (see Brillinger, 1972, Theorem 3.3). Expression (2.11.23) is the case $K = 2$.

Macchi (1975) discusses the fact that the product densities of a point process may have the form (2.11.14) without the process $p(t, \omega')$ being nonnegative. Asymptotic independence and mixing properties of the process $p(t, \omega')$ generally carry over to a doubly stochastic process N. See Westcott (1972) for the case of mixing.

A *cluster process* is a point process of point processes. It has two components, a process M' of cluster centers and a process $M''(\cdot\,|\,t)$ of cluster members (centered at t). Each point of M is assumed to initiate an independent process of cluster members. The cluster process itself then consists of the superposition of all the various cluster members. Suppose that the probability generating functional of M' is $G'[\xi]$ and that of $M''(\cdot\,|\,t)$ is $G''[\xi\,|\,t]$. Then the probability generating functional of the cluster process N is

$$G[\xi] = E \prod_j G''[\xi\,|\,\tau_j] = G'[G''[\xi\,|\,t]] \tag{2.11.25}$$

This relationship is discussed in the article by Moyal (1962). The densities of the process N may be determined in terms of those of the processes M', M'' using Faa de Bruno formulas. The cases of $K = 1, 2$ are

$$q_1(t_1) = \int q_1{}'(t) q_1{}''(t_1\,|\,t)\, dt \tag{2.11.26}$$

$$q_2(t_1, t_2) = \int q_1{}'(t) q_2{}''(t_1, t_2\,|\,t)\, dt + \iint q_2{}'(t, u) q_1{}''(t_1\,|\,t) q_1{}''(t_2\,|\,u)\, dt\, du \tag{2.11.27}$$

If $M''(\cdot\,|\,\tau_j)$, $j = 0, 1, \ldots,$ denote the successive clusters, then the process may be represented by

$$N(I) = \sum_j M''(I\,|\,\tau_j) = \int M''(I\,|\,t)\, dM'(t)$$

In many cases the cluster distribution is the same for all clusters with $M''(I\,|\,t) = M''(I - t)$ and

$$G''[\xi\,|\,t] = E \exp\left\{\int \log \xi(u + t)\, dM''(u)\right\}$$

If the clusters each have a single member with density $a(u)$, then

$$G''[\xi\,|\,t] = E\xi(u + t) = \int \xi(u + t) a(u)\, du$$

If M'' is Poisson with intensity $p''(u)$, then

$$G''[\xi\,|\,t] = \exp\left\{\int [\xi(u + t) - 1] p''(u)\, du\right\}$$

In the case in which the process M' is stationary as well, the process N is stationary. The relationships (2.11.25)–(2.11.27) then become

$$G[\xi] = G'[G''[\xi(t+\cdot)]]$$

$$q_1 = q_1' \int q_1''(v)\,dv = q_1' E\{M''(-\infty,\infty)\}$$

$$q_2(v) = q_1' \int q_2''(v-t,-t)\,dt + \iint q_2'(t-u) q_1''(v-t) q_1''(-u)\,dt\,du \qquad (2.11.28)$$

Taking the Fourier transform of (2.11.28), the power spectrum of the process N may be seen to be given by

$$(2\pi)^{-1} p_1' \operatorname{var}\left\{\int \exp\{-i\lambda u\}\,dM''(u)\right\} + f_2'(\lambda)\left|E\int \exp\{-i\lambda u\}\,dM''(u)\right|^2 \qquad (2.11.29)$$

where

$$E\int \exp\{-i\lambda u\}\,dM''(u) = \int \exp\{-i\lambda u\} q_1''(u)\,du$$

$$\operatorname{var}\left\{\int \exp\{-i\lambda u\}\,dM''(u)\right\} = \int q_1''(u)\,du + \iint \exp\{-i\lambda u + i\lambda v\} q_2''(u,v)\,du\,dv$$

The expression (2.11.29) may be found in the chapter by Daley and Vere-Jones (1972). Westcott (1971) shows that a stationary cluster process is mixing if the cluster center process is mixing.

In the *Poisson cluster process*, the process M' of cluster centers is Poisson. Suppose the intensity of the Poisson is $p'(t)$. Then from (2.11.25) N has probability generating functional

$$\exp\left\{\int [G''[\xi|t] - 1] p'(t)\,dt\right\} \qquad (2.11.30)$$

The cumulant densities may be determined as follows:

$$\log G[\xi+1] = \int \left(\sum \frac{1}{k!} \int \cdots \int q_k''(t_1,\ldots,t_k)\,\xi(t_1)\cdots\xi(t_k)\,dt_1\cdots dt_k\right) p'(t)\,dt$$

and so

$$q_K(t_1,\ldots,t_K) = \int q_K''(t_1,\ldots,t_K|t) p'(t)\,dt \qquad (2.11.31)$$

Some particular Poisson cluster processes may be distinguished. In the *Neyman–Scott process*, the cluster members are independent random variables with density $a_t(u)$ and the probability that the cluster is of size n is $\pi_t(n)$. If $g_t(z) = \sum_n \pi_t(n)z^n$, then

$$G''[\xi\,|\,t] = \underset{n}{E}\ \underset{u}{E} \prod_{j=1}^{n} \xi(t+u_j) = \underset{n}{E}\left[\int \xi(t+u)a_t(u)\,du\right]^n$$

$$= g_t\left[\int \xi(t+u)a_t(u)\,du\right] \tag{2.11.32}$$

Suppose that the variate n has factorial moments $\mu_t^{\langle k\rangle}$ given by

$$g_t(z+1) = \sum_{k=0}^{\infty} \frac{1}{k!} \mu_t^{\langle k\rangle} z^k$$

The cumulant densities $q_K''(\cdot\,|\,t)$ may then be determined as

$$G''[\xi+1\,|\,t] = g_t\left[\int \xi(t+u)a_t(u)\,du + 1\right]$$

$$= \sum_k \frac{1}{k!} \mu_t^{\langle k\rangle} \left[\int \xi(v)a_t(v-t)\,dv\right]^k$$

and so

$$q_K''(t_1, \ldots, t_K\,|\,t) = \mu_t^{\langle K\rangle} a_t(t_1 - t) \cdots a_t(t_K - t) \tag{2.11.33}$$

From (2.11.31), the cumulant densities of the Neyman–Scott process are given by

$$\int \mu_t^{\langle K\rangle} a_t(t_1 - t) \cdots a_t(t_K - t)p'(t)\,dt \tag{2.11.34}$$

In the stationary case $p'(t) = p'$, $\mu_t^{\langle K\rangle} = \mu^{\langle K\rangle}$, $a_t(u) = a(u)$, and, in particular, one has

$$q_1 = p'E\{n\}$$

$$q_2(u) = p'E\{n(n-1)\} \int a(u-t)a(t)\,dt$$

If the characteristic function of the density is given by $\phi(\lambda) = \int \exp\{i\lambda u\}a(u)\,du$, then

$$f_2(\lambda) = (2\pi)^{-1}p'E\{n\} + (2\pi)^{-1}p'E\{n(n-1)\}\,|\phi(\lambda)|^2$$

and

$$f_K(\lambda_1, \ldots, \lambda_{K-1}) = (2\pi)^{-K+1}p' \sum_{k=1}^{K} \phi\left(\sum_{j \in \nu_1} \lambda_j\right) \cdots$$

$$\times\, \phi\left(\sum_{j \in \nu_{k-1}} \lambda_j\right) \phi\left(\sum_{j \in \nu_k} \lambda_j\right) \mu^{\langle k\rangle}$$

where the summation extends over all partitions $(\nu_1, \ldots, \nu_k)$ of the set $(1, \ldots, K)$ and $\lambda_K = -\lambda_1 - \cdots - \lambda_{K-1}$.

The *Bartlett–Lewis process* is a Poisson cluster process with $p'(t) = p'$ and the individual cluster processes renewal processes (i.e., the points are located at $0 = u_0, u_1, u_2, \ldots,$ with the $u_{j+1} - u_j$, $j = 0, 1, \ldots,$ independent and identically distributed). The probability that the cluster is of size n is $\pi(n)$. The cumulant densities for this process may be determined from expression (2.11.31) once expressions have been determined for the cumulant densities of a renewal process. This is done later in this section.

One important problem concerning the analysis of cluster processes is that of identifying the points of the cluster center process M' given the points of the overall process N.

Consider next a Poisson process N_R on the plane $(-\infty, \infty)^2$, having parameter measure R satisfying $R(I \times (-\infty, \infty))$, $R((-\infty, \infty) \times I) < \infty$ for compact I in $B(-\infty, \infty)$. The probability generating functional of N_R is given by

$$E \exp\left\{\iint \log \mathcal{N}(u, v) N_R(du, dv)\right\} = \exp\left\{\iint [\mathcal{N}(u, v) - 1] R(du, dv)\right\} \tag{2.11.35}$$

Consider the point process N_R^* on the line determined by superposing the two marginals of N_R, namely

$$N_R^*(t) = N_R((-\infty, t] \times (-\infty, \infty)) + N_R((-\infty, \infty) \times (-\infty, t])$$

Its probability generating functional is determined from (2.11.35) by setting $\mathcal{N}(u, v) = \xi(u)\xi(v)$ and may be written

$$G[\xi] = \exp\left\{\int [\xi(u) - 1][R(du \times (-\infty, \infty)) + R((-\infty, \infty) \times du] + \iint [\xi(u) - 1][\xi(v) - 1] R(du, dv)\right\} \tag{2.11.36}$$

In the case in which R is absolutely continuous with density $r(u, v)$ and

$$r_1(u) = \int r(u, v)\, dv < \infty, \qquad r_2(v) = \int r(u, v)\, du < \infty$$

expression (2.11.36) shows that N_R^* has intensity function $r_1(t) + r_2(t)$ and autocovariance density $2r(t_1, t_2)$. This construction shows that any nonnegative function $q_2(t_1, t_2)$ which is integrable in either variable, may be the autocovariance density of a point process. In the stationary case $q_2(t_1, t_2) = q_2(t_1 - t_2)$, and it is seen that any nonnegative integrable function can be the autocovariance density of a stationary point process. The rate of the process, as constructed earlier, will be $q_1 = \int q_2(u)\, du$.

The *Gauss–Poisson process* (introduced by Newman, 1970, and discussed by Milne and Westcott, 1972) has probability generating functional

$$\exp\left\{\int [\xi(u) - 1]Q_1(du) + \tfrac{1}{2}\iint [\xi(u) - 1][\xi(v) - 1]Q_2(du, dv)\right\} \tag{2.11.37}$$

with $Q_2(du, dv) = Q_2(dv, du)$. It is seen that the Gauss–Poisson process may be represented as $N + N^*_{Q_2/2}$ where N is a Poisson process independent of N with intensity measure $Q_1(dt) - Q_2(dt \times (-\infty, \infty))$. It is clear that this procedure may be extended to construct point processes with given (integrable in $K - 1$ arguments) nonnegative cumulant density of order K. From expression (2.5.16) the index of dispersion of the process here will be

$$I(t) = 1 + \frac{\{\int_0^t \int_0^t q_2(t_1 - t_2)\, dt_1\, dt_2\}}{\{\int_0^t \int q_2(t_1 - t_2)\, dt_1\, dt_2\}}$$

and so $I(\infty) = 2$. Also, $f_2(0) = q_1/\pi$ and $|f_2(\lambda) - q_1/2\pi| \le |f_2(0) - q_1/2\pi|$, indicating that this process too is overdispersed compared to a Poisson with the same intensity.

Hawkes (1972) introduced the class of *self-exciting point processes*. These are defined on the interval $(-\infty, \infty)$ and have conditional intensity of the form

$$\begin{aligned}\gamma(t, \omega) &= \mu + \int_{-\infty}^{t} a(t - u)\, dN(u)\\ &= \mu + \sum_{\tau_j \le t} a(t - \tau_j)\end{aligned} \tag{2.11.38}$$

where $\mu, a(u) \ge 0$, and $\int_0^\infty a(u)\, du < 1$. (The first condition here ensures that the conditional intensity is nonnegative; the second ensures that the process has finite intensity.) The parameters of such a process satisfy the following relationships:

$$p_1 = \mu + p_1 \int_0^\infty a(u)\, du$$

$$q_2(u) = \int_0^\infty a(v) q_2(u - v)\, dv$$

$$f_2(\lambda) = p_1/[2\pi|1 - A(\lambda)|^2]$$

$u > 0$, where $A(\lambda) = \int_0^\infty \exp\{-i\lambda u\} a(u)\, du$. The process is stationary, and has the following interpretation as a cluster process: (i) Immigrants arrive in accordance with a Poisson process of intensity μ. (ii) The immigrant who arrived at time τ_j generates descendants in accordance with a Poisson process of rate $a(t - \tau_j)$. (iii) The descendants in turn generate descendants, and so on. (This representation is discussed by Hawkes and Oakes, 1974.)

The process may be modulated by $\beta(t, \omega) = \beta$ to produce a further self-exciting process, provided $\beta \int_0^\infty a(u)\,du < 1$. By arranging for $A(\lambda)$ to be a rational function of λ, the power spectrum may be a rational function and the process will be analogous to the stationary N-tuple Markov Gaussian series.

An explicit expression is not known for the probability generating functional of a self-exciting process; however, it may be represented as

$$G[\xi] = \exp\left\{-\mu \int_{-\infty}^{\infty} (1 - H[\xi(\cdot\ + u)])\,du\right\}$$

where $H[\xi]$ satisfies the integral equation

$$H[\xi] = \xi(0)\exp\left\{-\int (1 - H[\xi(\cdot\ + u)])a(u)\,du\right\}$$

$H[\xi]$ is here the probability generating functional of the cluster generated by an immigrant arriving at time 0 (see Hawkes and Oakes, 1974).

Using expressions (2.2.17) and (2.11.38) the likelihood function may be written down directly in terms of μ, $a(\cdot)$, and $N(t)$, $0 \le t \le T$. Also, from expression (2.2.16) one has, for example,

$$\begin{aligned}\mathrm{Prob}&\{\tau_J > t \mid \tau_j, j < J\}\\ &= \exp\left\{-\mu(t - \tau_{J-1}) - \int_{-\infty}^{t}\int_{\max\{v,\tau_{J-1}\}}^{t} a(u - v)\,du\,dN(v)\right\}\end{aligned}$$

Rice (1975) has considered the problem of determining $A(\lambda)$ and $a(u)$ from p_1 and $f_2(\lambda)$. The particular case of $a(u) = \alpha\exp\{-\beta u\}$ is investigated in some detail by Oakes (1975).

Suppose that the members of the sequence $\tau_0, y_1, y_2, \ldots$ are statistically independent variates. Then the corresponding point process is called a *renewal process*. The Poisson and Erlang processes are examples of renewal processes. Generally the variates $y_1, y_2, \ldots$ are assumed to be identically distributed. Suppose their c.d.f. is $A(y)$, their density $a(y)$, and their hazard function $h(y) = (d/dy)\log[1 - A(y)]$. Suppose $A_0(y)$, $a_0(y)$, $h_0(y)$ are the corresponding parameters for τ_0. Then the conditional intensity function is here given by

$$\begin{aligned}\gamma(t, \omega) &= h_0(t) &&\text{for}\quad N(t) = 0\\ &= h(t - \tau_{N(t)-1}) &&\text{for}\quad N(t) \ge 1\end{aligned} \tag{2.11.39}$$

By first principles, the intensity function of the process is given by

$$\begin{aligned}p_1(t) &= a_0(t) + \int a(t - u)a_0(u)\,du\\ &\quad + \iint a(t - u)a(u - v)a_0(v)\,du\,dv + \cdots\end{aligned} \tag{2.11.40}$$

Suppose that the Laplace transform of $p_1(t)$ is denoted, for Re $s > 0$, by $\hat{p}_1(t) = \int_0^\infty \exp\{-st\} p_1(t)\, dt$, with a similar definition for $\hat{a}_0(s)$, $\hat{a}(s)$. Taking the Laplace transform of each side of Eq. (2.11.40), we obtain

$$\begin{aligned} \hat{p}_1(s) &= \hat{a}_0(s) + \hat{a}_0(s)\hat{a}(s) + \hat{a}_0(s)\hat{a}(s)^2 + \cdots \\ &= \hat{a}_0(s)/[1 - \hat{a}(s)] \end{aligned} \tag{2.11.41}$$

If the process is to be stationary, then $p_1(t) = p_1$ and Eq. (2.11.41) reads

$$p_1/s = \hat{a}_0(s)/[1 - \hat{a}(s)]$$

and therefore the initial distribution should have Laplace transform $\hat{a}_0(s) = p_1[1 - a(s)]/s$. It may be verified that this is the Laplace transform of $a_0(y) = p_1[1 - A(y)]$, and it may be further verified that for the density to integrate to 1, it is necessary that $p_1 = 1/E\{y_j\}$. The inverse character of the relationship between the expected interval length and the intensity should have been anticipated. In Section 2.5 it was seen that the relationship applied to general stationary point processes. The general theory of Palm distributions may be applied to show that the point process is strictly stationary with the choice $a_0(y) = [1 - A(y)]/E\{y_j\}$.

The product density of order 2 of the process is given by

$$p_2(t_1, t_2) = \text{Prob}\{dN(t_2) = 1 \mid N\{t_1\} = 1\} p(t_1)/dt_2$$

Now for $t_2 > t_1$

$$\begin{aligned} &\text{Prob}\{dN(t) = 1 \mid N\{0\} = 1\}/dt_2 \\ &\qquad = a(t) + \int a(t-u)a(u)\, du + \iint a(t-u)a(u-v)a(v)\, du\, dv + \cdots \\ &\qquad = b(t) \end{aligned}$$

The Laplace transform of $b(t)$ is here given by

$$\begin{aligned} \hat{b}(s) &= \int_0^\infty \exp\{-st\} b(t)\, dt = \hat{a}(s) + \hat{a}(s)^2 + \cdots \\ &= \hat{a}(s)/[1 - \hat{a}(s)] \end{aligned} \tag{2.11.42}$$

for Re $s > 0$. In particular cases this relationship may be inverted to obtain $b(t)$, and then one has the product density of order 2 as

$$p_2(t_1, t_2) = p_1(t_1) b(t_2 - t_1)$$

for $t_2 > t_1$. In the stationary case $p_2(u) = p_1 b(|u|)$. This relationship may be used to determine the power spectrum. Except at $\lambda = 0$

$$f_2(\lambda) = (2\pi)^{-1}p_1 + (2\pi)^{-1}\int \exp\{-i\lambda u\}p_2(u)\,du$$

$$= (2\pi)^{-1}p_1 + (2\pi)^{-1}\int_0^\infty \exp\{-i\lambda u\}p_2(u)\,du$$

$$+ (2\pi)^{-1}\int_0^\infty \exp\{i\lambda u\}p_2(u)\,du$$

$$= (2\pi)^{-1}p_1 + (2\pi)^{-1}p_1\hat{b}(i\lambda) + (2\pi)^{-1}p_1\hat{b}(-i\lambda) \qquad (2.11.43)$$

(by continuity, the same formula applies at $\lambda = 0$). This, together with (2.11.42), gives the required expression. A similar expression may be set down for the general cumulant spectrum. First set

$$g_k(\lambda_1, \ldots, \lambda_k) = p_1 \sum_P \hat{b}(i\lambda_{P1})\hat{b}(i\lambda_{P1} + i\lambda_{P2}) \cdots \hat{b}(\lambda_{P1} + \cdots + i\lambda_{Pk-1})$$

where the summation is over all permutations P of the set $(1, \ldots, k)$. Then

$$f_K(\lambda_1, \ldots, \lambda_{K-1}) = (2\pi)^{-K+1} \sum_{k=1}^{K} g_k\left(\sum_{j \in \nu_1} \lambda_j, \ldots, \sum_{j \in \nu_k} \lambda_j\right) \qquad (2.11.44)$$

where the summation is over all partitions $(\nu_1, \ldots, \nu_k)$ of $(1, \ldots, K)$ and it is understood that $\lambda_K = -\lambda_1 - \cdots - \lambda_{K-1}$. The corresponding product density is given by

$$p_K(t_1, \ldots, t_K) = p_1 b(t_2 - t_1) \cdots b(t_K - t_{K-1}) \qquad (2.11.45)$$

for $t_1 < t_2 < \cdots < t_K$.

The likelihood function of a given set of data on $[0, T]$ may be written

$$h_0(\tau_0)h(y_1) \cdots h(y_J) \exp\left\{-\int_0^{\tau_0} h_0(u)\,du - \int_0^{y_1} h(u)\,du - \cdots\right.$$

$$\left. - \int_0^{y_J} h(u)\,du - \int_0^{T-\tau_J} h(u)\,du\right\} \qquad (2.11.46)$$

with $J = N(T) - 1$. Cox (1972) develops some properties of the modulated renewal process having conditional intensity function

$$\exp\{\beta_1 z_1(t) + \cdots + \beta_p z_p(t)\}\gamma(t, \omega)$$

where $\gamma(t, \omega)$ is the conditional intensity function of a renewal process, $\beta_1, \ldots, \beta_p$ are unknown parameters, and $z_1(t), \ldots, z_p(t)$ are known functions. For example, Cox suggests consideration of the likelihood conditional on the order statistics of $y_1, \ldots, y_{N(t)-1}$.

A variety of generalizations of the renewal process are possible. Expression (2.11.39) could be generalized to the case of nonidentical interpoint distributions by writing

$$\gamma(t, \omega) = h(t - \tau_{N(t)-1}, N(t))$$

where $h(u, j)$, $j = 0, 1, 2, \ldots$, are successive hazard functions. The renewal process may be generalized to the *random walk point process* by allowing the random variables $\tau_0, y_1, y_2, \ldots$ to take on negative values and then reordering the points obtained. Daley and Oakes (1974) develop a variety of properties of this process. Another form of extension allows a number of successive intervals to be dependent. Wold (1948) introduced the class of *Markov-dependent interval processes.* Vere-Jones (1975a) developed certain results for these processes.

P. A. W. Lewis and co-workers (see Jacobs and Lewis, 1976) have introduced classes of point processes in which the individual intervals have exponential distributions (as in the Poisson case), but the intervals are dependent. Their EARMA (1, 1) model takes the following form: (i) $\{\varepsilon_j\}$ is a sequence of independent exponential variates; (ii) $\{U_j\}$, $\{V_j\}$ are independent sequences of independent variates taking the values $\{0, 1\}$ with $\text{Prob}\{U_j = 0\} = \beta$, $\text{Prob}\{V_j = 0\} = \rho$; (iii) $A_j = \rho A_{j-1} + V_j \varepsilon_j$, $j = 1, 2, \ldots$, and $A_0 = \varepsilon_0$; (iv) the interval sequence of the process is given by

$$y_j = \beta\varepsilon_j + U_j A_{j-1}$$

$j = 1, 2, \ldots$; (v) $\tau_0 = 0$. This process is stationary. In the case $\rho = 0$, an "exponential moving average model" is obtained with power spectrum the ratio of fourth-order polynomials in λ.

Consider next a stationary Markov process $X(t)$, $0 \le t < \infty$, with state space $\{0, 1, 2, \ldots\}$ and transition intensities given by

$$\begin{aligned} \text{Prob}\{X(t + dt) = j \mid X(t) = k\} &= q_k Q_{kj}\, dt, \qquad j \ne k \\ &= 1 - q_k\, dt, \qquad j = k \end{aligned}$$

with $Q_{kk} = 0$, $\sum_j Q_{kj} = 1$. Rudemo (1973) considers the point process whose points correspond to the transition times of the process X. This process may be represented as follows: let $U(j)$, $j = 0, 1, \ldots$, be a Markov chain with transition matrix $Q = [Q_{kj}]$. Suppose the process began at its stationary distribution. Let $\{e(j)\}$, $j = 0, 1, \ldots$, be a sequence of independent exponentials with mean 1, and independent of the sequence $\{U(j)\}$. Then the points of the process may be represented as $\tau_0 = 0$,

$$\tau_{j+1} = \tau_j + \varepsilon(j)/q_{U(j)}$$

When the point process itself is Markov, it is referred to as a *pure birth process.* Its conditional intensity function is now given by

$$\gamma(t, \omega) = \beta(t, N(t)) \qquad (2.11.47)$$

and initial conditions, such as $N(0) = 1$, are assumed. When $\beta(t, N) = \beta(t)N$, $\beta(t)$ is referred to as the birthrate at time t. Expression (2.11.47) shows that the process corresponds to modulating a unit Poisson process by the function $\beta(t, N(t))$. The likelihood function may be written

$$\exp\left\{-\int_0^T \beta(t, N(t))\,dt + \int_0^T \log \beta(t, N(t))\,dN(t)\right\}$$

In the case in which $\beta(t, N) = \beta N$, $N(0) = 1$, the stochastic process is really $N(t) - 1$ and the likelihood given data on $[0, T]$ is

$$\exp\left\{-\beta \int_0^T N(t)\,dt\right\} \times \beta^{N(T)-1} \times 2 \times 3 \times \cdots \times (N(T) - 1) \qquad (2.11.48)$$

Details of this birth process have been developed by Keiding (1974).

A *stationary Markov point process* is provided by a doubly stochastic Poisson with time-independent intensity process $p(\omega')$. From expression (2.11.20), the conditional intensity is given by

$$\gamma(t, \omega) = \frac{E_{\omega'}\{p(\omega')^{N(t)+1} \exp\{-tp(\omega')\}\}}{E_{\omega'}\{p(\omega')^{N(t)} \exp\{-tp(\omega')\}\}}$$

showing that the process is in fact Markov. The intensity of the process is $\mu = Ep(\omega')$, the autocovariance density $\sigma^2 = \operatorname{var} p(\omega')$, and the power spectrum is

$$\frac{\mu}{2\pi} + \frac{\sigma^2}{2\pi}\delta(\lambda)$$

The delta function at the origin is indicative of the nonmixing character of the process. The product densities of this process are all constant.

A direct counterpart of the pure birth process is provided by the *pure death process*, wherein the points of the process correspond to times of death. One way of setting up such a process is to consider elements such that

$$\text{Prob}\{\text{failure in } (t, t + dt)\} = \exp\left\{-\int_0^t h(u)\,du\right\} h(t)$$

The conditional intensity function of the process N_j corresponding to the life of a single element is

$$\gamma_j(t, \omega) = h(t)[1 - N_j(t^-)]$$

If $N(t) = \sum_j N_j(t)$ corresponds to the times of failure of a population of J independent elements, then its conditional intensity function is

$$h(t)[J - N(t^-)]$$

A great deal of literature exists concerning *birth and death processes* (e.g., Bharucha-Reid, 1960). These may be considered bivariate point processes $\{N_1(t), N_2(t)\}$ with the process N_1 referring to the times of births and the process N_2 referring to the times of deaths. The number living at time t is then $L(t) = N_1(t) - N_2(t)$ (assuming that there have been no immigrants). If births and deaths cannot occur simultaneously, then the processes N_1 and N_2 may be determined directly from the process L. An equivalent notation could apply to a queueing system with N_1 referring to the arrival times of customers, N_2 to departure times, and L to the number of customers in the system.

An extensive literature also exists concerning *branching processes*. Here $N(t)$ may refer to the number of individuals alive at time t. The process evolves through the property that each individual has a probability of producing further individuals for the population. The individuals may be classified as to age or generation. A general reference to the theory of branching processes is provided by Athreya and Ney (1972).

It is perhaps sensible to conclude this section with a warning to the effect that a number of the processes discussed can lead to identical functional forms for certain of the parameters considered. Daley and Oakes (1974) mention that a particular power spectrum that is a ratio of quadratics in λ can arise as (i) a self-exciting process with exponential kernel, (ii) a Neyman–Scott process with exponential displacements and a Poisson number of cluster members, (iii) a random walk point process with the generating distribution having exponential tails. In a similar vein, a flat spectrum occurs for either a stationary Poisson process or a doubly stochastic Poisson with white noise random intensity $p(t, \omega')$.

3. INFERENCE

3.1. Linear Statistics for Ordinary Time Series

Suppose that the data $\{X(t), 0 < t < T\}$ are available for analysis. In many circumstances it is of interest to compute a linear statistic of the form

$$\int_0^T \phi(t)X(t)\,dt \tag{3.1.1}$$

for a given function ϕ. For example, suppose that the series X is given by

$$X(t) = \theta\phi(t) + \varepsilon(t) \tag{3.1.2}$$

$-\infty < t < \infty$, ϕ known, θ unknown, and ε a zero mean series. Then the least squares estimate of θ is

$$\hat{\theta} = \left\{\int_0^T \phi(t)X(t)\,dt\right\} \bigg/ \left\{\int_0^T |\phi(t)|^2\,dt\right\} \tag{3.1.3}$$

a statistic based on (3.1.1). A theorem is presently given indicating the large sample behavior of such statistics. First, an assumption is formulated concerning the coefficient functions ϕ.

Assumption 3.1. Given the (possibly complex-valued) functions $\phi_j{}^T(t)$, $0 \le t < \infty$, $j = 1, \ldots, J$, suppose that as $T \to \infty$

(i) $$\left\{\int_0^T \phi_j{}^T(t+u)\overline{\phi_k{}^T(t)}\,dt\right\} \bigg/ T \to b_{jk}(u)$$

for $0 \le u < \infty$ with $b_{jj}(u)$ finite and continuous at 0;

(ii) $$\left\{\sup_{0 \le t \le T} |\phi_j{}^T(t)|\right\} \big/ \sqrt{T} \to 0$$

for $j, k = 1, \ldots, J$.

The conditions of this assumption are similar to those of Grenander (1954) and Hannan (1970, p. 215). Examples of functions satisfying the conditions are given shortly. Under the assumption, there exist functions of bounded variation $G_{jk}(\lambda)$, $-\infty < \lambda < \infty$, such that

$$b_{jk}(u) = \int_{-\infty}^{\infty} \exp\{iu\lambda\}\,dG_{jk}(\lambda)$$

for $0 \le u < \infty$ and $j, k = 1, \ldots, J$. The $dG_{jk}(\lambda)$ may be viewed as the limit of

$$dG_{jk}^T(\lambda) = (2\pi T)^{-1} \left[\int_0^T \exp\{-i\lambda t\}\phi_j{}^T(t)\,dt\right] \left[\int_0^T \exp\{i\lambda t\}\overline{\phi_k{}^T(t)}\,dt\right] d\lambda$$

in the sense of weak convergence. Now one can state the following theorem:

Theorem. Let $X(t)$, $-\infty < t < \infty$, be a series of the form $X(t) = \mu(t) + \varepsilon(t)$, with $\mu(t)$ nonstochastic and $\varepsilon(t)$ a zero mean stationary series satisfying condition (2.4.18) and having power spectrum $f_{\varepsilon\varepsilon}(\lambda)$. Suppose $\phi_j{}^T$, $j = 1, \ldots, J$, are functions satisfying Assumption 3.1. Let

$$U_j{}^T = \int_0^T \phi_j{}^T(t)X(t)\,dt$$

Then, as $T \to \infty$, the variate $\{U_1{}^T, \ldots, U_J{}^T\}$ is asymptotically normal with

$$EU_j{}^T = \int_0^T \phi_j{}^T(t)\mu(t)\,dt$$

$$\operatorname{cov}\{U_j{}^T, U_k{}^T\} \sim 2\pi T \int_{-\infty}^{\infty} f_{\varepsilon\varepsilon}(\lambda)\,dG_{jk}(-\lambda) \qquad (3.1.4)$$

for $j, k = 1, \ldots, J$.

This theorem is proved quite simply by evaluating joint cumulants and showing that, when standardized, those of order greater than 2 tend to 0 as $T \to \infty$. The case of discrete time and of mixing conditions other than (2.4.18) has been considered by Hannan (1973a).

Example 1. Suppose $J = 1$, $\phi^T(t) = 1$. Then $b(u) = 1$, $dG(\lambda) = \delta(\lambda)\,d\lambda$, and

$$\int_0^T X(t)\,dt$$

is seen to be asymptotically normal with mean $\int_0^T \mu(t)\,dt$ and variance $2\pi T f_{\varepsilon\varepsilon}(0)$.

Example 2. Suppose $J = 1$, $\phi^T(t) = \exp\{-i\lambda_0 t\}$. Then $b(u) = \exp\{-i\lambda_0 u\}$, $dG(\lambda) = \delta(\lambda + \lambda_0)\,d\lambda$, and

$$\int_0^T \exp\{-i\lambda_0 t\}X(t)\,dt$$

is asymptotically (complex) normal with variance $2\pi T f_{\varepsilon\varepsilon}(\lambda_0)$.

Example 3. Suppose $J = 1$, $\phi^T(t) = (t/T)^n$. Then $b(u) = 1/(2n + 1)$, $dG(\lambda) = b(0)\,\delta(\lambda)\,d\lambda$, and

$$\left\{\int_0^T t^n X(t)\,dt\right\}\Big/T^n$$

is asymptotically normal with variance $2\pi T f_{\varepsilon\varepsilon}(0)/(2n + 1)$.

Example 4. Suppose $J = 1$, $\phi^T(t) = \rho^{t/T}$, with $0 < \rho \le 1$. Then $b(u) = (1 - \rho^2)/[2 \ln 1/\rho]$, $dG(\lambda) = b(0)\,\delta(\lambda)\,d\lambda$, and

$$\int_0^T \rho^{t/T} X(t)\,dt$$

is asymptotically normal with variance $2\pi T f_{\varepsilon\varepsilon}(0)b(0)$.

Example 5. Suppose the series is given by (3.1.2) with ϕ satisfying Assumption 3.1. Then $\hat{\theta}$ of (3.1.3) is asymptotically normal with mean θ and variance

$$2\pi T^{-1}\left[\int_{-\infty}^{\infty} f_{\varepsilon\varepsilon}(\lambda)\, dG(\lambda)\right]\Big/ b(0)^2$$

Example 6. Suppose the series X is given by (3.1.2) with ϕ satisfying Assumption 3.1. Consider choosing θ to minimize

$$E\left\{\int_0^T \int_0^T w(s-t)[X(s) - \theta\phi(s)][X(t) - \theta\phi(t)]\, ds\, dt\right\}$$

with $w(t) = \int \exp\{it\lambda\} W(\lambda)\, d\lambda$, $W(\lambda)$ being symmetric, nonnegative, bounded, and integrable. The extreme value of θ is given by

$$\left\{\int_0^T \int_0^T w(s-t)\overline{\phi(s)}X(t)\, ds\, dt\right\}\Big/\left\{\int_0^T \int_0^T w(s-t)\overline{\phi(s)}\phi(t)\, ds\, dt\right\} \tag{3.1.5}$$

From the theorem, this statistic is asymptotically normal with mean θ and variance

$$2\pi T^{-1}\left[\int f_{\varepsilon\varepsilon}(\lambda)|W(\lambda)|^2\, dG(\lambda)\right]\Big/\left|\int W(\lambda)\, dG(\lambda)\right|^2$$

By Schwarz's inequality

$$\int f_{\varepsilon\varepsilon}(\lambda)|W(\lambda)|^2\, dG(\lambda) \int f_{\varepsilon\varepsilon}(\lambda)^{-1}\, dG(\lambda) \geq \left|\int W(\lambda)\, dG(\lambda)\right|^2$$

and so the best choice of $W(\lambda)$ is $f_{\varepsilon\varepsilon}(\lambda)^{-1}$.

Results similar to those of Examples 5 and 6 may be developed for the case in which

$$X(t) = \theta_1\phi_1(t) + \cdots + \theta_P\phi_P(t) + \varepsilon(t)$$

Related references include Kholevo (1969) and Rozanov (1969). Central limit theorems may also be developed directly for the case of a series X such that $\mathscr{A}[X]$ is a series satisfying (2.4.18) for some linear filter $\mathscr{A}$.

3.2. Linear Statistics for Point Processes

In the case of a point process N, given the events of a realization on $(0, T]$, one may compute linear statistics

$$\int_0^T \phi(t)\, dN(t) = \phi(\tau_0) + \cdots + \phi(\tau_{N(T)-1}) \tag{3.2.1}$$

for given functions ϕ. For example, suppose that N is a process with rate function of the form $p_1(t) = \theta\phi(t)$, ϕ known and θ unknown. Then

$$\hat{\theta} = \left\{\int_0^T \phi(t)\, dN(t)\right\} \Big/ \left\{\int_0^T |\phi(t)|^2\, dt\right\}$$

is a linear statistic providing an unbiased estimate of θ. In particular, the choices $\phi(t) = \cos \lambda t$ and $\phi(t) = \sin \lambda t$ lead to the computation of the finite Fourier transform

$$\int_0^T \exp\{-i\lambda t\}\, dN(t) \tag{3.2.2}$$

Alternatively, taking $\phi(t)$ to be approximately the Dirac delta $\delta(t - t_0)$ leads to a statistic whose expected value is approximately $p_1(t_0)$, the rate of the process at time t_0. In the case of a Poisson process with rate function $p(t)$, from expression (2.11.6) the log likelihood function is

$$\int_0^T \log p(t)\, dN(t) - \int_0^T p(t)\, dt$$

and again a linear statistic appears in an important fashion. Grandell (1972) and Clevenson and Zidek (1975) consider the use of linear statistics in the estimation of the rate function in the case of doubly stochastic and time inhomogeneous Poisson processes, respectively.

Turning to the consideration of large sample properties of linear statistics based on point processes, suppose that for the process N of interest

$$\text{cum}\{dN(t_1), \ldots, dN(t_K)\} = C_K(dt_1, \ldots, dt_K)$$

$K = 1, 2, \ldots,$ with the C_K of bounded variation in finite intervals, in the manner of Brillinger (1972). Suppose

$$U_j^T = \int_0^T \phi_j^T(t)\, dN(t), \qquad j = 1, \ldots, J$$

Then one has the following theorem:

Theorem. Suppose that the

$$\int_0^T \int_0^T \phi_j^T(t_1)\phi_j^T(t_2) C_2(dt_1, dt_2)$$

$j = 1, \ldots, J$, are each of order of magnitude σ_T^2 as $T \to \infty$. Suppose in addition that

$$\int_0^T \cdots \int_0^T \phi_{j_1}^T(t_1) \cdots \phi_{j_K}^T(t_K) C_K(dt_1, \ldots, dt_K) = o(\sigma_T{}^K)$$

for $j_1, \ldots, j_K = 1, \ldots, J$ and $K = 3, 4, \ldots$. Then the variate $\{U_1{}^T, \ldots, U_J{}^T\}$ is asymptotically normal.

The proof of this theorem is immediate, as the standardized joint cumulants of order greater than 2 tend to 0 under the indicated assumptions.

In the case in which the process N is stationary, satisfying the mixing condition (2.5.29), and the $\phi_j{}^T$ satisfy Assumption 3.1, the conditions of the theorem are satisfied. The variate $\{U_1{}^T, \ldots, U_J{}^T\}$ is asymptotically normal with the covariance of $U_j{}^T$ and $U_k{}^T$ given by $2\pi T \int f_2(\lambda)\, dG_{jk}(-\lambda)$. In particular, the finite Fourier transform (3.2.2) is asymptotically (complex) normal with variance $2\pi T f_2(\lambda)$.

The discussion of this section has concentrated on the counts of the process. Clearly linear statistics may be formed in the intervals. The interval sequence is an ordinary discrete time series and results of the character of those of Section 3.1 may be set down directly.

3.3. Quadratic Statistics for Ordinary Time Series

A variety of the quantities computed in the analysis of ordinary time series are quadratic in the observations. Suppose that the data $\{X(t), 0 < t < T\}$ are available. Some examples are as follows:

(i) $$c_2{}^T(u) = T^{-1} \int_0^{T-|u|} [X(t+u) - c_1{}^T][X(t) - c_1{}^T]\, dt$$

with $c_1{}^T = T^{-1} \int_0^T X(t)\, dt$, computed as an estimate of the autocovariance function $c_2(u)$;

(ii) the periodogram

$$I_2{}^T(\lambda) = (2\pi T)^{-1} \left| \int_0^T \exp\{-i\lambda t\} X(t)\, dt \right|^2$$

(iii) The empirical spectral measure

$$F_2{}^T(\lambda) = \frac{2\pi}{T} \sum_{0 < 2\pi s/T \leq \lambda} I_2{}^T\left(\frac{2\pi s}{T}\right)$$

(iv) $$f_2{}^T(\lambda) = \frac{2\pi}{T} \sum_{s \neq 0} W^T\left(\lambda - \frac{2\pi s}{T}\right) I_2{}^T\left(\frac{2\pi s}{T}\right) \tag{3.3.1}$$

with W^T concentrated near 0 and integrating to 1, or

$$f_2{}^T(\lambda) = L^{-1} \sum_{j=1}^{L} I_2{}^T\left(\frac{2\pi s_j}{T}\right) \tag{3.3.2}$$

with $s_1, \ldots, s_J$ distinct positive integers near $T\lambda/2\pi$, as estimates of the power spectrum $f_2(\lambda)$

$$\text{(v)} \qquad \frac{(2\pi)^2}{T} \sum_s I_{ee}^T\left(\frac{2\pi s}{T}\right)\left|W\left(\frac{2\pi s}{T}\right)\right|^2 I_{\phi\phi}^T\left(\frac{2\pi s}{T}\right)$$

with $e(t)$ the residual series $X(t) - \hat{\theta}\phi(t)$, $\hat{\theta}$ given by (3.1.5), as an estimate of the numerator of the variance (3.1.6) and finally

$$\text{(vi)} \qquad \sum_{s \neq 0} \left[\log f\left(\frac{2\pi s}{T}\right) \Big/ g\left(\frac{2\pi s}{T}\right) + I_2{}^T\left(\frac{2\pi s}{T}\right)\left\{\left[1/f\left(\frac{2\pi s}{T}\right)\right] - \left[1/g\left(\frac{2\pi s}{T}\right)\right]\right\}\right]$$

computed for various functions f and g in the Gaussian estimation of time series parameters (see Dzhaparidze and Yaglom, 1973).

These examples lead to the consideration of quadratic expressions of the form

$$V_j{}^T = \frac{2\pi}{T} \sum_{s \neq 0} A_j{}^T\left(\frac{2\pi s}{T}\right) I_2{}^T\left(\frac{2\pi s}{T}\right)$$
$$= \iint a_j{}^T(s-t)X(s)X(t)\, ds\, dt \tag{3.3.3}$$

where

$$a_j{}^T(t) = \frac{2\pi}{T} \sum_{s \neq 0} \exp\left\{-i\frac{2\pi st}{T}\right\} A_j{}^T\left(\frac{2\pi s}{T}\right)$$

$j = 1, \ldots, J$. In connection with these variates, one has the following theorem:

Theorem. Suppose that the series X satisfies condition (2.4.18). Suppose $A_j{}^T(\alpha)$ of (3.3.3) $= A_j(\alpha)$ with A_j continuous, bounded, and absolutely integrable. Then the variate $\{V_1{}^T, \ldots, V_J{}^T\}$ is asymptotically normal with

$$EV_j{}^T \sim \int A_j(\alpha) f_2(\alpha)\, d\alpha \tag{3.3.4}$$

$$\operatorname{cov}\{V_j{}^T, V_k{}^T\} \sim \frac{2\pi}{T}\int A_j(\alpha)\overline{A_k(\alpha)} f_2(\alpha)^2\, d\alpha + \frac{2\pi}{T}\int A_j(\alpha) A_k(\alpha) f_2(\alpha)^2\, d\alpha$$
$$+ \frac{2\pi}{T}\iint A_j(\alpha)\overline{A_k(\beta)} f_4(\alpha, -\alpha, -\beta)\, d\alpha\, d\beta \tag{3.3.5}$$

as $T \to \infty$. Suppose $A_j{}^T(\alpha) = B_T^{-1} W(B_T^{-1}[\alpha - \lambda_j])$ in the manner of (3.3.1), with W continuous, bounded, absolutely integrable, and integrating to 1.

Then the variate $\{V_1^T, \ldots, V_J^T\}$ is asymptotically normal with

$$EV_j^T \sim f_2(\lambda_j)$$

$$\operatorname{cov}\{V_j^T, V_k^T\} \sim 2\pi \int W(\alpha)^2\, d\alpha\, (B_T T)^{-1} f_2(\lambda_j)^2 \qquad \text{if} \quad \lambda_j = \pm\lambda_k \neq 0$$

$$\sim 0 \qquad \text{if} \quad \lambda_j \neq \lambda_k$$

Suppose $J = 1$ and A^T is such as to give the statistic (3.3.2). Then the variate (3.3.2) is asymptotically $f_2(\lambda)\chi^2_{2L}/(2L)$.

The large sample distributions of the statistics at the beginning of this section may be obtained by choice of appropriate A_j^T.

Hannan (1973b) develops certain large sample results for quadratic statistics in discrete time series.

3.4. Quadratic Statistics for Point Processes

Important second-order parameters for stationary point processes include the *product density* $p_2(u)$ defined by

$$p_2(u)\, dt\, du = \operatorname{Prob}\{dN(t+u) = 1 \text{ and } dN(t) = 1\} \tag{3.4.1}$$

$u \neq 0$; the *covariance density* $q_2(u)$ of

$$q_2(u)\, dt\, du = \operatorname{cov}\{dN(t+u), dN(t)\} \tag{3.4.2}$$

$u \neq 0$, so that $q_2(u) = p_2(u) - p_1{}^2$; the *power spectrum*

$$f_2(\lambda) = (2\pi)^{-1}\left[p_1 + \int \exp\{-i\lambda u\} q_2(u)\, du\right] \tag{3.4.3}$$

$-\infty < \lambda < \infty$; the *spectral measure*

$$F_2(\lambda) = \int_0^\lambda f_2(\alpha)\, d\alpha \tag{3.4.4}$$

the *variance time curve*

$$V(t) = \operatorname{var} N(t)$$

$$= tp_1 + 2\int_0^t (t-u) q_2(u)\, du \tag{3.4.5}$$

$$= \int \left(\frac{\sin t\alpha/2}{\alpha/2}\right)^2 f_2(\alpha)\, d\alpha \tag{3.4.6}$$

and finally

$$\operatorname{var}\left\{\int \phi(t)\, dN(t)\right\} = p_1 \int \phi(t)^2\, dt + \iint \phi(s)\phi(t) q_2(s-t)\, ds\, dt$$

$$= \int |\Phi(\alpha)|^2 f_2(\alpha)\, d\alpha \tag{3.4.7}$$

where $\Phi(\alpha) = \int \exp\{i\alpha t\}\phi(t)\, dt$.

Given the stretch of data $\{N(t),\ 0 < t \le T\}$, with events at points $\tau_0 < \tau_1 < \cdots < \tau_{N(T)-1}$, estimates of each of these parameters may be constructed. To begin, suppose that β_T is a nonnegative scale factor tending to 0 as $T \to \infty$. As an estimate of $p_2(u)$ now consider

$$p_2{}^T(u) = \text{card}\{(j, k) \text{ such that } u - \beta_T < \tau_j - \tau_k < u + \beta_T\}/(2\beta_T T) \tag{3.4.8}$$

and

$$p_2{}^T(u) = \sum_{j,k} \frac{w([u - \tau_j + \tau_k]/\beta_T)}{(\beta_T T \int w(t)\, dt)} \tag{3.4.9}$$

with $w(t)$ nonnegative, as estimates of $p_2(u)$. Suppose the process N satisfies condition (2.5.29). Then when $\beta_T = L/T$, L fixed, the estimate (3.4.8) is distributed asymptotically as $\mathscr{P}/2L$ where $\mathscr{P}$ is Poisson with mean $2Lp_2(u)$. On the other hand, when $\beta_T \to 0$, but $\beta_T T \to \infty$, the estimate (3.4.9) is asymptotically normal with variance

$$\left[p_2(u) \int w(t)^2\, dt\right] \Big/ \left(\beta_T T \left|\int w(t)\, dt\right|^2\right)$$

These results may be found in the chapter by Brillinger (1975b). The covariance density may be estimated by

$$q_2{}^T(u) = p_2{}^T(u) - (N(T)/T)^2$$

involving either of the statistics (3.4.8) or (3.4.9).

The periodogram of the data under discussion is the quadratic statistic given by

$$I_2{}^T(\lambda) = (2\pi T)^{-1} \left|\int_0^T \exp\{-i\lambda t\}\, dN(t)\right|^2$$

Estimates of the power spectrum may be based on it through the formula (3.3.1) or (3.3.2). As was the case with ordinary time series, the first estimate will be asymptotically normal, the second distributed asymptotically as a multiple of a chi-squared.

A further situation in which a quadratic statistic based on point process data occurs is with the approximate log likelihood function

$$-\sum_{s=1}^{S} \left[\log f_2\left(\frac{2\pi s}{T}\right) + I_2{}^T\left(\frac{2\pi s}{T}\right) \Big/ f_2\left(\frac{2\pi s}{T}\right)\right] \tag{3.4.10}$$

derived by thinking of the periodogram ordinates as independent exponential variates (e.g., Hawkes and Adamopoulos, 1973).

The expressions (3.4.4), (3.4.6), (3.4.7), and (3.4.10) each suggest consideration of quadratic statistics of the form

$$V_j^T = \frac{2\pi}{T} \sum_{s=1}^{S} A_j\left(\frac{2\pi s}{T}\right) I_2{}^T\left(\frac{2\pi s}{T}\right)$$

$j = 1, \ldots, J$. Provided the $A_j(\alpha)$ are continuous, bounded, and absolutely integrable, the variate $\{V_1{}^T, \ldots, V_J{}^T\}$ is asymptotically normal with first- and second-order moments given by (3.3.4) and (3.3.5).

The construction of estimates of certain higher order "polynomial" parameters of time series and point processes by polynomial functions of the observations is discussed by Brillinger (1972, 1975b, c).

3.5. General Parameter Estimation for Ordinary Time Series

In a variety of circumstances, the probability distribution of an ordinary time series X depends on a finite-dimensional parameter θ. For example, the series might be defined by

$$X(t) = \alpha_0 \varepsilon(t) + \alpha_1 \varepsilon^{(1)}(t) + \cdots + \alpha_p \varepsilon^{(p)}(t) \tag{3.5.1}$$

or

$$\beta_0 X(t) + \beta_1 X^{(1)}(t) + \cdots + \beta_q X^{(q)}(t) = \varepsilon(t) \tag{3.5.2}$$

where ε is a white noise series of variance σ^2, and one might be interested in estimating the parameter $\theta = (\alpha_0, \ldots, \alpha_p, \sigma^2)$ in case (3.5.1) and $\theta = (\beta_0, \ldots, \beta_q, \sigma^2)$ in case (3.5.2). In other situations the series might have the form corresponding to linear regression on $0 < t < T$

$$X(t) = \theta_1 \phi_1{}^T(t) + \cdots + \theta_P \phi_P{}^T(t) + \varepsilon(t) \tag{3.5.3}$$

with ε a stationary error series, or the form corresponding to nonlinear regression

$$X(t) = \psi^T(t; \theta) + \varepsilon(t) \tag{3.5.4}$$

where the functional form of ψ^T is known, but not the actual value of θ.

A number of general estimation procedures take the following form: a loss function $Q^T(\theta; X)$ is given, and an estimate $\hat{\theta}$ taken as the value of θ giving the minimum of $Q^T(\theta; X)$. Examples of such procedures include *linear least squares* where $\theta = (\theta_1, \ldots, \theta_P)$ of (3.5.3) is estimated by minimizing

$$Q^T(\theta; X) = T^{-1} \int_0^T |X(t) - \theta_1 \phi_1{}^T(t) - \cdots - \theta_P \phi_P{}^T(t)|^2 \, dt \tag{3.5.5}$$

in the case of (3.5.3); *nonlinear least squares* where

$$Q^T(\theta; X) = T^{-1} \int_0^T |X(t) - \psi^T(t; \theta)|^2 \, dt \tag{3.5.6}$$

in the case of (3.5.4); *Gaussian estimation* where, as in the case of (3.5.1) and (3.5.2), the functional form $f_2(\lambda; \theta)$ of the power spectrum is known and

$$Q^T(\theta; X) = \frac{2\pi}{T} \sum_{s=1}^{S} \left[\log f_2\left(\frac{2\pi s}{T}; \theta\right) \Big/ g\left(\frac{2\pi s}{T}\right) + I_2{}^T\left(\frac{2\pi s}{T}\right)\left\{1 \Big/ f_2\left(\frac{2\pi s}{T}; \theta\right) - 1 \Big/ g\left(\frac{2\pi s}{T}\right)\right\}\right] \quad (3.5.7)$$

for some given function $g(\lambda)$; *maximum likelihood* where $Q^T(\theta; X)$ is taken as the negative of the likelihood ratio relative to some fixed measure, of the given data.

A lemma indicating conditions under which a $\hat{\theta}$ constructed in this manner is consistent, is presented by Brillinger (1975b). Among the conditions are the existence of $Q(\theta)$ with $Q(\theta) > Q(\theta_0)$ for $\theta \neq \theta_0$, the true parameter value, and $Q^T(\theta; X) \geq Q(\theta) + o_p(1)$, $Q^T(\theta_0; X) = Q(\theta_0) + o_p(1)$. Under further regularity conditions

$$\frac{\partial Q^T(\theta_0; X)}{\partial \theta_0}$$

is asymptotically $N_P(0, \mathbf{\Sigma}_T)$,

$$\frac{\partial^2 Q^T(\theta_0; X)}{\partial \theta_0 \, \partial \theta_0{}^*}$$

tends to $\mathbf{\Psi}$ in probability and $\hat{\theta}$ is asymptotically $N_P(\theta_0, \mathbf{\Psi}^{-1}\mathbf{\Sigma}_T\mathbf{\Psi}^{-1})$.

In the case of (3.5.3) and (3.5.5) and where the $\phi_j{}^T$ satisfy Assumption 3.1, one has

$$Q(\theta) = (\theta - \theta_0)[b_{jk}(0)](\theta - \theta_0)^* + \int f_{\varepsilon\varepsilon}(\alpha)\, d\alpha$$

$$\mathbf{\Psi} = 2[b_{jk}(0)]$$

$$\mathbf{\Sigma}_T = T^{-1} 8\pi \left[\int f_{\varepsilon\varepsilon}(\alpha)\, dG_{jk}(-\alpha)\right]$$

In the case of (3.5.7),

$$Q(\theta) = \int \left[\frac{\log f_2(\alpha; \theta)}{g(\alpha)} + f_2(\alpha; \theta_0)\left\{\frac{1}{f_2(\alpha; \theta)} - \frac{1}{g(\alpha)}\right\}\right] d\alpha$$

$$\mathbf{\Psi} = \left[\int \frac{\partial \log f_2(\alpha; \theta_0)}{\partial \theta_j} \frac{\partial \log f_2(\alpha; \theta_0)}{\partial \theta_k}\, d\alpha\right]$$

$$\mathbf{\Sigma}_T = T^{-1}\mathbf{\Psi} + T^{-1}\left[2\pi \iint \frac{\partial \log f_2(\alpha; \theta_0)}{\partial \theta_j} \frac{\partial \log f_2(\beta; \theta_0)}{\partial \theta_k} \times \frac{f_4(\alpha, -\alpha, -\beta; \theta_0)}{f_2(\alpha; \theta_0) f_2(\beta; \theta_0)}\, d\alpha\, d\beta\right]$$

As an example of (3.5.4) and (3.5.6), consider the model of a series as a sum of decaying cosines:

$$X(t) = \sum_{k=1}^{K} \alpha_k \exp\left\{-\frac{\phi_k t}{T}\right\} \cos\{\gamma_k t + \delta_k\} + \varepsilon(t)$$

for $0 < t < T$. The matrices $\mathbf{\Psi}, \mathbf{\Sigma}_T$ may be evaluated and the estimates found to be asymptotically normal with

$$\overrightarrow{\text{var}}\, \hat{\alpha} \sim T^{-1} 4\pi f_{\varepsilon\varepsilon}(\gamma) I_2(\phi) J(\phi)^{-1}$$

$$\overrightarrow{\text{var}}\, \hat{\phi} \sim T^{-1} 4\pi f_{\varepsilon\varepsilon}(\gamma) I_0(\phi) J(\phi)^{-1} \alpha^{-2}$$

$$\overrightarrow{\text{var}}\, \hat{\gamma} \sim T^{-3} 4\pi f_{\varepsilon\varepsilon}(\gamma) I_0(\phi) J(\phi)^{-1} \alpha^{-2}$$

$$\overrightarrow{\text{var}}\, \hat{\delta} \sim T^{-1} 4\pi f_{\varepsilon\varepsilon}(\gamma) I_2(\phi) J(\phi)^{-1} \alpha^{-2}$$

where $I_j(\phi) = \int_0^1 u^j \exp\{-2\phi u\}\, du$ and $J(\phi) = I_0(\phi) I_2(\phi) - I_1(\phi)^2$.

As an example of the use of the method of maximum likelihood, consider the diffusion process determined by

$$dX(t) = \mu(t, X(t))\, dt + dW(t)$$

where W is the Wiener process, $\mu(t, x) = \sum \theta_k \phi_k(t, x)$ with $\phi_k(t, x)$ known, $\theta_0 = 1$, and $X(0) = 0$. From (2.10.8), the logarithm of its likelihood ratio relative to W is given by

$$\sum_{k=0}^{K} \theta_k \int_0^T \phi_k(t, X(t))\, dX(t) - \tfrac{1}{2} \sum_{j,k=0}^{K} \theta_j \theta_k \int_0^T \phi_j(t, X(t)) \phi_k(t, X(t))\, dt$$

The maximum likelihood estimate of $\theta = (\theta_1, \ldots, \theta_K)$ is the solution of the system of equations

$$\sum_{k=1}^{K} \hat{\theta}_k \int_0^T \phi_j(t, X(t)) \phi_k(t, X(t))\, dt = \int_0^T \phi_j(t, X(t))[dX(t) - \phi_0(t, X(t))\, dt]$$

$j = 1, \ldots, K$. Under regularity conditions, including

$$\underset{T \to \infty}{p\text{-lim}}\; T^{-1} \int_0^T \phi_j(t, X(t)) \phi_k(t, X(t))\, dt = I_{jk}(\theta)$$

Taraskin (1974) shows that $\hat{\theta} = (\hat{\theta}_1, \ldots, \hat{\theta}_K)$ is asymptotically normal with mean θ and covariance matrix $T^{-1}[I_{jk}(\theta)]^{-1}$ as $T \to \infty$.

In the case in which $\phi_k(t, x) = \phi_k(x)$, the process X may be stationary. Suppose that it has stationary distribution $G_\theta(x)$; then Taraskin (1974) shows that

$$I_{jk}(\theta) = \int \phi_j(x) \phi_k(x)\, dG_\theta(x)$$

$j, k = 1, \ldots, K$. The stationary case is also considered by Brown and Hewitt (1975) and Kulinich (1975). When a diffusion process has known local variance $\sigma^2(t, x)$, it may be transformed so that $\sigma(t, x) = 1$ as earlier. Brown and Hewitt (1975) remark that in the case in which $\sigma(t, x) = \sigma(x)$, the latter value may be determined almost surely, as

$$\lim_{n\to\infty} \sum_{j=1}^{2^n} [X(jt2^{-n}) - X((j-1)t2^{-n})]^2 = \int_0^t \sigma^2(X(u))\, du$$

almost surely.

As a further use of the method of maximum likelihood consider the linear dynamic system (2.8.9) wherein $F, \ldots, L$ depend on an unknown parameter θ, but not on t, i.e., consider the system

$$\begin{aligned} dS(t) &= F_\theta S(t)\, dt + G_\theta X(t)\, dt + K_\theta\, dW_1(t) \\ dY(t) &= H_\theta S(t)\, dt + J_\theta X(t)\, dt + L_\theta\, dW_2(t) \end{aligned} \tag{3.5.8}$$

where $W_1(t)$, $W_2(t)$ are independent Wiener processes. Suppose that the dependence of $F_\theta, \ldots, L_\theta$ on θ is such that the system (3.5.8) is identifiable. (Conditions under which this occurs may be found, e.g., in Balakrishnan, 1973; Glover and Williams, 1974.) Suppose that the data $\{X(t), Y(t), 0 < t < T\}$ are available and that it is desired to estimate θ. If $L_\theta = L$, then the likelihood ratio of the process Y relative to the process LW_2 would be given by

$$\exp\left\{L^{-2} \int_0^T (H_\theta \hat{S}_\theta(t) + J_\theta X(t))\, dY(t) - L^{-2} \int_0^T (H_\theta \hat{S}_\theta(t) + J_\theta X(t))^2\, dt/2\right\} \tag{3.5.9}$$

where

$$\hat{S}_\theta(t) = F_\theta \int_0^t \hat{S}_\theta(u)\, du + G_\theta \int_0^t X(u)\, du + L^{-2} \int_0^t P_\theta(u) H_\theta{}^*\, dv_\theta(u)$$

$$v_\theta(t) = Y(t) - H_\theta \int_0^t \hat{S}_\theta(u)\, du - J_\theta \int_0^t X(u)\, du$$

and

$$\frac{dP_\theta(t)}{dt} = F_\theta P_\theta(t) + P_\theta(t) F_\theta{}^* + K_\theta K_\theta{}^* - L^{-2} P_\theta(t) H_\theta{}^* H_\theta P_\theta(t)$$

with $P_\theta(0) = 0$. As an estimate of θ, consider the value $\hat{\theta}$ that maximizes expression (3.5.9) for some L. Balakrishnan (1973) shows that under regularity conditions $\hat{\theta} \to \theta$ in probability provided $L_\theta = L$. Bagchi (1975) shows that this convergence continues to occur for other values of L, e.g., $L = 1$. Gupta and Mehra (1974) consider computational aspects of the problem.

Linear regression analysis of time series is discussed, for example, by Grenander (1954) and Hannan (1970). Nonlinear regression is discussed by

Hannan (1971) and Robinson (1972). The particular case of ψ^T a sum of sinusoids is investigated by Hannan (1973a). Gaussian estimation is considered by Munk and MacDonald (1960), Whittle (1962), Walker (1964), and Dzhaparidze and Yaglom (1973). Extension of the Gaussian procedure to the estimation of parameters of linear systems is carried out by Akisik (1975). Davies (1973) describes the construction of asymptotically optimal tests and estimators in the case of Gaussian time series. The use of the likelihood function in the estimation for Markov processes is indicated, for example, by Billingsley (1961) and Roussas (1972).

3.6. General Parameter Estimation for Point Processes

Consider now a point process N whose probability distribution depends on a P-dimensional parameter θ. Suppose that the data $\{N(t), 0 < t \le T\}$ of points located at $\tau_0 < \tau_1 < \cdots < \tau_{N(T)-1}$ are available for analysis. Let the conditional intensity function of the process be given by $\gamma(t; \theta)$. Following expression (2.2.17), the log likelihood function of the data may be written

$$-\int_0^T \gamma(t; \theta)\, dt + \int_0^T \log \gamma(t; \theta)\, dN(t) = -\int_0^T \gamma(t; \theta)\, dt + \sum_{j=0}^{N(T)-1} \log \gamma(\tau_j; \theta)$$

Define

$$Q^T(\theta; N) = T^{-1} \int_0^T \gamma(t; \theta)\, dt - T^{-1} \int_0^T \log \gamma(t; \theta)\, dN(t)$$

in the manner of the preceding section. The maximum likelihood estimate $\hat{\theta}$ is obtained by minimizing $Q^T(\theta; N)$. Under regularity conditions θ is the solution of the system of equations $\partial Q^T/\partial\hat{\theta} = 0$, i.e.,

$$\int_0^T \frac{\partial \gamma(t; \hat{\theta})}{\partial \hat{\theta}_k}\, dt - \sum_{j=0}^{N(T)-1} \frac{\log \gamma(\tau_j; \hat{\theta})}{\partial \hat{\theta}_k} = 0 \tag{3.6.1}$$

As in the preceding section, $\hat{\theta}$ will be consistent and asymptotically normal under regularity conditions. Here

$$EQ^T(\theta; N) = T^{-1} \int_0^T E\{\gamma(t; \theta)\}\, dt - T^{-1} \int_0^T E\{\gamma(t; \theta_0) \log \gamma(t; \theta)\}\, dt \tag{3.6.2}$$

for $\theta = \theta_0$, the entropy of McFadden (1965). In the stationary mixing case, expression (3.6.2) will tend to $E\{\gamma(t; \theta) - \gamma(t; \theta_0) \log \gamma(t; \theta)\}$. Let θ_0 denote the true value of the parameter. As in the preceding section, with

$$\frac{\partial Q^T(\theta_0; N)}{\partial \theta_0}$$

asymptotically $N_P(0, \Sigma_T)$ and

$$\frac{\partial^2 Q^T(\theta_0; N)}{\partial\theta_0\, \partial\theta_0^*} \sim \Psi_T$$

in probability, $\hat{\theta}$ will be asymptotically $N_P(\theta, \Psi_T^{-1}\Sigma_T\Psi_T^{-1})$. Here

$$\Psi_T = T^{-1} \int_0^T E\left\{\gamma(t; \theta)^{-1} \frac{\partial\gamma(t; \theta)}{\partial\theta} \frac{\partial\gamma(t; \theta)^*}{\partial\theta}\right\} dt$$

and

$$\Sigma_T = T^{-2} \int_0^T E\left\{\gamma(t; \theta)^{-1} \frac{\partial\gamma(t; \theta)}{\partial\theta} \frac{\partial\gamma(t; \theta)^*}{\partial\theta}\right\} dt$$

See Vere-Jones (1975b).

As a particular case of these results, consider the time inhomogeneous Poisson process with rate function $p^T(t; \theta)$ on $0 < t < T$. Suppose

$$\lim_{T\to\infty} T^{-1} \int_0^T [p^T(t; \theta) - p^T(t; \theta_0) \log p^T(t; \theta)]\, dt = Q(\theta)$$

where $Q(\theta) > Q(\theta_0)$ for $\theta \neq \theta_0$. Suppose also

$$\lim_{T\to\infty} T^{-1} \int_0^T p^T(t; \theta_0) \frac{\partial \log p^T(t; \theta_0)}{\partial\theta_0} \frac{\partial \log p^T(t; \theta_0)^*}{\partial\theta_0}\, dt = \Psi$$

Then the large sample distribution may be approximated by $N_P(\theta, T^{-1}\Psi)$. The particular case of

$$p^T(t; \theta) = \exp\left\{\sum_1^P \theta_k \phi_k(t)\right\}$$

seems important in practice, with the ϕ_k known functions (see Lewis, 1970; Cox, 1972; Keiding, 1974).

Keiding (1974) considers the case of the pure birth process where $\gamma(t; \theta) = \theta N(t)$ and $N(0) = 1$. The maximum likelihood estimate is here

$$\hat{\theta} = (N(T) - 1)\Big/\left\{\int_0^T N(t)\, dt\right\}$$

and $\{[\int_0^T N(t)\, dt]/\theta\}^{1/2}(\hat{\theta} - \theta)$ is asymptotically $N(0, 1)$.

Snyder (1975, p. 251) considers the case of a Poisson process of rate θ incident on a type I counter with dead time Δ. The conditional intensity function is here

$$\begin{aligned} \gamma(t; \theta) &= \theta \qquad \text{for} \quad \tau_{N(t)-1} + \Delta \le t \\ &= 0 \qquad \text{otherwise} \end{aligned}$$

from (2.7.13). Snyder shows that the maximum likelihood estimate of θ is

$$\hat{\theta} = \hat{p}\left[\frac{\tau_{N(t)-1} + \Delta}{T} - \hat{p}\Delta\right]^{-1} \qquad \text{if} \quad \tau_{N(T)-1} \leq T < \tau_{N(t)-1} + \Delta$$

$$= \hat{p}[1 - \hat{p}\Delta]^{-1} \qquad \text{if} \quad \tau_{N(T)-1} + \Delta \leq T$$

where $\hat{p} = N(T)/T$.

Vere-Jones (1975b) discusses a variety of aspects of maximum likelihood estimation for point processes. Aalen (1975) considers the case of $\gamma(t; \theta) = \theta(t)\phi(t, \omega)$ with ϕ given and possibly depending on the past of N. Aalen investigates the maximum likelihood estimation of θ in the case of constant $\theta(t)$, and considers the estimation of

$$\Theta(t) = \int_0^t \theta(u)J(u)\,du \qquad \text{by} \qquad \hat{\Theta}(t) = \int_0^t \phi(u)^{-1}J(u)\,dN(u)$$

where $J(t) = \lim_{h \downarrow 0} I\{\phi(t-h) > 0\}$, I here being the indicator function. He obtains a central limit theorem by assuming he is dealing with a sequence of such processes.

An analog of the Gaussian estimation procedure of the preceding section may be set down in this case of point process data. Suppose the process has mean rate $p(\theta)$ and power spectrum $f(\lambda; \theta)$. Set $g(\lambda; \theta) = f(\lambda; \theta)/p(\theta)$, $\hat{p} = N(\tau)/T$. As an estimate of θ, take the value $\hat{\theta}$ minimizing

$$Q^T(\theta; N) = \frac{2\pi}{T}\sum_{s=1}^{S}\left[\log g\left(\frac{2\pi s}{T}; \theta\right) + I_2{}^T\left(\frac{2\pi s}{T}\right)\Big/\left\{\hat{p}g\left(\frac{2\pi s}{T}; \theta\right)\right\} - 1\right]$$

In Brillinger (1975a), conditions are set down under which these estimates are consistent and asymptotically $N_P(\theta_0, \mathbf{\Psi}^{-1}\mathbf{\Sigma}_T\mathbf{\Psi}^{-1})$ where

$$\mathbf{\Psi} = \left[\int \frac{\partial \log g(\alpha; \theta)}{\partial \theta_j}\frac{\partial \log g(\alpha; \theta)}{\partial \theta_k}\,d\alpha\right]_{\theta=\theta_0}$$

$$\mathbf{\Sigma}_T = T^{-1}\mathbf{\Psi} + T^{-1}\left[2\pi \iint \frac{\partial \log g(\alpha; \theta)}{\partial \theta_j}\frac{\partial \log g(\beta; \theta)}{\partial \theta_k}\right.$$

$$\left. \times \frac{f_4(\alpha, -\alpha, -\beta; \theta)}{g(\alpha; \theta)g(\beta; \theta)}\,d\alpha\,d\beta\right]_{\theta=\theta_0}$$

Akisik (1975) extends this procedure to certain "linear" point process systems.

Davies (1977) investigates the problem of developing optimum tests of the hypothesis that a given point process is Poisson. Snyder (1975) discusses tests of hypotheses relevant to certain point processes.

REFERENCES

1. Aalen, O. O. (1975). Statistical Inference for a Family of Counting Processes. PhD. Thesis, Univ. of California, Berkeley, California.
2. Akisik, V. A. (1975). On the Estimation of Parametric Transfer Functions. PhD. Thesis, Univ. of California, Berkeley, California.
3. Anderson, T. W. (1971). "Statistical Analysis of Time Series." Wiley, New York.
4. Aronszajn, N. (1950). Theory of reproducing kernels, *Trans. Amer. Math. Soc.* **68**, 337–404.
5. Athreya, K. B., and Ney, P. E. (1972). "Branching Processes." Springer-Verlag, Berlin.
6. Bagchi, A. (1975). Continuous time systems identification with unknown noise covariance, *Automatica* **11**, 533–536.
7. Balakrishnan, A. V. (1973). Stochastic Differential Systems I, "Lecture Notes in Economics and Mathematical Systems," No. 84. Springer-Verlag, Berlin.
8. Bartlett, M. S. (1966). "Stochastic Processes." Cambridge Univ. Press, London and New York.
9. Benes, V. E. (1963). "General Stochastic Processes in the Theory of Queues." Addison-Wesley, Reading, Massachusetts.
10. Bharucha-Reid, A. T. (1960). "Elements of the Theory of Markov Processes and Their Applications." McGraw-Hill, New York.
11. Bhat, U. N. (1969). Sixty years of queueing theory, *Management Sci.* **15**, 280–294.
12. Billingsley, P. (1961). "Statistical Inference for Markov Processes." Univ. of Chicago Press, Chicago, Illinois.
13. Billingsley, P. (1968). "Convergence of Probability Measures." Wiley, New York.
14. Bochner, S. (1960). "Harmonic Analysis and the Theory of Probability." Univ. of California Press, Berkeley, California.
15. Box, G. E. P., and Jenkins, G. M. (1970). "Time Series Analysis, Forecasting and Control." Holden Day, San Francisco, California.
16. Breiman, L. (1968). "Probability." Addison-Wesley, Reading, Massachusetts.
17. Bremaud, P. M. (1974). The martingale theory of point processes over the real half line, "Lecture Notes in Economics and Mathematics Systems." Springer-Verlag, Berlin.
18. Brillinger, D. R. (1970). The identification of polynomial systems by means of higher order spectra, *J. Sound Vibrat.* **12**, 301–313.
19. Brillinger, D. R. (1972). The spectral analysis of stationary interval functions. 483–513 in *Proc. Berkeley Symp. 6th* (L. M. Le Cam, J. Neyman, and E. L. Scott, eds.), Vol. 1. Univ. of California Press, Berkeley, California.
20. Brillinger, D. R. (1973). Estimation of the mean of a stationary time series by sampling. *J. Appl. Probability* **10**, 419–431.
21. Brillinger, D. R. (1974). Fourier analysis of stationary processes. *Proc. IEEE* **62**, 1628–1643.
22. Brillinger, D. R. (1975a). "Time Series: Data Analysis and Theory." Holt, New York.
23. Brillinger, D. R. (1975b). Statistical inference for stationary point processes, *in* "Stochastic Processes and Related Topics" (M. L. Puri, ed.), Vol. 1, pp. 55–99. Academic Press, New York.
24. Brillinger, D. R. (1975c). Estimation of product densities, *in Comput. Sci. Statist., Annu. Symp. Interface, 8th UCLA, Los Angeles*, pp. 431–438.
25. Brillinger, D. R., Bryant, H. L., Jr., and Segundo, J. P. (1976). Identification of synaptic interactions, *Biol. Cybernet.* **22**, 213–228.
26. Brown, B. M., and Hewitt, J. I. (1975). Asymptotic likelihood theory for diffusion processes, *J. Appl. Probability* **12**, 228–238.

27. Bryant, H., and Segundo, J. P. (1975). How does the neuronal spike trigger read Gaussian white noise transmembrane current? *in Proc. Symp. Testing Identification Nonlinear Syst., 1st* (G. D. McCann and P. Z. Marmarelis, eds.), pp. 236–247. California Inst. of Technology, Pasadena, California.
28. Chou, C-S., and Meyer, P. A. (1975). Sur la representation des martingales comme integrales stochastiques dans les processus pontuels, "Lecture Notes in Mathematics," No. 465, pp. 226–236. Springer-Verlag, Berlin.
29. Cinlar, E. (1972). Superposition of point processes, *In* "Stochastic Point Processes" (P. A. W. Lewis, ed.), pp. 549–606. Wiley, New York.
30. Clevenson, M. L., and Zidek, J. V. (1975). Bayes linear estimates of the intensity function of the nonstationary Poisson process, *J. Amer. Stat. Assoc.* **72**, 112–120.
31. Cox, D. R. (1962). "Renewal Theory." Methuen, London.
32. Cox, D. R. (1972). The statistical analysis of dependencies in point processes, *in* "Stochastic Point Processes" (P. A. W. Lewis, ed.), pp. 55–66. Wiley, New York.
33. Cox, D. R., and Lewis, P. A. W. (1966). "The Statistical Analysis of Series of Events." Methuen, London.
34. Cox, D. R., and Smith, W. L. (1961). "Queues." Methuen, London.
35. Cramér, H. (1942). On harmonic analysis in certain functional spaces, *Ark. Math. Astr. Fysik* **28**, 1–7.
36. Cramér, H. (1976). Half a century with probability theory: some personal reflections, *Ann. Probability* **4**, 509–546.
37. Cramér, H., and Leadbetter, M. R. (1967). "Stationary and Related Stochastic Processes." Wiley, New York.
38. Cramér, H., Leadbetter, M. R., and Serfling, R. J. (1971). On distribution function—moment relationships in a stationary point process, *Z. Wahrschein.* **18**, 1–8.
39. Daley, D. J. (1969). Spectral properties of weakly stationary point processes, *Bull. Inst. Internat. Statist.* **37**, No. 2, 344–346.
40. Daley, D. J., and Oakes, D. (1974). Random walk point processes, *Z. Wahrschein.* **30**, 1–16.
41. Daley, D. J., and Vere-Jones, D. (1972). A summary of the theory of point processes, *in* "Stochastic Point Processes" (P. A. W. Lewis, ed.), pp. 299–383. Wiley, New York.
42. Davies, R. B. (1973). Asymptotic inference in stationary Gaussian time series, *Advances Appl. Probability* **5**, 469–497.
43. Davies, R. B. (1977). Testing the hypothesis that a point process is Poisson *Adv. Appl. Prob.* (to appear).
44. Doob, J. (1953). "Stochastic Processes." Wiley, New York.
45. Dzhaparidze, K. O., and Yaglom, A. M. (1973). Asymptotically efficient estimates of the spectrum parameters of stationary stochastic processes, *Proc. Prague Symp. Asympt. Statistics*, pp. 55–106. Karlova Univ., Prague.
46. Ephremides, A., and Thomas, J. B. (1973). "Random Processes." Dowden, Hutchinson and Ross, Stroudsburg, Pennsylvania.
47. Feinberg, S. (1974). Stochastic models for single neuron firing trains: a Survey, *Biometrics* **30**, 399–427.
48. Furstenberg, H. (1960). "Stationary Processes and Prediction Theory." Princeton Univ. Press, Princeton, New Jersey.
49. Gaver, D. P., and Lewis, P. A. W. (1977). First order autoregressive Gamma sequences and point processes (to appear).
50. Gelfand, I. M., and Vilenkin, N. Ya. (1964). "Generalized Functions," p. 4. Academic Press, New York.

51. Geman, D., and Horowitz, J. (1973). Remarks on Palm measures, *Ann. Inst. Henri Poincare B* **9**, 215–232.
52. Gihman, I. I., and Skorohod, A. V. (1972). "Stochastic Differential Equations." Springer-Verlag, Berlin.
53. Glover, K., and Williams, J. C. (1974). Parameterization of linear dynamical systems: canonical forms and identifiability, *IEEE Trans. Automatic Control* **AC-19**, 640–646.
54. Grandell, J. (1972). Statistical inference for doubly stochastic Poisson processes, *in* "Stochastic Point Processes" (P. A. W. Lewis, ed.), pp. 90–121. Wiley, New York.
55. Grenander, U. (1954). On the estimation of regression coefficients in the case of an autocorrelated disturbance, *Ann. Math. Statist.* **25**, 252–272.
56. Gupta, N. K., and Mehra, R. (1974). Computational aspects of maximum likelihood estimation and reduction in sensitivity function calculations, *IEEE Trans. Automatic Control* **AC-19**, December.
57. Haight, F. A. (1967). "Handbook of the Poisson Distribution." Wiley, New York.
58. Hannan, E. J. (1970). "Multiple Time Series." Wiley, New York.
59. Hannan, E. J. (1971). Nonlinear time series regression, *J. Appl. Probability* **8**, 767–780.
60. Hannan, E. J. (1973a). The estimation of frequency, *J. Appl. Probability* **10**, 510–519.
61. Hannan, E. J. (1973b). Multivariate time series analysis, *J. Multivariate Anal.* **3**, 395–407.
62. Hannan, E. J. (1973c). Central limit theorems for time series regression. *Z. Wahrschein.* **26**, 157–170.
63. Harris, T. E. (1963). "The Theory of Branching Processes." Springer-Verlag, Berlin.
64. Hawkes, A. G. (1972). Spectra of some mutually exciting point processes with associated variables, *in* "Stochastic Point Processes" (P. A. W. Lewis, ed.), pp. 261–271. Wiley, New York.
65. Hawkes, A. G., and Adamopoulos, L. (1973). Cluster models for earthquakes—regional comparisons, *Bull. Inst. Internat. Statist.* **45(3)**, 454–461.
66. Hawkes, A. G., and Oakes, D. (1974). A cluster process representation of a self-exciting process, *J. Appl. Probability* **11**, 493–504.
67. Hida, T. (1960). Canonical representations of Gaussian processes and their applications, *Mem. Coll. Sci. Kyoto A* **33**, 110–155.
68. Hida, T. (1970). "Stationary Stochastic Processes." Princeton Univ. Press, Princeton, New Jersey.
69. Ito, K. (1951). On stochastic differential equations, *Mem. Amer. Math. Soc.* **4**, 51.
70. Jacobs, P. A., and Lewis, P. A. W. (1977). A mixed autoregressive-moving average exponential sequence and point process *Adv. Appl. Prob.* **9**, 87–104.
71. Jacod, J. (1975). Multivariate point processes: predictable projection, Radon–Nikodym derivatives, representation of martingales, *Z. Wahrschein.* **31**, 235–253.
72. Jagers, P. (1974). Aspects of random measures and point processes, *Advan. Probability* **3**, 179–239.
73. Jowett, J. H., and Vere-Jones, D. (1972). The prediction of stationary point processes, *in* "Stochastic Point Processes" (P. A. W. Lewis, ed.), pp. 405–435. Wiley, New York.
74. Kailath, T. (1970). The innovations approach to detection and estimation theory, *Proc. IEEE* **58**, 680–695.
75. Kailath, T. (1971). RKHS approach to detection and estimation problems—Part I: deterministic signals in Gaussian noise, *IEEE Trans. Informat. Theory* **IT-17**, 530–549.
76. Kallenberg, O. (1975). Limits of compound and thinned point processes, *J. Appl. Probability* **12**, 269–278.
77. Keiding, N. (1974). Estimation in the birth process, *Biometrika* **61**, 71–80.
78. Kendall, D. G. (1975). The genealogy of genealogy branching processes before (and after) 1873, *Bull. London Math. Soc.* **7**, 225–253.

79. Kerstan, J., Matthes, K., and Mecke, J. (1974). "Unbegrenzt teilbare Punktprozesse." Akademie-Verlag, Berlin.
80. Khintchine, A. Ya. (1934). Korrelationstheories der stationaren Prozesse, *Math. Ann.* **109**, 604–615.
81. Kholevo, A. S. (1969). On estimation of regression coefficients, *Theory Probability Appl.* **14**, 79–104.
82. Knight, F. B. (1975). A predictive view of continuous time processes, *Ann. Probability* **3**, 573–596.
83. Kolmogorov, A. N. (1940). Curves in Hilbert space invariant with regard to a one parameter group of motions, *Dokl. Akad. Nauk. SSSR* **26**, 6–9.
84. Koopmans, L. H. (1974). "The Spectral Analysis of Time Series." Academic Press, New York.
85. Korenberg, M. J. (1973). Cross-correlation analysis of neural cascades, *Proc. Ann. Rocky Mountain Bioeng. Symp., 10th* 47–51.
86. Krickeberg, K. (1974). Moments of point processes, *in* "Stochastic Geometry" (E. F. Harding and D. G. Kendall, eds.), pp. 89–113. Wiley, New York.
87. Kulinich, G. L. (1975). On the estimation of the drift parameter of a stochastic diffusion equation, *Theory Probability Appl.* **20**, 384–387.
88. Kurtz, T. G. (1974). Point processes and completely monotone set functions, *Z. Wahrschein.* **31**, 51–67.
89. Leadbetter, M. R. (1972). Point processes generated by level crossings, *in* "Stochastic Point Processes" (P. A. W. Lewis, ed.). Wiley, New York.
90. Lewis, P. A. W. (1970). Remarks on the theory, computation and application of the spectral analysis of series of events, *J. Sound Vibrat.* **12**, 353–375.
90a. Lewis, P. A. W. (ed.) (1972). "Stochastic Point Processes." Wiley, New York.
90b. Loève, M. (1955). "Probability Theory." Van Nostrand–Reinhold, Princeton, New Jersey.
90c. Lotka, A. J. (1957). "Elements of Mathematical Biology." Dover, New York.
91. McCann, G. D., and Marmarelis, P. Z. (1975). *Proc. Symp. Testing and Identificat. of Nonlinear Syst., 1st*, California Inst. of Technol., Pasadena, California.
92. McFadden, J. A. (1965). The entropy of a point process, *SIAM J. Appl. Math.* **13**, 988–994.
93. Macchi, O. (1971). Stochastic point processes and multicoincidences, *Proc. IEEE Trans. Informat. Theory* **IT-17**, 2–7.
94. Macchi, O. (1975). The coincidence approach to stochastic point processes, *Advan. Appl. Probability* **7**, 83–122.
95. Masani, P. (1972). On helixes in Hilbert space. I, *Theory Probability Appl.* **17**, 1–19.
96. Meyer, P. A. (1971). Demonstration simplifiee d'un theoreme de Knight, "Lecture Notes in Math," No. 191, pp. 191–195. Springer-Verlag, Berlin.
97. Milne, R. K., and Westcott, M. (1972). Further results for Gauss-Poisson processes, *Advan. Appl. Probability* **4**, 151–176.
98. Mirasol, N. M. (1963). The output of an $M/G/\infty$ queueing system is Poisson, *Operations Res.* **11**, 282–284.
99. Moore, G. P., Perkel, D. H., and Segundo, J. P. (1966). Statistical analysis and functional interpretation of neuronal spike data, *Ann. Rev. Physiol.* **28**, 493–522.
100. Moyal, J. E. (1962). The general theory of stochastic population processes, *Acta Math.* **108**, 1–31.
101. Munk, W. H., and MacDonald, G. J. F. (1960). "The Rotation of the Earth." Cambridge Univ. Press, London and New York.
102. Murthy, V. K. (1974). "The General Point Process." Addison-Wesley, Reading, Massachusetts.

103. Nelson, E. (1959). Regular probability measures on function space, *Ann. Math.* **69**, 630–643.
104. Neveu, J. (1968). Sur la structure des processus ponctuels stationnaires, *C. R. Acad. Sci. Paris* **267**, 561–564.
105. Newman, D. S. (1970). A new family of point processes which are characterized by their second moment properties, *J. Appl. Probability* **7**, 338–358.
106. Oakes, D. (1975). The Markovian self-exciting process, *J. Appl. Probability* **12**, 69–77.
107. Papangelou, F. (1974). On the Palm probabilities of processes of points and processes of lines, *in* "Stochastic Geometry" (E. F. Harding and D. G. Kendall, eds.), pp. 114–147. Wiley, New York.
108. Parzen, E. (1959). Statistical inference on time series by Hilbert space methods, I, *in* "Time Series Analysis Papers" (1967, E. Parzen, ed.), pp. 251–382. Holden-Day, San Francisco, California.
109. Parzen, E. (1962). Extraction and detection problems and reproducing kernel Hilbert spaces, *SIAM J. Control* **1**, 35–62.
110. Pfaffelhuber, E. (1975). Generalized harmonic analysis for distributions, *IEEE Trans. Informat. Theory* **IT-21**, 605–611.
111. Prabhu, N. U. (1965). "Queues and Inventories." Wiley, New York.
112. Prohorov, Yu V. (1961). The method of characteristic functionals, *in Proc. Berkeley Symp. Math. Statist. Probability, 4th* (J. Neyman, ed.), Vol. 2, pp. 403–419. Univ. of California Press, Berkeley, California.
113. Prohorov, Yu V., and Rozanov, Yu A. (1969). "Probability Theory." Springer-Verlag, Berlin.
114. Quenouille, M. H. (1949). Problems in plane sampling, *Ann. Math. Statist.* **20**, 355–375.
115. Ramakrishnan, A. (1950). A stochastic process relating to particles distributed in a continuous infinity of states, *Proc. Cambridge Phil. Soc.* **46**, 596–602.
116. Renyi, A. (1967). Remarks on the Poisson process, *Studia Sci. Math. Hungar.* **2**, 119–123.
117. Rice, J. (1975). Estimated factorisation of the spectral density of a stationary point process, *Advan. Appl. Probability* **7**, 801–817.
118. Rice, S. O. (1945). Mathematical analysis of random noise, *Bell Syst. Tech. J.* **24**, 46–156.
119. Robinson, P. M. (1972). Non-linear regression for multiple time series, *J. Appl. Probability* **9**, 758–768.
119a. Roussas, G. G. (1972). "Contiguous Probability Measures." Cambridge Univ. Press, London.
120. Rozanov, Yu A. (1966). "Stationary Random Processes." Holden-Day, San Francisco, California.
121. Rozanov, Yu. A. (1969). On a new class of estimates, *in* "Multivariate Analysis" (P. R. Krishnaiah, ed.), Vol. II, pp. 437–441. Academic Press, New York.
122. Rubin, I. (1972). Regular point processes and their detection, *IEEE Trans. Informat. Theory* **IT-18**, 547–557.
123. Rudemo, M. (1973). Point processes generated by transitions of Markov chains, *Advan. Appl. Probability* **5**, 262–286.
124. Ruelle, D. (1969). "Statistical Mechanics." Benjamin, New York.
125. Ryll-Nardzewski, C. (1961). Remarks on the process of calls, *in Proc. Berkeley Symp., 4th* (J. Neyman, ed.), Vol. 2, pp. 455–465. Univ. of California Press, Berkeley, California.
126. de Sam Lazaro, J., and Meyer, P. A. (1975). Questions de theorie des flots, *in* "Lecture Notes in Mathematics," No. 465, pp. 2–96. Springer-Verlag, Berlin.
127. Schwartz, L. (1973). Surmartingales regulieres a valeurs mesures et desintegrations regulieres d'une mesure, *J. Anal. Math.* **26**, 1–168.

128. Segall, A., and Kailath, T. (1975). The modelling of randomly modulated jump processes, *IEEE Trans. Informat. Theory* **IT-21**, 135–143.
129. Shiryaev, A. N. (1963). Some problems in the spectral theory of higher-order moments, I, *Theory Probability Appl.* **5**, 265–284.
130. Snyder, D. L. (1975). "Random Point Processes." Wiley, New York.
131. Srinivasan, S. K. (1969). "Stochastic Theory and Cascade Processes." Amer. Elsevier, New York.
132. Srinivasan, S. K. (1974). "Stochastic Point Processes." Hafner, New York.
133. Taraskin, A. F. (1974). On the asymptotic normality of vector-valued stochastic integrals and estimates of drift parameters of a multidimensional diffusion process, *Theory Probability Math. Stat.* **2**, 209–224.
134. Udias, A., and Rice, J. (1975). Statistical analysis near San Andreas Observatory, Hollister, California, *Bull. Seismol. Soc. Amer.* **65**, 809–827.
135. Varayia, P. (1975). The martingale theory of jump processes, *IEEE Trans. Automatic Control* **AC-20**, 1.
136. Vere-Jones, D. (1968). Some applications of probability generating functionals to the study of input-output streams, *J. Roy. Statist. Soc. B* **30**, 321–333.
137. Vere-Jones, D. (1974). An elementary approach to the spectral theory of stationary random measures, *in* "Stochastic Geometry" (E. F. Harding and D. G. Kendall, eds.), pp. 307–321. Wiley, New York.
138. Vere-Jones, D. (1975a). A renewal equation for point processes with Markov dependent intervals, *Math. Nach.* **68**, 133–139.
139. Vere-Jones, D. (1975b). On updating algorithms for inference for stochastic processes, *in* "Perspectives in Probability and Statistics" (J. Gani, ed.), pp. 239–259. Academic Press, New York.
139a. Valkonskii, V. A., and Rozanov, Yu. A. (1959). Some limit theorems for random functions, I. *Theory Probability Appl.* **4**, 178–197.
140. Walker, A. M. (1964). Asymptotic properties of least squares estimates of parameters of the spectrum of stationary non-deterministic time series, *J. Austral. Math. Soc.* **4**, 363–384.
141. Wang, G. L. (1968). Contagion in stochastic models for epidemics, *Ann. Math. Statist.* **39**, 1863–1889.
142. Westcott, M. (1970). Identifiability in linear processes, *Z. Wahrschein.* **16**, 39–46.
143. Westcott, M. (1971). On existence and mixing results for cluster point processes, *J. Roy. Statist. Soc. B* **33**, 290–300.
144. Westcott, M. (1972). The probability generating functional, *J. Austral. Math. Soc.* **14**, 448–466.
145. Westergaard, H. (1968). "Contributions to the History of Statistics." Agathon, New York.
146. Whittle, P. (1962). Gaussian estimation in stationary time series, *Bull. Inst. Internat. Statist.* **39**, 105–129.
147. Whittle, P. (1963). "Prediction and Regulation." English Univ. Press, London.
148. Wiener, N. (1930). Generalized harmonic analysis, *Acta Math.* **55**, 117–258.
149. Wiener, N. (1949). "Time Series." MIT Press, Cambridge, Massachusetts.
150. Wiener, N. (1958). "Nonlinear Problems in Random Theory." MIT Press, Cambridge, Massachusetts.
151. Wold, H. (1948). On stationary point processes and Markov chains, *Skand. Akt.* **31**, 299–340.
152. Yaglom, A. M. (1962). "An Introduction to the Theory of Stationary Random Functions." Prentice-Hall, Englewood Cliffs, New Jersey.
153. Yvon, J. (1935). "La Theories Statistique des Fluids et l'Equation d'Etat." Herman, Paris.

Some Recent Developments on Real Multivariate Distributions*

P. R. KRISHNAIAH

DEPARTMENT OF MATHEMATICS AND STATISTICS
UNIVERSITY OF PITTSBURGH
PITTSBURGH, PENNSYLVANIA

1. INTRODUCTION

Multivariate distribution theory plays a very important role in drawing inference from multivariate data. In particular, distributions of certain functions of the eigenvalues of various random matrices are quite useful in applying various multivariate test procedures. These test procedures arise in the analysis of the data that arise in social, behavioral, medical, physical, engineering, and other sciences as well as other disciplines. In this chapter, we review some of the developments on the distribution problems connected with the eigenvalues or certain functions of the eigenvalues of real random

* Part of this work was sponsored by the Air Force Flight Dynamics Laboratory, Air Force Systems Command, United States Air Force under Grant AFOSR 77-3239.

ISBN 0-12-4266 01-0

matrices and discuss their applications in areas such as pattern recognition, principal component analysis, canonical correlation analysis, covariance structure analysis, and simultaneous test procedures. In a companion paper (Krishnaiah, 1976), the author gave a review of some developments on complex random matrices.

Since the time Roy (1939), Fisher (1939), and Hsu (1939) made the important contribution of deriving the joint distribution of the roots of the multivariate beta matrix, numerous developments took place on the distributions of the real random matrices. These contributions are made not only by mathematical statisticians but also by nuclear physicists. Nuclear physicists have been interested in these problems since some empirical evidence indicated that the statistical behavior of the energy levels of nuclei in high excitation can be investigated by studying the distributions of the eigenvalues of certain random matrices. For various contributions made by nuclear physicists to the area of distribution theory of eigenvalues of random matrices, the reader is referred to Carmeli (1974), Mehta (1967), Porter (1965), Garg (1972), and Wigner (1967). For some applications of the distributions of the eigenvalues of random matrices in probability theory, the reader is referred to Boyce (1968) and Bharucha-Reid (1970). It is not possible to give a thorough review of the literature on multivariate distributions in this chapter since numerous developments have taken place in this area since 1939. In this chapter we review some developments on the exact multivariate distribution theory as well as a few aspects of approximations. A discussion of the asymptotic distribution theory will be presented in a separate publication. For a review of the literature on some aspects of asymptotic multivariate distribution theory, the reader is referred to Siotani (1975).

In Section 2 of this chapter we discuss the evaluation of some integrals which arise in multivariate distribution theory. Section 3 is devoted to a review of the literature on the joint distributions of the eigenvalues of random matrices. The distributions of the individual roots of a wide class of random matrices are described in Section 4 whereas the distributions of the ratios of the roots of the Wishart matrix and multivariate beta matrix are given in Section 5. In Section 6, we review the literature on the distributions of the traces of multivariate F and multivariate beta matrices. Some discussion is also presented in Section 6 on the evaluation of the moments of elementary symmetric functions. In Section 7 we review some recent developments on approximations to the distributions of the likelihood ratio statistics with suitable Pearson-type distributions. Some applications of the multivariate distribution theory in inference on covariance structures and simultaneous tests on mean vectors are described in Section 8. In Section 9, we discuss how multivariate distributions arise in simultaneous tests for the equality of the eigenvalues. For various other applications of multivariate distributions in multivariate statistical inference, the reader is referred to Anderson (1958), Krishnaiah (1966, 1969a, 1973, 1977), Kshirsagar (1972), Rao (1965), Roy (1957), Roy *et al.* (1971), Timm (1975), and other books in multivariate statistical analysis. For a bibliography of multivariate statistical analysis, the reader is referred to Anderson *et al.* (1972), or Subrahmaniam and Subrahmaniam (1973).

2. PRELIMINARIES AND EVALUATION OF SOME INTEGRALS

The following notation is needed in the sequel. The transpose of a matrix A is denoted by A' whereas the inverse and determinant of a square matrix B are denoted by B^{-1} and $|B|$, respectively. The exponential of the trace of a matrix B is denoted by etr B. If A is a skew-symmetric matrix (i.e., $A = -A'$) and A is of even order (say $2m$), then the Pfaffian of A is defined (see Cullis, 1918, p. 521) as

$$\mathrm{Pf}(A) = \frac{1}{m!} \sum \pm a_{i_1 i_2} a_{i_3 i_4} \cdots a_{i_{2m-1} i_{2m}} \tag{2.1}$$

where the summation is over all permutations $i_1, \ldots, i_{2m}$ of $1, 2, \ldots, 2m$ subject to the restrictions $i_1 < i_2, i_3 < i_4, \ldots, i_{2m-1} < i_{2m}$. The sign is positive or negative according to whether the permutation is even or odd. The square of $\mathrm{Pf}(A)$ is equal to $|A|$. The number of distinct terms on the right-hand side of (2.1) is equal to $2m!/m!\,2$.

Now let

$$I(\psi; q, c, d) = \int_{c \le x_1 \le \cdots \le x_q \le d} \cdots \int V(x_1, \ldots, x_q) \prod_{i=1}^{q} \{\psi(x_i)\, dx_i\} \tag{2.2}$$

where $V(x_1, \ldots, x_q) = |(g_i(x_j))|$. Also, let

$$F_s^t(c, d) = \int_c^d F_s(c, \theta) g_t(\theta) \psi(\theta)\, d\theta, \qquad F_s(c, \theta) = \int_c^\theta g_s(x) \psi(x)\, dx$$

and

$$f_s^t(c, d) = F_s^t(c, d) - F_t^s(c, d)$$

In addition, let

$$\Delta(\psi; 2m, c, d) = |(f_i^j(c, d))_{i,j=1,\ldots,2m}|^{1/2} \tag{2.3}$$

and

$$G_t(\psi; 2m+1, c, d) = |(f_i^j(c, d))_{i,j=1,\ldots,t-1,t+1,\ldots,2m}|^{1/2} \tag{2.4}$$

with the understanding that $G_t(\psi; 1, c, d) = 1$. The following lemma is needed in the sequel.

Lemma 2.1. Let $\psi(x)$ be a function of x such that the integral in (2.2) exists. Then

$$I(\psi; q, c, d) = \Delta(\psi; q, c, d) \tag{2.5}$$

when $q = 2m$, and

$$I(\psi; q, c, d) = \sum_{i=1}^{2m+1} (-1)^{i+1} F_i(c, d) G_i(\psi; q, a, b) \tag{2.6}$$

when $q = 2m + 1$.

When $g_i(x_j) = x_j^{i-1}$, $\psi(x) = \exp(-x^k)$, $q = 2m$, and k is a positive integer, Mehta (1960) proved Eq. (2.5) by integrating out alternate variables $x_1, x_3, \ldots, x_{2m-1}$. By using the method of integration over alternate variables, it was shown (see Krishnaiah and Chang, 1971a; Krishnaiah and Waikar, 1971a) that Lemma 2.1 holds good in general. DeBruijn (1951) proved Lemma 2.1 when $q = 2m$, and also obtained an alternative expression for $I(\psi; 2m + 1, c, d)$. The proof of deBruijn was based upon noting that the value of the integral

$$\int \cdots \int_{a \le x_1 \le \cdots \le x_q \le b} E(x_1, \ldots, x_q) V(x_1, \ldots, x_q) \prod_{i=1}^{q} \psi(x_i)\, dx_1 \cdots dx_q \tag{2.7}$$

is the same as that of (2.2) and the integrand in (2.7) is a symmetric function of $x_1, \ldots, x_q$; here $E(x_1, \ldots, x_q) = \prod_{i>j}^{q} \mathrm{Sgn}(x_i - x_j)$ where $\mathrm{Sgn}(a)$ is equal to 1 or -1 accordingly as a is positive or negative. The work of Mehta (1960), Krishnaiah and Chang (1971a), and Krishnaiah and Waikar (1971a) was done independent of the work by deBruijn (1951).

Next, consider the following:

Lemma 2.2. Let $\eta(x_1, \ldots, x_q)$ be a symmetric function of $x_1, \ldots, x_q$. Then

$$\int \cdots \int_{c \le x_1 \le \cdots \le x_q \le d} \eta(x_1, \ldots, x_q) \left| (g_i(x_j)) \right| dx_1 \cdots dx_q$$

$$= \int_c^d \cdots \int_c^d E(x_1, \ldots, x_q) \eta(x_1, \ldots, x_q) \prod_{i=1}^{q} \{g_i(x_i)\, dx_i\} \tag{2.8}$$

In addition, if

$$\eta(x_1, \ldots, x_q) = \sum_{\kappa} c_\kappa x_1^{k_1} \cdots x_q^{k_q} \tag{2.9}$$

where c_κ depends on $k_1, \ldots, k_q$ and $\sum_\kappa$ denotes the summation over different elements of $\kappa = (k_1, \ldots, k_q)$ subject to suitable restrictions, we have

$$\int_{a \le x_1 \le \cdots \le x_q \le b} \cdots \int \eta(x_1, \ldots, x_q)|(g_i(x_j))|\, dx_1 \cdots dx_q = \sum_\kappa c_\kappa \operatorname{Pf}(A_\kappa) \tag{2.10}$$

where $(A_\kappa) = (a_{ij}^\kappa)$. In (2.10)

$$a_{ij}^\kappa = \int_c^d \int_c^d x^{k_i} y^{k_j} g_i(x) g_j(y) \operatorname{Sgn}(y - x)\, dx\, dy \tag{2.11}$$

for $i, j = 1, \ldots, 2q$ when $p = 2q$; if $p = 2q + 1$, we have the following terms in addition to the terms given by (2.11) for $i, j = 1, \ldots, 2q$:

$$a_{2q+2,2q+2}^\kappa = 0 \tag{2.12}$$

$$a_{i,2q+2}^\kappa = -a_{i,2q+2}^\kappa = \int_c^d x^{k_i} g_i(x)\, dx, \qquad i = 1, \ldots, 2q + 1$$

DeBruijn (1951) proved this lemma when $\eta(x_1, \ldots, x_q) = 1$. Lemma 2.2 can be shown (see Krishnaiah and Chattopadhyay, 1975) to be true in the general case following the same lines as deBruijn (1951).

The following definitions of hypergeometric functions with matrix argument are given by Herz (1955):

$$_0F_0(\Lambda) = \operatorname{etr} \Lambda \tag{2.13}$$

$$\begin{aligned} &{}_{p+1}F_q(\alpha_1, \ldots, \alpha_p, \gamma; \beta_1, \ldots, \beta_q; Z) \\ &= \frac{1}{\Gamma_m(\gamma)} \int_{\Lambda > 0} \operatorname{etr}(-\Lambda)\, {}_pF_q(\alpha_1, \ldots, \alpha_p; \beta_1, \ldots, \beta_q; \Lambda Z)|\Lambda|^{\gamma - (m+1)/2}\, d\Lambda \end{aligned} \tag{2.14}$$

$$\begin{aligned} &{}_pF_{q+1}(\alpha_1, \ldots, \alpha_p; \beta_1, \ldots, \beta_q, \gamma; \Lambda) \\ &\quad = \Gamma_m(\gamma) \frac{2^{m(m-1)/2}}{(2\pi i)^{m(m+1)/2}} \int_{\operatorname{Re} Z = X_0 > 0} \operatorname{etr}(Z) \\ &\quad \times {}_pF_q(\alpha_1, \ldots, \alpha_p; \beta_1, \ldots, \beta_q; \Lambda Z^{-1})|Z|^{-\gamma}\, dZ \end{aligned} \tag{2.15}$$

where Z and Λ are symmetric matrices of order $m \times m$ and

$$\Gamma_p(a) = \pi^{p(p-1)/4} \prod_{i=1}^p \Gamma(a - \tfrac{1}{2}(i - 1))$$

Let $\kappa = (k_1, \ldots, k_p)$ be a partition of k such that $k_p \geq \cdots \geq k_1 \geq 0$ and $k_1 + \cdots + k_p = k$. Also, let $C_\kappa(A)$ denote the zonal polynomial of the matrix A of order $p \times p$. The hypergeometric functions with matrix arguments are expressed in terms of zonal polynomials as follows by Constantine (1963):

$$_qF_r(a_1, \ldots, a_q; b_1, \ldots, b_r; S) = \sum_{k=0}^{\infty} \sum_{\kappa} \frac{(a_1)_\kappa \cdots (a_q)_\kappa}{(b_1)_\kappa \cdots (b_r)_\kappa} \frac{C_\kappa(S)}{k!} \tag{2.16}$$

$$_qF_r(a_1, \ldots, a_q; b_1, \ldots, b_r; S, T) = \sum_{k=0}^{\infty} \sum_{\kappa} \frac{(a_1)_\kappa \cdots (a_q)_\kappa}{(b_1)_\kappa \cdots (b_r)_\kappa} \frac{C_\kappa(S)C_\kappa(T)}{k!\, C_\kappa(I_p)} \tag{2.17}$$

where

$$(a)_\kappa = \prod_{i=1}^{p} (a - \tfrac{1}{2}(i-1))_{k_i}, \qquad (a)_k = a(a+1) \cdots (a+k-1)$$

and I_p is an identity matrix of order $p \times p$. In the above equations, $a_1, \ldots, a_q$ and $b_1, \ldots, b_r$ are real or complex constants whereas S and T are symmetric matrices of order $p \times p$.

For a review of the literature on zonal polynomials, the reader is referred to Subrahmaniam (1974). Zonal polynomials were introduced by Hua (1959) and James (1961b) independently.

3. JOINT DISTRIBUTIONS OF THE EIGENVALUES OF CERTAIN RANDOM MATRICES

Let X: $p \times m$ be a random matrix whose columns are distributed independently as multivariate normal with means given by $E(X) = M$ and covariance matrix Σ. Then the distribution of $S = XX'$ is known to be a central or noncentral Wishart distribution with m degrees of freedom according to whether $M = 0$ or $M \neq 0$. Next, let $S_1 = YY'$ where the columns of Y: $p \times n$ are distributed independently as multivariate normal with mean vector 0 and covariance Σ_1. Then, $S(S + S_1)^{-1}$ is known to be a central or noncentral multivariate beta matrix according to whether $M = 0$ or $M \neq 0$. Similarly, $F = SS_1^{-1}$ is known to be a central or noncentral multivariate F matrix according to whether $M = 0$ or $M \neq 0$.

Next, let $A = (a_{ij})$ be a $p \times p$ symmetric random matrix whose elements are distributed independently with means given by $E(A) = N$. Also, the variances of diagonal elements a_{ii} $(i = 1, \ldots, p)$ are equal to 2 whereas the variances of the off-diagonal elements a_{ij} $(i \neq j = 1, \ldots, p)$ are equal to 1. Then A is known to be a central or noncentral Gaussian matrix according to whether $N = 0$ or $N \neq 0$.

Let $l_1 \leq \cdots \leq l_p$ be the eigenvalues of the Wishart matrix S. When $M = 0$ and $\Sigma = I_p$, the joint density of $l_1, \ldots, l_p$ is known (see Nanda, 1948b) to be

$$f_1(l_1, \ldots, l_p) = C_1 |L|^{(m-p-1)/2} \operatorname{etr}(-\tfrac{1}{2}L) \prod_{i>j} (l_i - l_j)$$

$$0 < l_1 \leq \cdots \leq l_p < \infty \tag{3.1}$$

where $L = \operatorname{diag}(l_1, \ldots, l_p)$ and

$$C_1 = \frac{\pi^{p^2/2}}{2^{pm/2}\Gamma_p(m/2)\Gamma_p(p/2)} \tag{3.2}$$

The distribution of $l_1, \ldots, l_p$ when $\Sigma = I_p$ and $M \neq 0$ is given by

$$f_1(l_1, \ldots, l_p) = C_1 \operatorname{etr}(-\tfrac{1}{2}\Omega)\, {}_0F_1(m/2; \tfrac{1}{4}\Omega; L) \times \operatorname{etr}(-\tfrac{1}{2}L)|L|^{(m-p-1)/2} \prod_{i>j} (l_i - l_j) \tag{3.3}$$

where $\Omega = \operatorname{diag}(\omega_1, \ldots, \omega_p)$ and $\omega_1, \ldots, \omega_p$ are the latent roots of $MM'\Sigma^{-1}$. When $M = 0$, and $\Sigma \neq I_p$, the joint density of roots of S is

$$f_1(l_1, \ldots, l_p) = C_1 |\Sigma|^{-m/2}\, {}_0F_0(-\tfrac{1}{2}\Sigma^{-1}, L)|L|^{(m-p-1)/2} \prod_{i>j} (l_i - l_j) \tag{3.4}$$

The expressions in (3.3) and (3.4) were derived by James (1961a).

Next, let $\theta_1 \leq \cdots \leq \theta_p$ be the eigenvalues of SS_1^{-1}. When $\Sigma = \Sigma_1$ and $M = 0$, the joint density of these roots is given by

$$f_2(\theta_1, \ldots, \theta_p) = C_2 \prod_{i=1}^{p} \{\theta_i^r (1 - \theta_i)^{-r-s-p-1}\} \prod_{i>j} (\theta_i - \theta_j)$$

$$0 \leq \theta_1 \leq \cdots \leq \theta_p \leq \infty \tag{3.5}$$

where $r = \frac{1}{2}(m - p - 1)$, $s = \frac{1}{2}(n - p - 1)$, and

$$C_2 = \frac{\pi^{p^2/2}\Gamma_p(r + s + p + 1)}{\Gamma_p(m/2)\Gamma_p(n/2)\Gamma_p(p/2)}$$

This density was derived by Fisher (1939), Hsu (1939), and Roy (1939) independently.

When $\Sigma = \Sigma_1$, $E(X) = M$, and $m \geq p$ the joint density of $\theta_1, \ldots, \theta_p$ is known (see Constantine, 1963) to be

$$f_2(\theta_1, \ldots, \theta_p) = \operatorname{etr}(-\tfrac{1}{2}\Omega_0)\, {}_1F_1(\tfrac{1}{2}(m + n); \tfrac{1}{2}m; \tfrac{1}{2}\Omega_0, (I + \Phi^{-1})^{-1}) \times C_2 \frac{|\Phi|^r}{|I + \Phi|^{(m+n)/2}} \prod_{i>j} (\theta_i - \theta_j) \tag{3.6}$$

where $\Phi = \text{diag}(\theta_1, \ldots, \theta_p)$, $\Omega_0 = \text{diag}(\omega_{01}, \ldots, \omega_{0p})$, and $\omega_{01}, \ldots, \omega_{0p}$ are the eigenvalues of $M'\Sigma^{-1}M$. Roy (1942) derived this distribution when the rank of Ω_0 is 1 whereas Anderson (1946) derived it when the rank of Ω_0 is 2. When $n \geq p \geq m$, the joint density of the roots of SS_1^{-1} is known to be

$$f_2(\theta_1, \ldots, \theta_m) = \text{etr}(-\tfrac{1}{2}\Omega_0)\ {}_1F_1(\tfrac{1}{2}(m+n); \tfrac{1}{2}p; \tfrac{1}{2}\Omega_0, (I+\Phi^{-1})^{-1})$$
$$\times \frac{\Gamma_m(\frac{1}{2}(m+n))\pi^{m^2/2}}{\Gamma_m(p/2)\Gamma_m((m+n-p)/2)\Gamma_m(m/2)}$$
$$\times \frac{|\Phi|^{(p-m-1)/2}}{|I+\Phi|^{(m+n)/2}} \prod_{i>j}^{m} (\theta_i - \theta_j) \tag{3.7}$$

where $\Phi = \text{diag}(\theta_1, \ldots, \theta_m)$ and Ω_0 was defined earlier.

Suppose $u_1 \leq \cdots \leq u_p$ are the latent roots of $S(S+S_1)^{-1}$. Then the joint density of these roots can be obtained from the joint density of $\theta_1, \ldots, \theta_p$ by making the transformations $u_i = \theta_i/(1+\theta_i)$ $(i = 1, \ldots, p)$.

Next, let the m rows of $(Z_1'|Z_2')$ be distributed independently as $(p+q)$-variate normal with mean vector 0 and covariance matrix

$$\begin{pmatrix} \Sigma_{11} & \Sigma_{12} \\ \Sigma_{21} & \Sigma_{22} \end{pmatrix}$$

and let $p \leq q$. Also, let $r_1^2 \leq \cdots \leq r_p^2$ be the roots of $R^2 = (Z_1Z_1')^{-1}(Z_1Z_2')(Z_2Z_2')^{-1}(Z_2Z_1')$ and let $P^2 = \Sigma_{11}^{-1}\Sigma_{12}\Sigma_{22}^{-1}\Sigma_{21}$. When $P^2 = 0$, the joint distribution of $r_1^2, \ldots, r_p^2$ is known (see Fisher, 1939; Hsu, 1939; Roy, 1939) to be

$$f_3(r_1^2, \ldots, r_p^2) = C_3|R^2|^{(q-p-1)/2}|I-R^2|^{(m-p-q-1)/2} \prod_{i>j}^{p} (r_i^2 - r_j^2) \tag{3.8}$$

where

$$C_3 = \pi^{p/2} \prod_{i=1}^{p} \frac{\Gamma((m-i+1)/2)}{\Gamma((m+1-q-i)/2)\Gamma((p+1-i)/2)\Gamma((q+1-i)/2)}$$

In the noncentral case, it is known (see Constantine, 1963) that the joint density of $r_1^2, \ldots, r_p^2$ is given by

$$f_3(r_1^2, \ldots, r_p^2) = |I-P^2|^{m/2}\ {}_2F_1(m/2, m/2; q/2, P^2, R^2)$$
$$\times C_3|R^2|^{(q-p-1)/2}|I-R^2|^{(m-q-p-1)/2}$$
$$\times \prod_{i>j} (r_i^2 - r_j^2) \tag{3.9}$$

Next, let $a_1, \ldots, a_p$ be the latent roots of the Gaussian matrix A defined earlier. When $N = 0$, the joint density of the roots of A is known (see Hsu, 1941) to be

$$f_4(a_1, \ldots, a_p) = C_4 \prod_{i=1}^{p} \exp(-a_i^2/2) \prod_{i>j} (a_i - a_j)$$

$$-\infty < a_1 \leq \cdots \leq a_p < \infty \qquad (3.10)$$

where

$$C_4 = \left[2^{p/2} \prod_{i=1}^{p} \Gamma((p+1-i)/2)\right]^{-1}$$

The joint densities of the eigenvalues of a class of random matrices in the central cases are of the form

$$f_5(\delta_1, \ldots, \delta_p) = C \prod_{i=1}^{p} \psi(\delta_i)\phi(\delta_1, \ldots, \delta_p)$$

$$a \leq \delta_1 \leq \cdots \leq \delta_p \leq b \qquad (3.11)$$

where $\phi(x_1, \ldots, x_p) = (\phi_i(x_j))$. In the noncentral cases, the joint densities of the eigenvalues of a class of random matrices are of the form

$$f(\delta_1, \ldots, \delta_p) = C \prod_{i=1}^{p} \psi(\delta_i)\phi(\delta_1, \ldots, \delta_p) \sum_{k=0}^{\infty} \sum_{\kappa} a(\kappa)\eta_\kappa(\Delta)$$

$$a \leq \delta_1 \leq \cdots \leq \delta_p \leq b \qquad (3.12)$$

where $\Delta = \operatorname{diag}(\delta_1, \ldots, \delta_p)$, C is a constant, $\kappa = (k_1, \ldots, k_p)$ is a partition of k subject to suitable restrictions, $\eta_\kappa(\Delta)$ is a symmetric function of $\delta_1, \ldots, \delta_p$, and the second summation is over all partitions κ of k. The joint densities of the roots of the Wishart matrix, multivariate beta matrix, multivariate F matrix, and Gaussian matrix in the central cases are special cases of Eq. (3.11). The joint densities of the roots of these random matrices under various noncentral situations are special cases of Eq. (3.12).

4. DISTRIBUTIONS OF THE INDIVIDUAL ROOTS OF A CLASS OF RANDOM MATRICES

In this section, we review some of the literature on the distributions of the individual roots of a class of random matrices. Let the joint density of the eigenvalues $\delta_1 \leq \cdots \leq \delta_p$ of a random matrix be of the form given by Eq. (3.11). Then $P[c \leq \delta_1 \leq \delta_p \leq d]$, $P[\delta_p \leq d]$, and $P[\delta_1 \leq c]$ can be evaluated by applying Lemma 2.1. Krishnaiah and Waikar (1971a) derived expressions for the joint density of any few consecutive roots $\delta_{r+1}, \ldots, \delta_{r+s}$

$(0 \leq r \leq p-1, 1 \leq s \leq p-r)$, probability integrals associated with any pair of roots δ_r, δ_s $(1 \leq r < s \leq p)$, and the cummulative distribution function (c.d.f.) of any intermediate root. The c.d.f. of any intermediate root δ_r $(1 < r < p)$ is given by

$$P[\delta_r \leq x] = P[\delta_{r+1} \leq x] + P[\delta_1 \leq \cdots \leq \delta_r \leq x \leq \delta_{r+1} \leq \cdots \leq \delta_p] \tag{4.1}$$

The second term on the right-hand side of (4.1) can be evaluated by expanding the determinant $\phi(\delta_1, \ldots, \delta_p)$ by the first r columns using Laplace expansion and applying Lemma 2.1. Repeated application of (4.1) gives the c.d.f. of intermediate roots in terms of the c.d.f. of the largest or smallest root and terms which are of the same form as the second term in (4.1).

Next, let the joint density of $\delta_1, \ldots, \delta_p$ be of the form given by Eq. (3.12) where

$$\eta_\kappa(\Delta) = \sum_{\mathbf{r}} b_{\mathbf{r}}^\kappa \delta_1^{r_1} \cdots \delta_p^{r_p} \tag{4.2}$$

and $\sum_{\mathbf{r}}$ is the summation over $r_1, r_2, \ldots, r_p$ subject to some suitable restrictions and $b_{\mathbf{r}}^\kappa$ depends on $r_1, \ldots, r_p$ and κ. Krishnaiah and Chattopadhyay (1975) gave expressions for the moments of the elementary symmetric functions of the roots $\delta_1, \ldots, \delta_p$, joint density of any few consecutive roots, as well as the probability integral of any individual root.

Expressions for the cumulative density functions of the extreme roots as well as the intermediate roots of the central multivariate beta matrix were given by Roy (1945) in terms of "pseudodeterminants" and reduction formulas for these pseudodeterminants. Further work on the reduction formulas of these pseudodeterminants was done in the literature (e.g., see Pillai, 1956). Expressions for the roots of the central Wishart matrix S with $\Sigma = I_p$ can be obtained (see Nanda, 1948b) as limiting cases of the corresponding expressions for the roots of the central multivariate beta matrix. Gnanadesikan (1956) gave a reduction formula for the probability integral associated with the joint density of extreme roots of S in the central case when $\Sigma = I$. The expressions for the roots in terms of the pseudodeterminants and the associated reduction formulas are very complicated. They are not of much practical use except when the number of variables is small.

Exact expressions for the densities of the extreme roots of the central multivariate beta matrix as well as the largest root of the central Wishart matrix are available in the literature (e.g., see Sugiyama, 1966, 1967) in terms of zonal polynomials. An exact finite series expression for the density of the smallest root of the Wishart matrix S with $\Sigma = I_p$ in the central case was given by Krishnaiah and Chang (1971b) when $\frac{1}{2}(m - p - 1)$ is an integer whereas Sugiyama (1970) gave an expression for the joint distribution of the extreme roots of this matrix; both of these expressions are in terms of zonal polynomials. Khatri (1972) obtained finite series expressions for the densities of the extreme roots of the Wishart matrix S when $E(S/m) = \Sigma$, noncentral multivariate beta matrix, and canonical correlation matrix when $\frac{1}{2}(m - p - 1)$ is an integer; these expressions were derived independently by A. G. Constantine (unpublished).

Davis (1972b) showed that the marginal densities of the individual roots of the central multivariate beta matrix and central Wishart matrix S with $\Sigma = I_p$ satisfy certain differential equation. Applying this result, Davis (1972c) obtained recurrence relations for the densities of the individual roots of these matrices.

Upper percentage points of the largest root of the central multivariate beta matrix were given by Nanda (1951) and Foster and Rees (1957) for $p = 2$ whereas they were given for $p = 3, 4$ by Foster (1957, 1958). Heck (1960) gave charts for these percentage points for $p = 2(1)5$ whereas Pillai (1960) gave percentage points for $p = 2(1)6$. Pillai (1965) also suggested an approximation for the c.d.f. of the largest root of the central multivariate beta matrix. Using this approximation, percentage points of this statistic were computed for values of p up to 20 in a series of papers by Pillai (e.g., see Pillai, 1967). Using the expressions given by Krishnaiah and Chang (1971a), Schuurmann *et al.* (1973b) computed exact values of c for $p = 2(1)10$, $\alpha = 0.100, 0.050, 0.025, 0.010$, where

$$P[1 - c \leq u_1 \leq \cdots \leq u_p \leq c] = 1 - \alpha$$

Pillai and Dotson (1969) computed upper 5 and 1% points of the median root for $p = 3$. Krishnaiah and co-workers (1973) computed exact upper 5 and 1% of the distribution of u_s, $s = 2, 3, \ldots, p - 1$, for $p = 4, 5, 6, 7$, and of the distributions of u_2 and u_7 for $p = 8$ by using the results of Krishnaiah and Waikar (1971a). Using the expressions given by Krishnaiah and Chang (1971a), Chang (1974) computed the exact upper percentage points of u_1 and u_p whereas Schuurmann and Waikar (1974) computed the exact upper percentage points of u_1. Hanumara and Thompson (1968) constructed tables for approximate values of c_1 and c_2 for $p = 3(1)10$ where

$$P[l_1 \geq c_1] = 1 - \alpha, \qquad P[l_p \leq c_2] = 1 - \alpha$$

Pillai and Chang (1970) also computed approximate upper percentage points of l_p for values of p up to 20. Using the results of Krishnaiah and Waikar (1971a), Clemm *et al.* (1973a) constructed upper 10, 5, 2.5, and 1% points of the distributions of l_i ($i = 1, \ldots, p - 1$) for $p = 2(1)10$.

5. DISTRIBUTIONS OF THE RATIOS OF THE ROOTS

Throughout this section, $l_1 \le \cdots \le l_p$ denote the eigenvalues of the Wishart matrix S when $M = 0$ and $\Sigma = I_p$ whereas $u_1 \le \cdots \le u_p$ denote the eigenvalues of $S(S + S_1)^{-1}$ when $\Sigma = \Sigma_1$ and $M = 0$.

Let $v_j = l_j/\sum_i l_i$ for $j = 1, \ldots, p$. Also, let $f_{p,m}^{(j)}$ and $g_{p,m}^{(j)}$, respectively, denote the density functions of v_j and l_j. Davis (1972a) showed that

$$\mathscr{L}\{(1 + w)^{(pm-4)/2} f_{p,m}^{(j)}(1/1 + w)\} = 2\Gamma(pm/2) e^s s^{(2-pm)/2} g_{p,m}^{(j)}(2s) \quad (5.1)$$

where $\mathscr{L}(h(w))$ denotes the Laplace transformation of $h(w)$. Making use of the expressions in Krishnaiah and Chang (1971a) for the densities of l_1 and l_p and Eq. (5.1), Schuurmann *et al.* (1973a) obtained the densities of v_1 and v_p; the latter authors also computed the upper percentage points of v_p and lower percentage points of v_1. Krishnaiah and Schuurmann (1974b) obtained the marginal densities of $v_2, \ldots, v_{p-1}$ using Eq. (5.1) and the expressions given by Krishnaiah and Waikar (1971a) for the densities of $l_2, \ldots, l_{p-1}$. Percentage points of $v_2, \ldots, v_{p-1}$ were also computed by Krishnaiah and Schuurmann (1974b).

Sugiyama (1970) and Krishnaiah and Waikar (1971b) gave expressions for the density of l_1/l_p but their forms are complicated from a computational point of view. Waikar and Schuurmann (1973) and Krishnaiah and Schuurmann (1974b) gave alternative expressions for the distribution function of l_1/l_p and computed percentage points of this statistic.

6. DISTRIBUTIONS OF ELEMENTARY SYMMETRIC FUNCTIONS

In this section, we give a review of the literature on the distributions of the elementary symmetric functions of a class of random matrices. We first discuss the distribution of the trace of the Wishart matrix.

When $\Sigma = I_p$, it is known that the distribution of tr S is a central or noncentral chi-square distribution with pm degrees of freedom according to whether $M = 0$ or $M \ne 0$. When $\Sigma \ne I$, the distribution of tr S is the same as the distribution of the sum of chi-square variables whose joint distribution is a multivariate chi-square distribution in the sense of Krishnamoorthy and Parthasarathy (1951). This distribution is infinitely divisible (see Krishnaiah and Rao, 1961; Krishnaiah and Waikar, 1973). When $\Sigma \ne I$, exact expressions for the density of tr S in the noncentral case are given by Krishnaiah and Waikar (1973) whereas Krishnaiah (1977b) gives an asymptotic expression in terms of normal density and Hermite polynomials in the central case. If S is partitioned as $S = (S_{ij})$ where S_{ij}: $p_i \times p_j$, then the joint distribution of tr $S_{11}, \ldots,$ tr S_{pp} is the same as the joint distribution of correlated quadratic forms considered by Krishnaiah (1977b) and Khatri *et al.* (1977). We will now review the literature on the distributions of the traces of a multivariate F matrix SS_1^{-1} and a multivariate beta matrix $S(S + S_1)^{-1}$ when $\Sigma = \Sigma_1$.

The statistic $T_0^2 = \text{tr } SS_1^{-1}$ was proposed by Lawley (1938) for testing certain hypotheses in multivariate statistical analysis. Hotelling (1951) derived the exact null distribution of T_0^2 for $p = 2$. Grubbs (1954) computed the percentage points of T_0^2 for $p = 2$. This statistic is known in the literature as the Lawley–Hotelling trace statistic. Using the first four moments, Pillai (1954) approximated the distribution of T_0^2 with F distribution and computed (see Pillai, 1960) approximate percentage points of this distribution using this approximation. For other approximations which are essentially based on the moments, the reader is referred to Tiku (1971) and Hughes and Saw (1972). Constantine (1966) derived the nonnull distribution of T_0^2 for $|T_0^2| < 1$. Davis (1968) showed that T_0^2 in the central case satisfies an ordinary differential equation of order p, and constructed some tables (see Davis, 1970b) for this statistic by using analytic continuation of the series obtained by Constantine (1966) for $|T_0^2| < 1$. Pillai and Young (1971) expressed the distribution of T_0^2 in the central case as a linear combination of the inverse Laplace transformations of certain pseudodeterminants when $\frac{1}{2}(m - p - 1)$ is an integer. Krishnaiah and Chang (1972) expressed the Laplace transformation $L(t;\, T_0^2)$ of T_0^2 as a linear combination of terms of the form $\prod_{i=1}^{q} g(t;\, \alpha_i, \beta_i)$ where $q = [\frac{1}{2}(p + 1)]$ denotes the integral part of $\frac{1}{2}(p + 1)$, α_i and β_i are certain constants, and

$$g(t;\, c, d) = \int_0^{\infty} \exp(-tz) \bigg/ \left(1 + \frac{z}{b}\right)^a dz$$

Then they took inverse Laplace transformation and expressed the density $h(x)$ of T_0^2 as a linear combination of terms of the form

$$g_0(x;\, \alpha_1, \beta_1) * g_0(x;\, \alpha_2, \beta_2) * \cdots * g_0(x;\, \alpha_q, \beta_q)$$

where $*$ denotes convolution and

$$g_0(x;\, \alpha, \beta) = \beta^{\alpha}/(\beta + x)^{\alpha}, \qquad 0 < x < \infty$$

Using the expression of Krishnaiah and Chang (1972), one can compute exact percentage points of the distribution of T_0^2. However, the complexity of these computations increases substantially as p increases. But the evaluation of the expression for the density of T_0^2 given by Krishnaiah and Chang (1972) is much easier than the corresponding expression given by Pillai and Young (1971) when $\frac{1}{2}(m - p - 1)$ is an integer.

The statistic $T = \text{tr } S(S + S_1)^{-1}$ was proposed by Bartlett (1939) for testing certain hypotheses in multivariate analysis. The exact distribution of this statistic was first considered by Nanda (1950) who derived the distribution for the special cases $p = 2, 3, 4$ and $m = p + 1$. Later, several workers in the field considered various aspects of this test statistic. Using the first four moments of T, Pillai (1960) approximated the distribution of T with beta distribution and computed percentage points of this distribution. Mijares (1964a) extended the tables of Pillai using the same approximation. Pillai and Jayachandran (1967) gave the noncentral distributions of T_0^2 and T for $p = 2$. Khatri and Pillai (1968b) obtained an expression for the nonnull distribution of T for $|T| < 1$. Davis (1968) showed that the distribution of T satisfies a differential equation. Krishnaiah and Chang (1972) obtained an expression for the Laplace transformation $L(t;\, T)$ of T as a linear combination of terms of the form $\prod_{i=1}^{q} \exp(-c_i t) g(a_i t;\, b_i)$ where $q = [\frac{1}{2}(p+1)]$ denotes the integral part of $\frac{1}{2}(p+1)$, a_i is equal to 1 or 2, $b_i \geq 0$, c_i $(0 \leq c_i \leq q)$ are integers, and

$$g(a_i t;\, b_j) \exp(-c_k t) = \left(\frac{1}{a_i}\right)^{b_j + 1} \int_{c_k}^{c_k + a_i} \exp(-ty)(y - c_k)^{b_j}\, dy$$

Then, taking the inverse Laplace transformation Krishnaiah and Chang (1972) expressed the density $h(x)$ of T as a linear combination of terms of the form

$$g_0(x;\, a_1, b_1, c_1) * \cdots * g_0(x;\, a_q, b_q, c_q)$$

where $*$ denotes convolution and

$$g_0(x;\, a_i, b_i, c_i) = (1/a_i)^{b_j + 1}(x - c_i)^{b_j}$$

where $c_i < x < c_i + a_i$. This expression may be used to compute percentage points of the distribution of T. Schuurmann *et al.* (1975) simplified this expression further and also computed exact percentage points of the distribution of T for $p = 2(1)5$; for more extensive exact percentage points. the reader is referred to a technical report (ARL 73-0008) by the same authors.

Next, let the joint density of the roots $l_1 \leq \cdots \leq l_p$ be of the form (3.12) where $\eta_\kappa(L)$ is given by Eq. (4.2). Krishnaiah and Chattopadhyay (1975) expressed the moments of elementary symmetric functions of the roots $l_1, \ldots, l_p$ as linear combinations of certain double integrals by making use of Lemma 2.2. These double integrals can be evaluated explicitly for the cases when the joint density of the roots is any of the forms given by Eqs. (3.1)–(3.10). Here we note that the moments of the elementary symmetric functions for some special cases are evaluated explicitly in the literature; for details, the reader is referred to Mijares (1964b) Khatri and Pillai (1968a), Pillai and Jouris (1969), and deWaal (1973), and some references cited therein.

When the joint density of the roots is of the form (3.12) and $\eta_\kappa(L)$ is given by (4.2), the Laplace transformation of $T = \sum_{i=1}^{p} l_i$ is known (see Krishnaiah and Chattopadhyay, 1975) to be

$$\mathscr{L}(t;\, T) = \sum_{k=0}^{\infty} \sum_{\kappa} \sum_{\mathbf{r}} a(\kappa) b_{\mathbf{r}} \operatorname{Pf}(B_r) \tag{6.1}$$

where $B_r = (b_{rij})$. When $p = 2k$, we have

$$b_{rij} = \int_a^b \int_a^b x^{r_i + i - 1} y^{r_j + j - 1} \psi(x) \psi(y) \exp(-tx - ty)\, dx\, dy \qquad i, j = 1, \ldots, 2k \tag{6.2}$$

When p is odd, we have, in addition,

$$b_{r,2k+2,2k+2} = 0$$

$$b_{r,i,2k+2} = -b_{r,i,2k+2} = \int_a^b x^{r_i + i - 1} \psi(x) \exp(-tx)\, dx \qquad i = 1, \ldots, 2k + 1 \tag{6.3}$$

The distribution of T may be obtained by taking the inverse Laplace transformation of the right-hand side of (6.1). We can use the methods discussed by Krishnaiah and Chang (1972) to invert the right-hand side of (6.1) for the following cases: (i) $a = 0$, $b = \infty$, and $\psi(x)$ is of the form $x^c(1 - x)^{-c-d}$; (ii) $a = 0$, $b = 1$, and $\psi(x)$ is of the form $x^c(1 - x)^d$.

7. DISTRIBUTIONS OF THE LIKELIHOOD RATIO STATISTICS

There has been a considerable amount of activity in deriving the distributions of the likelihood ratio statistics for testing various hypotheses on the mean vectors and covariance matrices of multivariate normal populations. A review of these developments until 1957 was given by Anderson (1958). In the last few years, several papers were written by Consul, Mathai, and others on deriving the exact distributions of likelihood ratio statistics for testing the hypotheses on mean vectors and covariance matrices of the multivariate normal populations. Most of these developments were reviewed by Mathai (1973) and Mathai and Saxena (1973). The expressions that are available in the literature for these exact distributions are complicated for computational purposes except in some special cases.

Box (1949) derived an asymptotic expression for the distributions of a class of statistics W $(0 \le W \le 1)$ whose moments are of the form

$$E\{W^h\} = K \left(\frac{\prod_{j=1}^{c} y_j^{y_j}}{\prod_{k=1}^{a} x_k^{x_k}} \right)^h \frac{\prod_{k=1}^{a} \Gamma(x_k(1 + h) + \xi_k)}{\prod_{j=1}^{c} \Gamma(y_j(1 + h) + \eta_j)} \tag{7.1}$$

for $h = 0, 1, \ldots$, where K is a normalizing constant such that $E\{W^0\} = 1$, and $\sum_{k=1}^{a} x_k = \sum_{j=1}^{c} y_j$. Some likelihood ratio test statistics used in multivariate analysis belong to this class of statistics. The asymptotic series obtained by Box involves a linear combination of chi-square variates. There are many situations in which we have to take several terms in this series to get the desired degree of accuracy. In these situations, it is complicated to use the above asymptotic expression.

Recently, Krishnaiah and others in some of their papers approximated the distributions of certain powers of some likelihood ratio test statistics with Pearson's type I distribution. The accuracy of this approximation was found to be sufficient for practical purposes. A review of this work is now given.

Let $\mathbf{X}' = (\mathbf{X}_1', \ldots, \mathbf{X}_q')$ be distributed as multivariate normal with mean vector $\boldsymbol{\mu}' = (\boldsymbol{\mu}_1', \ldots, \boldsymbol{\mu}_q')$ and covariance matrix Σ. Also, let $\mathbf{X}_i$ be of order $p_i \times 1$ and $E\{(\mathbf{X}_i - \boldsymbol{\mu}_i)(\mathbf{X}_j' - \boldsymbol{\mu}_j') = \Sigma_{ij}$. In addition, let H_1, H_2, and H_3 denote the following hypotheses:

$$
\begin{aligned}
&H_1\colon\ \Sigma_{ij} = 0 \qquad (i \neq j = 1, \ldots, q)\\
&H_2\colon\ \Sigma = \sigma^2\Sigma_0 \qquad (\sigma^2 \text{ unknown, } \Sigma_0 \text{ known})\\
&H_3\colon\ \Sigma = \Sigma_0 \qquad (\Sigma_0 \text{ known})
\end{aligned}
\tag{7.2}
$$

Next, let $(\mathbf{x}_{1j}', \ldots, \mathbf{x}_{qj}')$ $(j = 1, \ldots, N)$ be N independent observations on X' and let $A = (A_{tu})$ where

$$A_{tu} = \sum_{j=1}^{N} (\mathbf{x}_{tj} - \bar{\mathbf{x}}_{t\cdot})(\mathbf{x}_{uj} - \bar{\mathbf{x}}_{u\cdot})', \qquad \bar{\mathbf{x}}_{t\cdot} = \frac{1}{N}\sum_{j=1}^{N} \mathbf{x}_{tj}$$

for $t, u = 1, \ldots, q$. Also, let $s = p_1 + \cdots + p_q$ and $n = N - 1$.

The likelihood ratio statistic V_1 for testing H_1 is known (see Wilks, 1935) to be

$$V_1 = \frac{|A|}{\prod_{j=1}^{q} |A_{jj}|} \tag{7.3}$$

Consul (1969) had expressed the density of V_1 in terms of Meijer's G function (Meijer, 1946) whereas Mathai and Saxena (1973) gave an expression in terms of a series by expanding the G function. Davis and Field (1971) computed the percentage points of $-2\rho \log V_1$ by using a Cornish–Fisher type (see Davis, 1971) of inversion of Box's asymptotic series; here ρ is a correction factor. Lee *et al.* (1977a) approximated the distribution of $V_1^{1/4}$ with Pearson's type I distribution by equating the first four moments and computed the percentage points of the distribution of $V_1^* = -2 \log V_1$ using this approximation. They compared some of the entries in their tables with the corresponding values obtained by using Box's asymptotic series of order n^{-13} and the corresponding entries in the tables of Davis and Field (1971). These comparisons indicate that the accuracy of the approximation used by Lee and associates is sufficient for practical purposes.

Mauchly (1940) obtained the following expression for the likelihood ratio statistic V_2 for testing H_2:

$$V_2 = \frac{|A\Sigma_0^{-1}|}{\{\text{tr } A\Sigma_0^{-1}/s\}} \tag{7.4}$$

An expression for the density of V_2 was given by Consul (1969) in terms of the G function whereas Mathai and Rathie (1970b) and Nagarsenker and Pillai (1973a) expressed this density in the form of a series. Nagarsenker and Pillai (1973a) computed the lower percentage points of the distribution of V_2 for some values of the parameters by using their expression for the density of V_2. Lee *et al.* (1977a) approximated the distribution of $V_2^{1/4}$ with Pearson's type I distribution by equating the first four moments. Using this approximation, they computed the lower percentage points of the distribution of V_2 for a few values of the parameters. These values compare favorably with the corresponding entries in the tables of Nagarsenker and Pillai (1973a).

Anderson (1958) derived the likelihood ratio statistic for testing H_3. The modified likelihood ratio test statistic (obtained by changing N to n in the likelihood ratio statistic) is given by

$$V_3 = (e/n)^{sn/2} |A\Sigma_0^{-1}|^{n/2} \operatorname{etr}(-\tfrac{1}{2}A\Sigma_0^{-1}) \tag{7.5}$$

Korin (1968) obtained an asymptotic expression of order n^{-15} for the distribution of V_3^* and computed the upper percentage points of the distribution of V_3^* for some values of the parameters; here V_3^* denotes $-2 \log V_3$. Nagarsenker and Pillai (1973b) obtained an expression for the distribution of V_3^* and computed the upper percentage points of V_3^* by using expression (7.5). Lee *et al.* (1977a) approximated the distribution of $V_3^{1/34}$ with the Pearson type I distribution. Using this approximation, they computed the percentage points of the distribution of V_3^* for a few values of the parameters and compared these values with the corresponding entries in the tables of Korin (1968) and Nagarsenker and Pillai (1973b). The accuracy of the approximation used by Lee *et al.* (1977a) is sufficient for practical purposes.

Next, let $\mathbf{X}_1, \ldots, \mathbf{X}_q$ be distributed independently as p-variate normal with mean vectors $\boldsymbol{\mu}_1, \ldots, \boldsymbol{\mu}_q$ and covariance matrices $\Sigma_1, \ldots, \Sigma_q$, respectively. Also, let $\mathbf{x}_{ij}$ ($j = 1, \ldots, N_i$) be jth independent observation on $\mathbf{X}_i$. In addition, let

$$A_{ii} = \sum_{j=1}^{N_i} (\mathbf{x}_{ij} - \bar{\mathbf{x}}_{i\cdot})(\mathbf{x}_{ij} - \bar{\mathbf{x}}_{i\cdot})', \qquad N_i\bar{\mathbf{x}}_{i\cdot} = \sum_{j=1}^{N_i} \mathbf{x}_{ij}$$

$n_g = N_g - 1$ and $n = \sum_{g=1}^{q} n_g$. Wilks (1932) derived the likelihood ratio statistic for testing H_4 where

$$H_4: \ \Sigma_1 = \cdots = \Sigma_q \tag{7.6}$$

The modified likelihood ratio test statistic for H_4 (obtained by changing N_g to n_g in the likelihood ratio test statistic for H_4) is given by

$$V_4 = \frac{\prod_{g=1}^{q} |A_{gg}|^{n_g/2} n^{pn/2}}{|\sum_{g=1}^{q} A_{gg}| \prod_{g=1}^{q} n_g^{pn_g/2}} \tag{7.7}$$

Bishop (1939) investigated the accuracy of approximating the distribution of the $(2/m)$th power of the likelihood ratio test statistic with beta distribution for a few cases when $m = q(n_0 + 1)$ and $n_i = n_0$ $(i = 1, \ldots, q)$. Korin (1969) computed the upper percentage points of the distribution of $V_4^* = -2 \log V_4$ using Box's asymptotic expression with terms up to order n^{-15}. Davis and Field (1971) computed the percentage points of V_4^* using Cornish–Fisher-type inversion (see Davis, 1971) of Box's asymptotic series. Lee *et al.* (1977a) computed the percentage points of V_4^* by approximating the distribution of $V_4^{1/b}$ with Pearson type I distribution. The constant b was chosen to be 30 when $p = 2, 3$, whereas it was chosen to be 45 when $p = 4, 5$. When $p = 6$, b was chosen to be equal to 70. A comparison of some of the entries in the tables of Lee *et al.* (1977a) with the corresponding entries in the tables of Korin (1969) and Davis and Field (1971) indicated that the accuracy of the approximation used by Lee *et al.* (1977a) is sufficient for practical purposes.

Next, let H_5 denote the hypothesis

$$H_5: \begin{cases} \Sigma_{11} & = \cdots = \Sigma_{q_1, q_1} \\ \Sigma_{q_1+1, q_1+1} & = \cdots = \Sigma_{q_2^*, q_2^*} \\ \vdots & \quad \vdots \\ \Sigma_{q_{k-1}^*+1, q_{k-1}^*+1} & = \cdots = \Sigma_{q, q} \end{cases} \tag{7.8}$$

where $q_0^* = 0$, $q_j^* = \sum_{i=1}^{j} q_i$, and $q_k^* = q$. The problem of testing H_5 arises in testing certain linear structures (see Krishnaiah and Lee, 1976) on the covariance matrices. The likelihood ratio statistic for testing H_5 is known (see Lee *et al.*, 1977a) to be

$$V_5 = \frac{\prod_{i=1}^{q} |A_{ii}/n_i|^{n_i/2}}{\prod_{j=1}^{k} |\sum_{i=q_{j-1}^*+1}^{q_j} A_{ii}/n_j^*|^{n_j^*/2}} \tag{7.9}$$

where $n_j^* = n_{q_{j-1}+1} + \cdots + n_{q_j^*}$. Lee *et al.* (1977a) approximated the distribution of $V_5^{1/b}$ with Pearson's type I distribution by equating the first four moments. The constant b was chosen to be equal to 6, 15, 20, and 30 accordingly to whether p equals 1, 2, 3, and 4, respectively. Using this approximation, these authors computed upper percentage points of the distribution of V_5^* where $V_5^* = -2 \log V_5$ and compared some of these values with the corresponding values obtained by using Box's asymptotic series of order n^{-13}. This comparison indicated that the accuracy of the approximation used by Lee and co-workers is sufficient for practical purposes.

Next, consider the hypothesis H_6 where

$$H_6: \begin{cases} \Sigma_1 = \cdots = \Sigma_k \\ \boldsymbol{\mu}_1 = \cdots = \boldsymbol{\mu}_k \end{cases}$$

Wilks (1932) derived the likelihood ratio statistic for testing H_6 and the moments of this statistic. Let λ_6 denote the modified likelihood ratio statistic (see Anderson, 1958) obtained by replacing N_i with n_i $(i = 1, \ldots, q)$ in the likelihood ratio statistic. Chang *et al.* (1977) computed the percentage points of the distribution of λ_6 by approximating $\lambda_6^{1/b}$ with Pearson's type I distribution where b is a suitably chosen integer. The accuracy of this approximation is found to be satisfactory for practical purposes. These authors also used a similar approximation to the distribution of the likelihood ratio statistic λ_7 for testing the hypothesis that $\Sigma_1 = \Sigma_0$ and $\boldsymbol{\mu} = \boldsymbol{\mu}_0$ simultaneously where Σ_0 and $\boldsymbol{\mu}_0$ are known. Nagarsenker and Pillai (1974) computed the percentage points of λ_7 using a different method. Lee *et al.* (1976) computed percentage points for the likelihood ratio statistics for testing some hypotheses of compound symmetry by approximating certain powers of the likelihood ratio statistics with Pearson's type I distribution.

8. APPLICATIONS IN INFERENCE ON COVARIANCE STRUCTURES AND SIMULTANEOUS TEST PROCEDURES

Let $\mathbf{X}$: $pq \times 1$ be distributed as a multivariate normal with mean vector $\boldsymbol{\mu}$ and covariance matrix Σ. Then it is of interest to test the hypothesis H_1 where

$$H_1: \quad \Sigma = G_1 \otimes \Sigma_1 + \cdots + G_k \otimes \Sigma_k \tag{8.1}$$

and the G_i are known $p \times p$ matrices which commute with each other and $k(q+1) \leq p(pq+1)$. This problem is considered by Bock and Bargmann (1966), Srivastava (1966), and Anderson (1969) when $q = 1$. Special cases of this problem have been considered by several workers in the field since the mid-1930s. The structure (8.1) is known to be the block version of sphericity when $k = 1$ and $G_1 = I_p$ whereas it is known to be the block version of the intraclass correlation model when $k = 2$, $G_1 = I_p$, and $G_2 = J - I_p$ where J is a matrix whose elements are equal to unity. Similarly, if $k = q$, $G_i = W_i$, $W_1 = I_p$, and

$$W_j = \begin{pmatrix} 0 & I_{q-j+1} \\ I_{j-1} & 0 \end{pmatrix}$$

for $j = 2, 3, \ldots, q$, the structure (8.1) is known to be the block version of circular symmetry. Olkin (1973) considered the problem of testing the hypotheses that the structure of Σ has block versions of sphericity, intraclass correlation model, and circular symmetry. Krishnaiah and Lee (1976) considered the problem of testing H in the general case. We will now briefly review the procedures of Krishnaiah and Lee (1976) and mention how the multivariate distributions discussed in this chapter are useful in the application of these procedures.

Since the G_i commute with each other, there exists an orthogonal matrix Γ such that $\Gamma G_i \Gamma' = \text{diag}(\lambda_{i1}, \ldots, \lambda_{ip})$ for $i = 1, \ldots, k$. We know that $\mathbf{Y} = (\Gamma \otimes I_q)\mathbf{X}$ is distributed as a multivariate normal with covariance matrix Σ^* where

$$\Sigma^* = \begin{bmatrix} \Sigma_{11}^* & \cdots & \Sigma_{1p}^* \\ \vdots & \vdots & \vdots \\ \Sigma_{p1}^* & \cdots & \Sigma_{pp}^* \end{bmatrix} \tag{8.2}$$

If Σ is of the structure given by (8.1), then $\Sigma^* = \text{diag}(\psi_1, \ldots, \psi_p)$ where

$$\begin{bmatrix} \psi_1 \\ \vdots \\ \psi_p \end{bmatrix} = \Lambda \begin{bmatrix} \Sigma_1 \\ \vdots \\ \Sigma_k \end{bmatrix} \tag{8.3}$$

and

$$\Lambda = \begin{bmatrix} \lambda_{11} I_q & \cdots & \lambda_{k1} I_q \\ \vdots & \vdots & \vdots \\ \lambda_{1p} I_q & \cdots & \lambda_{kp} I_q \end{bmatrix} \tag{8.4}$$

If $k = p$, the problem of testing H_1 is equivalent to the problem of testing H_2 where H_2: $\Sigma_{ij}^* = 0$ $(i \neq j = 1, \ldots, p)$. This is equivalent to the problem of testing for the multiple independence of several sets of variables when their joint distribution is a multivariate normal. The distribution problem associated with the likelihood ratio statistic for testing H_2 was discussed in Section 7. Next, let $k < p$ and $E\Lambda = 0$ where E: $q(p - k) \times pq$ is a known matrix of rank $q(p - k)$. Also, let

$$H_3\colon \quad E\psi = 0 \tag{8.5}$$

where $\psi' = (\psi_1', \ldots, \psi_p')$. Then the problem of testing H_1 is equivalent to testing $H_2 \cap H_3$. A special case of testing H_3 is equivalent to the problem of testing the hypothesis that the covariance matrices within different groups of multivariate normal populations are equal. An approximation to the distribution of the likelihood ratio statistic associated with testing this hypothesis was discussed in Section 7.

Next, consider the growth curve model*

$$E(X) = A\mu B \tag{8.6}$$

* This model was considered by Potthoff and Roy (1964).

where the rows of X: $n \times p$ are distributed independently as multivariate normal with means given by Eq. (8.6) and covariance matrix Σ. Here A: $n \times m$ and B: $r \times p$ $(m < n;\ r \leq p)$ are known and μ is unknown. Next, let $F = (F_1 \mid F_2)$ where F_1: $p \times r$ and F_2: $p \times (p - r)$ are such that $BF_2 = 0$ and $BF_1 = I$. Also, let $Y = XF_1$ and $Z = XF_2$. Then the rows of $(Y \mid Z)$ are distributed independently as multivariate normal with covariance matrix

$$W = \begin{pmatrix} W_{11} & W_{12} \\ W_{21} & W_{22} \end{pmatrix}$$

where $W_{ij} = F_i' \Sigma F_j$ $(i, j = 1, 2)$, and the means are given by

$$E(Y \mid Z) = (A\mu \mid 0) \tag{8.7}$$

When Z is held fixed, the rows of Y are distributed independently as multivariate normal with covariance matrix $\Sigma_0 = W_{11} - W_{12} W_{22}^{-1} W_{21}$ and means

$$E(Y/Z) = (A/Z)\begin{pmatrix} \mu \\ \zeta \end{pmatrix} \tag{8.8}$$

where $\zeta' = W_{12}' W_{22}^{-1}$. Rao (1966) reduced model (8.6) to the conditional model (8.8). Khatri (1966) derived the likelihood ratio test for a linear hypothesis on location parameters under model (8.6). Krishnaiah (1969b,c) discussed certain simultaneous test procedures on location parameters under model (8.6). We now discuss the problem of testing the hypotheses $H_1, \ldots, H_q$ simultaneously where

$$H_i\colon\quad C_i \mu = 0 \tag{8.9}$$

and C_i: $u_i \times m$ is known and is of rank u_i for $i = 1, \ldots, q$. Also, let S_{H_i} denote the sum of the squares and cross products (SP) matrix associated with H_i and let S_e denote the error SP matrix. Then, it is known (see Khatri, 1966) that

$$S_{H_i} = (C_i \hat{\mu})' R^{-1} (C_i \hat{\mu}) \tag{8.10}$$

and

$$S = (BE^{-1}B')^{-1} \tag{8.11}$$

where

$$\hat{\mu} = (A'A)^{-1} A'XE^{-1}B'(BE^{-1}B')^{-1}, \qquad E = X'[I - A(A'A)^{-1}A']X$$

$$R = C_i[(A'A)^{-1} + (A'A)^{-1}A' \times X\{E^{-1} - E^{-1}B'(BE^{-1}B')^{-1}BE^{-1}\}X'A(A'A)^{-1}]C_i$$

Also, let $l_{i1} \leq \cdots \leq l_{ip}$ denote the eigenvalues of $S_{H_i} S^{-1}$ and let $\psi(l_{i1}, \ldots, l_{ip})$ be a suitable function of these eigenvalues. Then we accept or reject H_i accordingly as

$$\psi(l_{i1}, \ldots, l_{ip}) \lesseqgtr c_\alpha \tag{8.12}$$

where

$$P\left[\psi(l_{i1}, \ldots, l_{ip}) \leq c_\alpha;\ i = 1, \ldots, q \,\middle|\, \bigcap_{i=1}^{q} H_i\right] = 1 - \alpha \tag{8.13}$$

The exact evaluation of c_α is very complicated. But using Bonferroni's inequality and using some results on distributions reviewed in this chapter, we can obtain bounds on c_α when $\psi(y_1, \ldots, y_p)$ is equal to $\sum_{j=1}^{p} y_j$, y_p or some other suitable functions.

9. APPLICATIONS IN SIMULTANEOUS TESTS FOR THE EQUALITY OF THE EIGENVALUES

In this section, we discuss the applications of some multivariate distributions in the area of simultaneous tests on the equality of the eigenvalues of the covariance matrix, canonical correlation matrix, and a matrix associated with the multivariate analysis of variance. The following notation is needed in the sequel.

Let $\mathbf{x}$: $p \times 1$ be distributed as multivariate normal with mean vector $\boldsymbol{\mu}$ and covariance matrix Σ and let $\mathbf{x}_j$ $(j = 1, \ldots, N)$ denote the jth independent observation on $\mathbf{x}$. Also, let $\lambda_1 \leq \cdots \leq \lambda_p$ be the eigenvalues of Σ, whereas the eigenvalues of S are denoted by $l_1 \leq \cdots \leq l_p$ where $S = \sum_{j=1}^{N} (\mathbf{x}_j - \bar{\mathbf{x}}.) \times (\mathbf{x}_j - \bar{\mathbf{x}}.)'$ and $N\bar{\mathbf{x}}. = \sum_{j=1}^{N} \mathbf{x}_j$. Next, let H: $\lambda_1 = \cdots = \lambda_p$, H_{ij}: $\lambda_i = \lambda_j$, A_{ij}: $\lambda_i > \lambda_j$ $(i > j)$, H_i: $\lambda_i = \lambda$, A_i: $\lambda_i > \lambda$, and $p\lambda = \sum_{j=1}^{p} \lambda_j$.

If we are testing the hypotheses H_{ij} $(i > j = 1, \ldots, q;\ q \leq p)$ simultaneously against A_{ij}, we accept or reject H_{ij} accordingly as

$$l_i/l_j \lesseqgtr c_1 \tag{9.1}$$

where

$$P[l_q/l_1 \leq c_1 \,|\, \lambda_1 = \cdots = \lambda_q] = 1 - \alpha \tag{9.2}$$

Next, consider the problem of testing $H_{i+1,i}$ $(i = 1, \ldots, q - 1;\ q \leq p)$ simultaneously against $A_{i+1,i}$. In this case, we accept or reject $H_{i+1,i}$ accordingly as

$$l_{i+1}/l_i \lesseqgtr c_2 \tag{9.3}$$

where

$$P[l_{i+1}/l_i \leq c_2;\ i = 1, \ldots, q - 1 \,|\, \lambda_1 = \cdots = \lambda_q] = 1 - \alpha \tag{9.4}$$

Next, consider the problem of testing the hypotheses $H_2, \ldots, H_q$ simultaneously against $A_2, \ldots, A_q$. In this case, we accept or reject H_i accordingly as

$$l_i/\bar{l} \lesseqgtr c_3 \tag{9.5}$$

where $p\bar{l} = \sum_{j=1}^{p} l_j$ and c_3 is chosen such that

$$P\left[l_q/\bar{l} \leq c_3 \,\middle|\, \bigcap_{j=2}^{q} H_j\right] = 1 - \alpha \tag{9.6}$$

When $q = p$, these procedures are equivalent to the procedures proposed by Krishnaiah and Waikar (1971a).

Sometimes we know in advance that $\lambda_1 = \cdots = \lambda_q = \lambda_0$ where λ_0 is unknown. In this case, we test H_{iq} $(i = q + 1, \ldots, q + t; q + t \leq p)$ as follows. Accept or reject H_{iq} accordingly as

$$l_i/(l_1 + \cdots + l_q) \lesseqgtr c_4 \tag{9.7}$$

where

$$P[l_{q+t}/(l_1 + \cdots + l_q) \leq c_4 \,|\, \lambda_1 = \cdots = \lambda_{q+t}] = 1 - \alpha \tag{9.8}$$

Next, let H_{i0}: $\lambda_i = \lambda_0$ $(i = r, \ldots, t; r \geq 1, t \leq p)$ where λ_0 is known. In this case, we accept or reject H_{i0} accordingly as

$$l_i \lesseqgtr c_5 \tag{9.9}$$

where c_5 is chosen such that

$$P[l_t \leq c_5 \,|\, \lambda_r = \cdots = \lambda_t = \lambda_0] = 1 - \alpha \tag{9.10}$$

Next, consider the problem of testing H: $\lambda_1 = \cdots = \lambda_p = \lambda_0$ where λ_0 is known against the alternative $\bigcup_{i=1}^{p-1} [\lambda_{i+1} > \lambda_i]$. In this case, we accept H if

$$\max_i \, (l_{i+1} - l_i) < c_6 \tag{9.11}$$

and reject otherwise where

$$P[\max_i \, (l_{i+1} - l_i) \leq c_6 \,|\, H] = 1 - \alpha \tag{9.12}$$

Similarly, we can propose a procedure based on $l_p - l_1$ to test H.

The probability integrals in (9.2), (9.4), (9.6), and (9.8) involve nuisance parameters when $q < p$. However, we can obtain bounds on them which are free from nuisance parameters.

Next, let $l_1 \leq \cdots \leq l_p$ be the eigenvalues of the central Wishart matrix with n degrees of freedom, $E(S/n) = \mathrm{diag}(\lambda_1, \ldots, \lambda_p)$, and $\lambda_1 = \cdots = \lambda_s$. Also, let $\mu_s \geq \cdots \geq \mu_1$ be the eigenvalues of U where U is a Wishart matrix of order $s \times s$ with n degrees of freedom and $E(U/n) = \lambda_1 I_s$. If $\phi(x_1, \ldots, x_s)$ is a monotonic nondecreasing function of each argument $x_1, \ldots, x_s$, it is easily seen (see Fujikoshi and Isogai, 1976), by applying Poincaré's separation theorem, that

$$P[\phi(l_p, \ldots, l_{p-s+1}) \leq c_7] \leq P[\phi(\mu_s, \ldots, \mu_1) \leq c_7] \leq P[\phi(l_s, \ldots, l_1) \leq c_7] \tag{9.13}$$

where $\phi(x_s, \ldots, x_1)$ is a monotonic nondecreasing function of each x_j. Now, let $\psi(x_1, \ldots, x_s)$ be another monotonic nondecreasing function of each of the arguments $x_1, \ldots, x_s$. Then,

$$P\left[\frac{\phi(l_p, \ldots, l_{p-s+1})}{\psi(l_s, \ldots, l_1)} \leq c_8\right] \leq P\left[\frac{\phi(\mu_s, \ldots, \mu_1)}{\psi(\mu_s, \ldots, \mu_1)} \leq c_8\right] \tag{9.14}$$

Inequalities (9.13) and (9.14) can be used to obtain bounds on the distribution functions of the individual roots and certain ratios of the roots of the Wishart matrix.

Krishnaiah and Lee (1977) discussed various simultaneous test procedures based on the linear combinations of the roots of S as well as the ratios of the linear combinations of these roots. Asymptotic joint distributions of certain functions of the eigenvalues of the MANOVA matrix, canonical correlation matrix, Wishart matrix, and correlation matrix are derived by Krishnaiah and Lee (1978) when the population roots have multiplicity.

We will now discuss some applications of the procedures just discussed. In the area of principal component analysis, the experimenter is interested in finding out the number of significant principal components which explain the variation among the experimental units. Each principal component is a linear combination of the original variables and the coefficients of these variables in the ith principal component are the elements of the eigenvector corresponding to the ith largest root. The principal components which correspond to the nonsignificant eigenvalues do not contribute much in explaining the variation among experimental units. By using the procedures described earlier, we can find out which of the eigenvalues are not significant.

Next, consider the factor analysis model

$$\mathbf{y} = F\mathbf{x} + \boldsymbol{\varepsilon} \tag{9.15}$$

where $\mathbf{x}: k \times 1$ and $\boldsymbol{\varepsilon}: p \times 1$ are distributed independently and normally with a common mean vector $\mathbf{0}$ and covariance matrices given by $E(\mathbf{x}\mathbf{x}') = I_k$ and $E(\boldsymbol{\varepsilon}\boldsymbol{\varepsilon}') = \sigma^2 I_p$. The covariance matrix of $\mathbf{y}$ is given by $\Sigma = \sigma^2 I_p + FF'$. In this model $\mathbf{x}$ denotes a vector of unobservable variables and $\boldsymbol{\varepsilon}$ denotes the vector of errors of measurement; the rank of FF' is $k < p$ and it is usually unknown. In the area of factor analysis, it is of interest to test the hypothesis that k is a specified value. This is equivalent to testing the hypotheses $H_2, \ldots, H_a$

$(a < p)$ simultaneously where H_i: $\lambda_i = \sigma^2$ and $\lambda_1 \leq \cdots \leq \lambda_p$ are the eigenvalues of Σ. If we know in advance that the rank of FF' is at most t, then the problem of testing the hypothesis that k is a specified value may be viewed as the problem of testing the hypotheses $H_{p-t+1}, \ldots, H_a$ $(a < p)$ simultaneously given that $\lambda_{p-t} = \cdots = \lambda_1 = \sigma^2$. These hypotheses can be tested by using the methods discussed in this section.

Next consider the following two-way classification model with one observation per cell:

$$y_{ij} = \mu + \alpha_i + \beta_j + \eta_{ij} + \varepsilon_{ij} \tag{9.16}$$

where y_{ij} $(i = 1, \ldots, t;\ j = 1, \ldots, b)$ is the observed value in the jth block due to the ith treatment, μ the general mean, α_i the effect due to the ith treatment, β_j the effect due to the jth block, η_{ij} the interaction term, and the ε_{ij} are random errors which are distributed independently and normally with zero means and variances σ^2. Also $\sum_i \alpha_i = \sum_i \eta_{ij} = \sum_j \eta_{ij} = \sum_j \beta_j = 0$. Without loss of generality let us assume that $t \leq b$. Then it is of interest to test for the structure of interaction terms. We denote the roots of $\Gamma\Gamma'$ by $\theta_1{}^2 \leq \cdots \leq \theta_t{}^2$ where $\theta_1 = 0$ and $\Gamma = (\eta_{ij})$. The eigenvectors of $\Gamma\Gamma'$ and $\Gamma'\Gamma$ associated with $\theta_2{}^2, \ldots, \theta_t{}^2$ are denoted by $\mathbf{u}_2, \ldots, \mathbf{u}_t$ and $\mathbf{v}_2, \ldots, \mathbf{v}_t$, respectively. Then Γ can be decomposed as

$$\Gamma = \sum_{k=2}^{t} \theta_k \mathbf{u}_k \mathbf{v}_k{}' \tag{9.17}$$

It is realistic to assume, in several situations, that the last few roots $\theta_2, \ldots, \theta_r$ are equal to zero or negligible. So, we assume that

$$\Gamma = \sum_{k=r+1}^{t} \theta_k \mathbf{u}_k \mathbf{v}_k{}', \qquad 1 < r \leq t - 1 \tag{9.18}$$

It is known (e.g., see Mandel, 1969) that the least squares estimates of θ_k, $\mathbf{u}_k$, and $\mathbf{v}_k$ are given by $\hat{\theta}_k$, $\hat{\mathbf{u}}_k$, and $\hat{\mathbf{v}}_k$, respectively, where $\hat{\theta}_2{}^2 \leq \cdots \leq \hat{\theta}_t{}^2$ are the nonzero eigenvalues of the matrix ZZ', $Z = (d_{ij})$, $d_{ij} = y_{ij} - \bar{y}_{i\cdot} - \bar{y}_{\cdot j} + \bar{y}_{\cdot\cdot}$, and

$$\bar{y}_{i\cdot} = \frac{1}{b} \sum_{j=1}^{b} y_{ij}, \qquad \bar{y}_{\cdot j} = \frac{1}{t} \sum_{i} y_{ij}, \qquad \bar{y}_{\cdot\cdot} = \frac{1}{bt} \sum_{i} \sum_{j} y_{ij}$$

Also, $\hat{\mathbf{u}}_k$ and $\hat{\mathbf{v}}_k$ are the eigenvectors of ZZ' and $Z'Z$ corresponding to $\hat{\theta}_k$. Now, let C_t: $t \times (t-1)$ be such that $C_t C_t' = I_t - (1/t)J_t$ and $C_t'C_t = I_{t-1}$. The nonzero characteristic roots of ZZ' are equal to the roots of W where

$$W = C_t'Y(I_b - (1/b)J_b)Y'C_t \tag{9.19}$$

It is known that the distribution of W is Wishart distribution with $b - 1$ degrees of freedom and

$$E(W/(b-1)) = \sigma^2 I_{t-1} + (\Omega/(b-1)) = \Sigma^* \tag{9.20}$$

where $\Omega = C_t'\Gamma\Gamma'C_t$. Here we note that $\theta_{r+1}^2 \le \cdots \le \theta_t^2$ are the nonzero roots of Ω. Let $\lambda_1^* \le \cdots \le \lambda_{t-1}^*$ be the roots of Σ^*. Then $\lambda_1^* = \cdots = \lambda_{r-1}^* = \sigma^2$ and $\lambda_{r+j}^* = \sigma^2 + \theta_{r+j+1}^2$ $(j = 0, 1, \ldots, t - r - 1)$. So the problem of testing the hypothesis $\lambda_{r+j}^* = \lambda_1^*$ $(j = 0, 1, \ldots, t - 1 - r)$ is equivalent to the problem of testing the hypothesis that $\theta_{r+j+1} = 0$. Procedures discussed in this section can be used to test simultaneously the hypotheses that $\theta_j = 0$ $(r + 1 \le j \le t)$.

The problem of testing the hypothesis of "no interaction" in two-way classification with one observation per cell was considered by Tukey (1949) from a different point of view. Gollob (1968) and Mandel (1969) considered the problem of testing the hypothesis of "no interaction" under the model (9.18). They partitioned the sums of squares, i.e., tr ZZ', into various sums of squares associated with the interaction terms $\theta_j \mathbf{u}_j \mathbf{v}_j'$ as follows. The sum of squares associated with the source of variation due to $\theta_j \mathbf{u}_j \mathbf{v}_j'$ is $\hat{\theta}_j^2$ and the error sum of squares is $\hat{\theta}_2^2 + \cdots + \hat{\theta}_r^2$. Then these authors proposed to test the hypotheses $\theta_j = 0$ individually by using the statistics $F_j = \hat{\theta}_j^2/(\hat{\theta}_2^2 + \cdots + \hat{\theta}_r^2)$ and treating $\hat{\theta}_2^2, \ldots, \hat{\theta}_t^2$ as independent chi-square variables with certain degrees of freedom. But $\hat{\theta}_2^2, \ldots, \hat{\theta}_{t-1}^2$ are distributed neither as chi-square variates nor independently. Corsten and Van Eijnsbergen (1972) showed that the likelihood ratio statistic for testing the hypothesis that $\theta_j = \cdots = \theta_t = 0$ is given by $(\hat{\theta}_{r+1}^2 + \cdots + \hat{\theta}_t^2)/(\hat{\theta}_2^2 + \cdots + \hat{\theta}_r^2)$. Johnson and Graybill (1972) independently obtained this statistic for the special case when $r = t - 1$. Recently, Yochmowitz and Cornell (1978) derived the likelihood ratio statistic for testing the hypothesis that $\theta_{r+j} = 0$ against the alternative that $\theta_{r+j} \ne 0$ and $\theta_{r+j-1} = 0$. It can be shown that the likelihood ratio statistic for testing $\theta_{r+j} = 0$ against $\theta_{r+j} \ne 0, \ldots, \theta_{r+j-c} \ne 0, \theta_{r+j-c-1} = 0$ is given by

$$\Lambda = \left\{\frac{\hat{\theta}_2^2 + \cdots + \hat{\theta}_{r+j-c-1}^2}{\hat{\theta}_2^2 + \cdots + \hat{\theta}_{r+j}^2}\right\}^{bt/2}$$

Next, let $\mathbf{x}_1, \ldots, \mathbf{x}_N$ be N independent random samples drawn from a p-variate population with the distribution function $G(\mathbf{x})$ where $G(\mathbf{x}) = \pi_1 F_1(\mathbf{x}) + \cdots + \pi_k F_k(\mathbf{x})$ and $\pi_j > 0$ $(j = 1, \ldots, k)$, $k - 1 < p$, $p < N$, $\pi_1 + \cdots + \pi_k = 1$, and $F_j(\mathbf{x})$ denotes the distribution function of multivariate normal with mean vector $\boldsymbol{\mu}_j$ and covariance matrix $\sigma^2 I_p$. Let S be the sample SP matrix. Then,

$$E(S/N - 1) = \sigma^2 I_p + \Omega \tag{9.21}$$

where

$$\Omega = \sum_{j=1}^{k} \pi_j(\boldsymbol{\mu}_j - \boldsymbol{\mu})(\boldsymbol{\mu}_j - \boldsymbol{\mu})' \qquad \text{and} \qquad \boldsymbol{\mu} = \pi_1\boldsymbol{\mu}_1 + \cdots + \pi_k\boldsymbol{\mu}_k$$

If $\Omega = \mathrm{diag}(\delta_p, \ldots, \delta_{p-k+2}, 0, \ldots, 0)$, then we can simultaneously test the hypotheses $\delta_j = 0$ for $j = p - k + 2, \ldots, t$ $(t \leq p)$, by using the methods discussed in this section. The problem of testing the hypothesis that $G(x) = F_1(x)$ was discussed by Bryant (1975) and Krishnaiah and Schuurmann (1974b).

When the number of variables is large, it is of interest in the area of pattern recognition to find the number of significant discriminant functions which will adequately discriminate between the patterns. We now discuss how some of the multivariate distributions are useful in tackling this problem.

Let $\mathbf{x}_1, \ldots, \mathbf{x}_k$ be distributed independently as p-variate normal with mean vectors $\boldsymbol{\mu}_1, \ldots, \boldsymbol{\mu}_k$ and a common covariance matrix Σ. Also, let $\mathbf{x}_{ij}$ $(j = 1, \ldots, n_i)$ denote the jth independent observation on $\mathbf{x}_i$. Here $\mathbf{x}_{ij}$ may denote the observation on the jth individual in the ith pattern. Also, let $S_1 = (s_{1tu})$ and $S_2 = (s_{2tu})$, denote the SP matrices associated with variation between groups and variation within groups, respectively. Let $\theta_1 \leq \cdots \leq \theta_p$ be the latent roots of $S_1 S_2^{-1}$ and let $\lambda_1 \leq \cdots \leq \lambda_p$ be the roots of $I + \Omega$ where

$$\Omega = \frac{1}{k-1} \sum_{i=1}^{k} n_i(\boldsymbol{\mu}_i - \bar{\boldsymbol{\mu}})(\boldsymbol{\mu}_i - \bar{\boldsymbol{\mu}})'\Sigma^{-1}, \qquad \bar{\boldsymbol{\mu}} = \frac{1}{k} \sum_{i=1}^{k} \boldsymbol{\mu}_i$$

We assume that $k - 1 < p$. The rank of Ω is $k - 1$ and so $\lambda_1 = \cdots = \lambda_{p-k+1} = 1$. Now, let H_j: $\lambda_j = 1$ and A_j: $\lambda_j > 1$ for $j = p - k + 2, \ldots, q$ $(q \leq p)$. The hypotheses H_j $(j = p - k + 2, \ldots, q)$ can be tested simultaneously as follows. Accept or reject H_j accordingly as

$$\theta_j \lesseqgtr c_9 \tag{9.22}$$

where

$$P[\theta_q \leq c_9 \,|\, \lambda_1 = \cdots = \lambda_q = 1] = 1 - \alpha \tag{9.23}$$

If $\lambda_1 = \cdots = \lambda_{q*} = 1$, we conclude that the discriminant functions associated with $\lambda_{q*+1}, \ldots, \lambda_p$ are sufficient to discriminate between the patterns. The discriminant function associated with λ_j is given by $a_{j1}x_1 + \cdots + a_{jp}x_p$ where $(a_{j1}, \ldots, a_{jp})$ is the eigenvector of $I + \Omega$ associated with λ_j and x_j is the observation on the jth variate. An estimate of this discriminant function is $b_{j1}x_1 + \cdots + b_{jp}x_p$ where $(b_{j1}, \ldots, b_{jp})$ is the eigenvector of $S_1 S_2^{-1}$ associated with θ_j. The discriminant functions which are important can be used to classify an observation into one of the pattern. Sometimes, it is of interest to find out which of the significant discriminant functions are of equal importance besides finding out the number of nonsignificant discriminant functions. In these situations, it is of interest to test the hypotheses H_{ij}

$(p-k+1 \leq j < i; i \leq q \leq p)$ simultaneously where H_{ij}: $\lambda_i = \lambda_j$. In this case, we accept or reject H_{ij} accordingly as

$$l_i/l_j \lesseqgtr c_{10} \tag{9.24}$$

where

$$P[l_q/l_{p-k+1} \leq c_{10} | \lambda_1 = \cdots = \lambda_q = 1] = 1 - \alpha \tag{9.25}$$

Let us now discuss how some of the distributions reviewed in this chapter are useful in the reduction of dimensionality in canonical correlation analysis.

Let $(\mathbf{x}', \mathbf{y}')$ be distributed as a $(p+q)$-variate normal with mean vector $(\boldsymbol{\mu}', \mathbf{v}')$ and covariance matrix Σ where

$$\Sigma = \begin{pmatrix} \Sigma_{11} & \Sigma_{12} \\ \Sigma_{21} & \Sigma_{22} \end{pmatrix}$$

Here Σ_{11}: $p \times p$ is the covariance matrix of $\mathbf{x}$ and Σ_{22}: $q \times q$ is the covariance matrix of $\mathbf{y}$. Without loss of generality, we assume that $p \leq q$. Now, let $(\mathbf{x}_j, \mathbf{y}_j)$ $(j = 1, \ldots, N)$ be the jth independent observation on $(\mathbf{x}, \mathbf{y})$ and let

$$S = \sum_{j=1}^{N} \begin{pmatrix} \mathbf{x}_j - \bar{\mathbf{x}}. \\ \mathbf{y}_j - \bar{\mathbf{y}}. \end{pmatrix} (\mathbf{x}_j' - \bar{\mathbf{x}}.', \mathbf{y}_j' - \bar{\mathbf{y}}.') = \begin{pmatrix} S_{11} & S_{12} \\ S_{21} & S_{22} \end{pmatrix}$$

where $N\bar{\mathbf{x}}. = \sum_{j=1}^{N} \mathbf{x}_j$ and $N\bar{\mathbf{y}}. = \sum_{j=1}^{N} \mathbf{y}_j$. Also, let $\rho_1^2 \leq \cdots \leq \rho_p^2$ be the eigenvalues of $\Sigma_{11}^{-1}\Sigma_{12}\Sigma_{22}^{-1}\Sigma_{21}$ and let $r_1^2 \leq \cdots \leq r_p^2$ be the eigenvalues of $S_{11}^{-1}S_{12}S_{22}^{-1}S_{21}$. Now, let H_j: $\rho_j^2 = 0$ for $j = 1, \ldots, p$. The hypotheses $H_1, \ldots, H_t$ are tested simultaneously as follows. We accept or reject H_j accordingly as

$$r_j^2 \lesseqgtr c_{11}$$

where

$$P[r_t^2 \leq c_{11} | \rho_1^2 = \cdots = \rho_t^2 = 0] = 1 - \alpha \tag{9.26}$$

The probability integrals in (9.23), (9.25), and (9.26) involve nuisance parameters but we can obtain bounds (e.g., see Fujikoshi and Isogai, 1976) which are free from nuisance parameters. We can also use procedures based on various ratios of the roots.

REFERENCES

Anderson, T. W. (1946). The non-central Wishart distribution and certain problems of multivariate statistics, *Ann. Math. Statist.* **17**, 409–431.

Anderson, T. W. (1958). "An Introduction to Multivariate Statistical Analysis." Wiley, New York.

Anderson, T. W. (1969). Statistical inference for covariance matrices with linear structure. *In* "Multivariate Analysis-II" (P. R. Krishnaiah, ed.). Academic Press, New York.

Anderson, T. W., Das Gupta, S., and Styan, G. P. H. (1972). "A Bibliography of Multivariate Statistical Analysis." Oliver and Boyd, Edinburgh.

Bartlett, M. S. (1939). A note on tests of significance in multivariate analysis, *Proc. Cambridge Phil. Soc.* **35**, 180–185.

Bartlett, M. S. (1950). Tests of significance in factor analysis, *Brit. J. Psychol.* (Statist. Section). **3**, 77–85.

Bharucha-Reid, A. T. (1970). Random algebraic equations. *In* "Probabilistic Methods in Applied Mathematics" (A. T. Bharucha-Reid, ed.), Vol. 1, pp. 1–52. Academic Press, New York.

Bishop, D. J. (1939). On a comprehensive test for the homogeneity of variances and covariances in multivariate problems, *Biometrika* **31**, 31–55.

Bock, R. D., and Bargmann, R. E. (1966). Analysis of covariance structures, *Psychometrika* **31**, 507–534.

Box, G. E. P. (1949). A general distribution theory for a class of likelihood criteria, *Biometrika* **36**, 317–346.

Boyce, W. E. (1968). Random eigenvalue problems. *In* "Probabilistic Methods in Applied Mathematics" (A. T. Bharucha-Reid, ed.), Vol. 2, pp. 1–73. Academic Press, New York.

Bryant, P. G. (1975). On testing for clusters using the sample covariance, *J. Multivariate Anal.* **5**, 96–105.

Carmeli, M. (1974). Statistical theory of energy levels and random matrices in physics, *J. Statist. Phys.* **10**, 259–297.

Chang, T. C. (1974). Upper percentage points of the extreme roots of the MANOVA matrix, *Ann. Inst. Statist. Math.*, Supple. **8**.

Chang, T. C., Krishnaiah, P. R., and Lee, J. C. (1977). Approximations to the distributions of the likelihood ratio statistics for testing the hypotheses on covariance matrices and mean vectors simultaneously. *In* "Applications of Statistics". (P. R. Krishnaiah, ed.). North-Holland Publ., Amsterdam.

Clemm, D. S., Chattopadhyay, A. K., and Krishnaiah, P. R. (1973a). Upper percentage points of the individual roots of the Wishart matrix, *Sankhyā, Ser. B* **35**, 325–338.

Clemm, D. S., Krishnaiah, P. R., and Waikar, V. B. (1973b). Tables for the extreme roots of the Wishart matrix, *J. Statist. Comp. Simulation* **2**, 65–92.

Constantine, A. G. (1963). Some non-central distribution problems in multivariate analysis, *Ann. Math. Statist.* **34**, 1270–1285.

Constantine, A. G. (1966). The distribution of Hotelling's generalized T_0^2, *Ann. Math. Statist.* **37**, 215–225.

Constantine, A. G., unpublished manuscript.

Consul, P. C. (1969). The exact distributions of likelihood criteria for different hypotheses. *In* "Multivariate Analysis-II" (P. R. Krishnaiah, ed.). Academic Press, New York.

Corsten, L. C. A., and van Eijbergen, A. C. (1972). Multiplicative effects in two-way analysis of variance, *Statistica Neerlandica* **26**, 61–68.

Cullis, C. E. (1918). "Matrices and Determinoids, II." Cambridge Univ. Press, London and New York.

Davis, A. W. (1968). A system of linear differential equations for the distribution of Hotelling's generalized T_0^2, *Ann. Math. Statist.* **39**, 815–832.

Davis, A. W. (1970a). Further applications of a differential equation for Hotelling's generalized T_0^2, *Ann. Inst. Statist. Math.* **22**, 77–87.

Davis, A. W. (1970b). Exact distribution of Hotelling's generalized T_0^2, *Biometrika* **57**, 187–191.

Davis, A. W. (1970c). On the null distribution of sum of the roots of a multivariate beta distribution, *Ann. Math. Statist.* **41**, 1557–1562.

Davis, A. W. (1971). Percentile approximations for a class of likelihood criteria, *Biometrika* **58**, 349–356.

Davis, A. W. (1972a). On the ratios of the individual latent roots to the trace of a Wishart matrix, *J. Multivariate Anal.* **2**, 440–443.

Davis, A. W. (1972b). On the marginal distributions of the latent roots of the multivariate beta matrix, *Ann. Math. Statist.* **43**, 1664–1670.

Davis, A. W. (1972c). On the distributions of the latent roots and traces of certain random matrices, *J. Multivariate Anal.* **2**, 189–200.

Davis, A. W., and Field, J. B. F. (1971). Tables of some multivariate test criteria, Tech. Rep. No. 32, Div. of Math. Statist., CSRIO, Australia.

DeBruijn, N. G. (1955). On some multiple integrals involving determinants, *J. Indian Math.* **19**, 133–151.

DeWaal, D. J. (1973). On the elementary symmetric functions of the Wishart and correlation matrices, *South African Statist. J.* **7**, 47–60.

Fisher, R. A. (1939). The sampling distribution of some statistics obtained from non-linear equations, *Ann. Eugenics.* **9**, 238–249.

Foster, F. G. (1957). Upper percentage points of the generalized beta distribution, II, *Biometrika* **44**, 441–443.

Foster, F. G. (1958). Upper percentage points of the generalized beta distribution, III, *Biometrika* **45**, 492–493.

Foster, F. G., and Rees, D. H. (1957). Upper percentage points of the generalized beta distribution, I, *Biometrika* **44**, 237–247.

Fujikoshi, Y., and Isogai, T. (1976). Lower bounds for the distributions of certain multivariate test statistics, *J. Multivariate Anal.* **6**, 250–255.

Garg, J. B. (ed.) (1972). "Statistical Properties of Nuclei." Plenum Press, New York.

Gnanadesikan, R. (1956). Contributions to multivariate analysis including univariate and multivariate component analysis and factor analysis, Mimeo Ser. No. 158, Inst. of Statist., Univ. of North Carolina, Chapel Hill, North Carolina.

Gollob, H. F. (1968). A statistical model which combines features of factor analytic and analysis of variance techniques, *Psychometrika* **33**, 73–116.

Grubbs, F. E. (1954). Tables of 1% and 5% probability levels of Hotelling's generalized T^2 statistics, Tech. Note No. 926, Ballistic Res. Lab., Aberdeen Proving Ground, Maryland.

Hanumara, R. C., and Thompson, W. A., Jr. (1968). Percentage points of the extreme roots of a Wishart matrix, *Biometrika* **55**, 505–512.

Hayakawa, T. (1967). On the distribution of a quadratic form in a multivariate normal sample, *Ann. Inst. Statist. Math.* **18**, 191–201.

Hayakawa, T. (1969). On the distribution of the latent roots of a positive definite random symmetric matrix, I, *Ann. Inst. Statist. Math.* **21**, 1–21.

Heck, D. L. (1960). Charts on some upper percentage points of the distribution of the largest characteristic root, *Ann. Math. Statist.* **31**, 625–642.

Herz, C. S. (1955). Bessel functions of matrix argument, *Ann. Math. Ser. 2*, **61**, 474–523.

Hotelling, H. (1951). A generalized T-test and measure of multivariate dispersion, *Proc. Berkeley Symp., 2nd* **1**, 23–42.

Hsu, P. L. (1939). On the distribution of roots of certain determinantal equations, *Ann. Eugenics* **9**, 250–258.

Hsu, P. L. (1941). On the limiting distribution of roots of a determinantal equation, *J. London Math. Soc.* **16**, 183–194.

Hua, L. K. (1959). "Harmonic Analysis of Functions of Several Complex Variables in the Classical Domains." Moscow.

Hughes, D. T., and Saw, J. G. (1972). Approximating the percentage points of Hotelling's generalized T_0^2 statistic, *Biometrika* **59**, 224–226.

James, A. T. (1960). Distribution of the latent roots of the covariance matrix, *Ann. Math. Statist.* **31**, 151–158.

James, A. T. (1961a). The distribution of noncentral means with known covariance, *Ann. Math. Statist.* **32**, 874–882.

James, A. T. (1961b). Zonal polynomials of the real positive definite symmetric matrices, *Ann. Math. Ser. 2* **74**, 456–469.

James, A. T. (1964). Distributions of matrix variates and latent roots derived from normal samples, *Ann. Math. Statist.* **35**, 475–501.

James, A. T. (1966). Inference on latent roots by calculation of hypergeometric functions of matrix arguments. *In* "Multivariate Analysis" (P. R. Krishnaiah, ed.). Academic Press, New York.

James, A. T. (1969). Tests of equality of latent roots of the covariance matrix. *In* "Multivariate Analysis-II" (P. R. Krishnaiah, ed.). Academic Press, New York.

Johnson, D. E., and Graybill, F. A. (1972). An analysis of a two-way model with interaction and no replication, *J. Amer. Statist. Assoc.* **67**, 862–868.

Khatri, C. G. (1966). A note on a MANOVA model applied to problems in growth curves, *Ann. Inst. Statist. Math.* **18**, 75–86.

Khatri, C. G. (1972). On the exact finite series distribution of the smallest or largest root of matrices in three situations, *J. Multivariate Anal.* **2**, 201–207.

Khatri, C. G., and Pillai, K. C. S. (1968a). On the moments of elementary symmetric functions of the roots of two matrices and approximations to a distribution, *Ann. Math. Statist.* **39**, 1274–1281.

Khatri, C. G., and Pillai, K. C. S. (1968b). On the non-central distributions of two test criteria in multivariate analysis of variance, *Ann. Math. Statist.* **39**, 215–226.

Khatri, C. G., and Srivastava, M. S. (1971). On exact non-null distributions of likelihood ratio criteria for sphericity test and equality of two covariance matrices, *Sankhyā Ser. A* **33**, 201–206.

Khatri, C. G., Krishnaiah, P. R., and Sen, P. K. (1977). A note on the joint distribution of correlated quadratic forms, *J. Statist. Planning Inference* **1**, 299–307.

Korin, B. P. (1968). On the distribution of a statistic used for testing a covariance matrix, *Biometrika* **55**, 171–178.

Korin, B. P. (1969). On testing of equality of k covariance matrices. *Biometrika* **56**, 216–217.

Korin, B. P., and Stevens, E. H. (1973). Some approximations for the distribution of a multivariate likelihood ratio criterion, *J. Roy. Statist. Soc. Ser. B*, **35**, 24–27.

Krishnaiah, P. R. (ed.) (1966). "Multivariate Analysis" Academic Press, New York.

Krishnaiah, P. R. (ed.) (1969a). "Multivariate Analysis-II" Academic Press, New York.

Krishnaiah, P. R. (1969b). Simultaneous test procedures under general MANOVA models. *In* "Multivariate Analysis-II" (P. R. Krishnaiah, ed.), pp. 121–143. Academic Press, New York.

Krishnaiah, P. R. (1969c). Further results on "Simultaneous test procedures under general MANOVA models," *Bulletin of the International Statistical Institute*, **43**, 288–289.

Krishnaiah, P. R. (ed.) (1973). "Multivariate Analysis-III" Academic Press, New York.

Krishnaiah, P. R. (1975). Tests for the equality of the covariance matrices of correlated multivariate normal populations. *In* "A Survey of Statistical Design and Linear Models" (J. N. Srivastava, ed.). North-Holland Publ., Amsterdam.

Krishnaiah, P. R. (1976). Some recent developments on complex multivariate distributions, *J. Multivariate Anal.* **6**, 1–30.

Krishnaiah, P. R. (ed.) (1977a). "Multivariate Analysis-IV" North-Holland Publ., Amsterdam.

Krishnaiah, P. R. (1977b). On generalized multivariate gamma type distributions and their applications in reliability, *Proc. Conf. Theory and Appl. of Reliability with Emphasis on Bayesian and Nonparametric Methods* (I. N. Shimi and C. P. Tsokos, eds.). Academic Press, New York.

Krishnaiah, P. R., and Chang, T. C. (1971a). On the exact distributions of the extreme roots of the Wishart and MANOVA matrices, *J. Multivariate Anal.* **1**, 108–117.

Krishnaiah, P. R., and Chang, T. C. (1971b). On the exact distributions of the smallest root of the Wishart matrix using zonal polynomials. *Ann. Inst. Statist. Math.* **23**, 293–295.

Krishnaiah, P. R., and Chang, T. C. (1972). On the exact distributions of the traces of $S_1(S_1 + S_2)^{-1}$ and $S_1 S_2^{-1}$, *Sankhyā Ser. A* **34**, 153–160.

Krishnaiah, P. R., and Chattopadhyay, A. K. (1975). On some noncentral distributions in multivariate analysis, *South African Statist. J.* **9**, 37–46.

Krishnaiah, P. R., and Lee, J. C. (1976). On covariance structures, *Sankhyā* **38**.

Krishnaiah, P. R., and Lee, J. C. (1977). Inference on the eigenvalues of the covariance matrices of real and complex multivariate normal populations. *In* "Multivariate Analysis-IV" (P. R. Krishnaiah, ed.). North-Holland Publ., Amsterdam.

Krishnaiah, P. R., and Lee, J. C. (1978). On the asymptotic joint distributions of certain functions of the eigenvalues of some random matrices, (abstract) *Bulletin of the International Statistical Institute* (in press).

Krishnaiah, P. R., and Rao, M. M. (1961). Remarks on a multivariate gamma distribution, *Amer. Math. Monthly* **68**, 342–346.

Krishnaiah, P. R., and Schuurmann, F. J. (1974a). On the distributions of the ratios of the extreme roots of the real and complex multivariate beta matrices, ARL 74-0122, Wright-Patterson Air Force Base, Ohio.

Krishnaiah, P. R., and Schuurmann, F. J. (1974b). On the evaluation of some distributions that arise in simultaneous tests for the equality of the latent roots of the covariance matrix, *J. Multivariate Anal.* **4**, 265–282.

Krishnaiah, P. R., and Waiker, V. B. (1971a). Exact joint distributions of any few ordered roots of a class of random matrices, *J. Multivariate Anal.* **1**, 308–315.

Krishnaiah, P. R., and Waikar, V. B. (1971b). Simultaneous tests for equality of latent roots against certain alternatives—I, *Ann. Inst. Statist. Math.* **23**, 451–468; also ARL 69–119 (1969).

Krishnaiah, P. R., and Waikar, V. B. (1972). Simultaneous tests for the equality of latent roots against certain alternatives—II, *Ann. Inst. Statist. Math.* **24**, 81–85.

Krishnaiah, P. R., and Waikar, V. B. (1973). On the distribution of a linear combination of correlated quadratic forms, *Commun. Statist.* **1**, 371–380.

Krishnaiah, P. R., Schuurmann, F. J., and Waikar, V. B. (1973). Upper percentage points of the intermediate roots of the MANOVA matrix, *Sankhyā Ser. B* **35**, 339–358.

Krishnamoorthy, A. S., and Parthasarathy, M. (1951). A multivariate gamma distribution, *Ann. Math. Statist.* **22**, 549–557; correction, *ibid.* **31**, 229 (1960).

Kshirsagar, A. M. (1972). "Multivariate Analysis." Dekker, New York.

Lawley, D. N. (1938). A generalization of Fisher's z test, *Biometrika* **30**, 180–187; correction, *ibid.* 467–469.

Lawley, D. N. (1956). Tests of significance for the latent roots of covariance and correlation matrices, *Biometrika* **43**, 128–136.

Lee, J. C., Chang, T. C., and Krishnaiah, P. R. (1977a). Approximations to the distributions of the likelihood ratio statistics for testing certain structures on the covariance matrices of real multivariate normal populations. *In* "Multivariate Analysis-IV" (P. R. Krishnaiah, ed.). North-Holland Publ., Amsterdam.

Lee, J. C., Krishnaiah, P. R., and Chang, T. C. (1976). On the distribution of the likelihood ratio test statistic for compound symmetry. *South African Statist. J.* **10**, 49–62.

Lee, J. C., Krishnaiah, P. R., and Chang, T. C. (1977b). Approximations to the distributions of the determinants of real and complex multivariate beta matrices, *South African Statist. J.* **11**, 13–26.

Lee, Y. S. (1972). Some results on the distribution of Wilks' likelihood-ratio criterion, *Biometrika* **59**, 649–664.

Mandel, J. (1969). The partitioning of interaction in analysis of variance, *J. Res. Nat. Bur. Standards* **73B**, 309–328.

Mathai, A. M. (1971). On the distribution of the likelihood-ratio criterion for testing linear hypotheses on regression coefficients, *Ann. Inst. Statist. Math.* **23**, 181–197.

Mathai, A. M. (1973). A review of the different techniques used for deriving the exact distributions of multivariate test criteria, *Sankhyā Ser. A* **35**, 39–60.

Mathai, A. M., and Rathie, P. N. (1970a). The exact distribution of Votaw's criteria, *Ann. Inst. Statist. Math.* **22**, 89–116.

Mathai, A. M., and Rathie, P. N. (1970b). The exact distribution for the sphericity test, *J. Statist. Res. (Dacca)* **4**, 140–159.

Mathai, A. M., and Rathie, P. N. (1971). The problem of testing independence, *Statistica* **31**, 673–688.

Mathai, A. M., and Saxena, R. K. (1973). "Generalized Hypergeometric Functions with Applications in Statistics and Physical Sciences." Springer-Verlag, Berlin.

Mauchly, J. W. (1940). Significant test for sphericity of a normal n-variate distribution, *Ann. Math. Statist.* **11**, 204–209.

Mehta, M. L. (1960). On the statistical properties of the level spacings in nuclear spectra, *Nucl. Phys.* **18**, 395–419.

Mehta, M. L. (1967). "Random Matrices and the Statistical Theory of Energy Levels." Academic Press, New York.

Mehta, M. L., and Gaudin, M. (1960). On the density of eigenvalues of a random matrix, *Nucl. Phys.* **18**, 420–427.

Meijer, C. S. (1946). On the G-function, I, *Indag. Math.* **8**, 124–134.

Mijares, T. A. (1964a). Percentage Points of the Sum $V_1^{(s)}$ of s Roots ($s = 1 - 50$). The Statistical Center, Univ. of the Philippines, Manila.

Mijares, T. A. (1964b). On elementary symmetric functions of the roots of a multivariate matrix distributions, *Ann. Math. Statist.* **35**, 1186–1198.

Nagarsenker, B. N., and Pillai, K. C. S. (1973a). The distribution of the sphericity test criterion, *J. Multivariate Anal.* **3**, 226–235.

Nagarsenker, B. N., and Pillai, K. C. S. (1973b). Distribution of the likelihood ratio criterion for testing a hypothesis specifying a covariance matrix, *Biometrika* **60**, 359–364.

Nagarsenker, B. N., and Pillai, K. C. S. (1974). Distribution of the likelihood ratio criterion for testing $\Sigma = \Sigma_0$ and $\mu = \mu_0$, *J. Multivariate Anal.* **4**, 114–122.

Nanda, D. N. (1948a). Distribution of a root of a determinantal equation, *Ann. Math. Statist.* **19**, 47–57.

Nanda, D. N. (1948b). Limiting distribution of a root of a determinantal equation, *Ann. Math. Statist.* **19**, 340–350.

Nanda, D. N. (1950). Distribution of the sum of roots of a determinantal equation under a certain condition, *Ann. Math. Statist.* **21**, 432–439.

Nanda, D. N. (1951). Probability distribution tables of the larger root of a determinantal equation with two roots, *J. Indian Soc. Agr. Statist.* **3**, 175–177.

Olkin, I. (1973). Testing and estimation for structures which are circularly symmetric in blocks. *In* "Multivariate Statistical Inference" (D. G. Kabe and R. P. Gupta, eds.). North-Holland Publ., Amsterdam.
Pillai, K. C. S. (1954). On some distribution problems in multivariate analysis, Mimeo Ser. No. 54, Inst. of Statist., Univ. of North Carolina, Chapel Hill, North Carolina.
Pillai, K. C. S. (1956). On the distribution of the largest or the smallest root of a matrix in multivariate analysis, *Biometrika* **43**, 122–127.
Pillai, K. C. S. (1960). "Statistical Tables for Tests of Multivariate Hypotheses." Statistical Center, Univ. of the Philippines, Manila.
Pillai, K. C. S. (1965). On the distribution of the largest characteristic root of a matrix in multivariate analysis, *Biometrika* **52**, 405–414.
Pillai, K. C. S. (1967). Upper percentage points of the largest root of a matrix in multivariate analysis, *Biometrika* **54**, 189–194.
Pillai, K. C. S., and Chang, T. C. (1970). An approximation to the cdf of the largest root of a covariance matrix, *Ann. Inst. Statist. Math.* (*Supple.*) **6**, 115–124.
Pillai, K. C. S., and Dotson, C. O. (1969). Power comparisons of tests of two multivariate hypotheses based on individual characteristic roots, *Ann. Inst. Statist. Math.* **21**, 49–66.
Pillai, K. C. S., and Gupta, A. K. (1969). On the exact distribution of Wilks' criterion, *Biometrika* **56**, 109–118.
Pillai, K. C. S., and Jayachandran, T. (1967). Power comparisons of tests of two multivariate hypotheses based on four criteria, *Biometrika* **54**, 195–210.
Pillai, K. C. S., and Jouris, G. M. (1969). On the moments of elementary symmetric functions of the roots of two matrices, *Ann. Inst. Statist. Math.* **21**, 309–320.
Pillai, K. C. S., and Young, D. L. (1971). On the exact distribution of Hotelling's generalized T_0^2. *J. Multivariate Anal.* **1**, 90–107.
Porter, C. E. (ed.) (1965). "Statistical Theories of Spectra: Fluctuations." Academic Press, New York.
Potthoff, R. F., and Roy, S. N. (1964). A generalized multivariate analysis of variance model useful especially for growth curve problems, *Biometrika* **51**, 313–326.
Rao, C. R. (1955). Estimation and tests of significance in factor analysis, *Psychometrika* **20**, 93–111.
Rao, C. R. (1965). "Linear Statistical Inference and Its Applications." Wiley, New York.
Rao, C. R. (1966). Covariance adjustment and related problems in multivariate analysis. *In* "Multivariate Analysis" (P. R. Krishnaiah, ed.), pp. 87–103. Academic Press, New York.
Roy, S. N. (1939). *p*-statistics or some generalizations in analysis of variance appropriate to multivariate problems, *Sankhyā* **4**, 381–396.
Roy, S. N. (1942). The sampling distribution of *p*-statistics and certain allied statistics on the nonnull hypothesis, *Sankhyā* **6**, 15–34.
Roy, S. N. (1945). The individual sampling distribution of the maximum, the minimum, and any intermediate of the *p*-statistics on the null hypothesis, *Sankhyā* **7**, 133–158.
Roy, S. N. (1957). "Some Aspects of Multivariate Analysis." Wiley, New York.
Roy, S. N., Gnanadesikan, R., and Srivastava, J. N. (1971). "Analysis and Design of Certain Multiresponse Experiments." Pergamon, New York.
Schatzoff, M. (1966). Exact distribution of Wilks' likelihood-ratio criterion, *Biometrika* **53**, 347–358.
Schuurmann, F. J., and Waikar, V. B. (1973). Tables for the power function of Roy's two-sided test for testing the hypothesis $\Sigma = I$ in the bivariate case, *Comm. in Statist.* **1**, 271–280.
Schuurmann, F. J., and Waikar, V. B. (1974). Upper percentage points of the smallest root of the MANOVA matrix, *Ann. Inst. Statist. Math. Supple.* **8**, 79–94.
Schuurmann, F. J., Krishnaiah, P. R., and Chattopadhyay, A. K. (1973a). On the distributions of the ratios of the extreme roots to the trace of the Wishart matrix, *J. Multivariate Anal.* **3**, 445–453.

Schuurmann, F. J., Waikar, V. B., and Krishnaiah, P. R. (1973b). Percentage points of the joint distribution of the extreme roots of the random matrix $S_1(S_1 + S_2)^{-1}$, *J. Statist. Comput. Simulation* **2**, 17–38.

Schuurmann, F. J., Krishnaiah, P. R., and Chattopadhyay, A. K. (1975). Exact percentage points of the distribution of the trace of a multivariate beta matrix, *J. Statist. Comp. Simulation* **3**, 331–343.

Siotani, M. (1975). Asymptotic expansions for the non-null distributions of multivariate test statistics. *In* "A Modern Course on Statistical Distributions in Scientific Work I" (G. P. Patil, *et al.*, eds.), pp. 299–317. Dordrect, Boston, Massachusetts.

Srivastava, J. N. (1966). On testing hypotheses regarding a class of covariance structures, *Psychometrika* **31**, 147–164.

Subrahmaniam, K. (1974). Recent trends in multivariate normal distribution theory: on the zonal polynomials and other functions of matrix argument. Part I: zonal polynomials, Tech. Rept. No. 69, Dept. of Statistics, University of Manitoba.

Subrahmaniam, K., and Subrahmaniam, K. (1973). "Multivariate Analysis: A Selected and Abstracted Bibliography, 1957–1972." Marcel Dekker, New York.

Sugiyama, T. (1966). On the distribution of the largest latent root and the corresponding latent vector for principal component analysis, *Ann. Math. Statist.* **37**, 995–1001.

Sugiyama, T. (1967). Distribution of the latent root and the smallest latent root of the generalized B statistics and F statistics in multivariate analysis, *Ann. Math. Statist.* **38**, 1152–1159.

Sugiyama, T. (1970). Joint distribution of the extreme roots of a covariance matrix, *Ann. Math. Statist.* **41**, 655–657.

Tiku, M. L. (1971). A note on the distribution of Hotelling's T^2 generalized $T_0{}^2$, *Biometrika* **58**, 237–241.

Timm, N. H. (1975). "Multivariate Analysis with Applications in Education and Psychology." Wadsworth Publ., Belmont, California.

Tukey, J. W. (1949). One degree of freedom for non-additivity, *Biometrics* **5**, 232–242.

Votaw, D. G., Jr. (1948). Testing compound symmetry in a normal multivariate distribution, *Ann. Math. Statist.* **19**, 447–473.

Waikar, V. B., and Schuurmann, F. J. (1973). Exact joint density of the largest and smallest roots of the Wishart and MANOVA matrices, *Utilitas Math.* **4**, 253–260.

Wigner, E. P. (1967). Random matrices in physics, *SIAM Rev.* **9**, 1–23.

Wilks, S. S. (1932). Certain generalizations in the analysis of variance, *Biometrika* **24**, 471–494.

Wilks, S. S. (1935). On the independence of k sets of normally distributed statistical variables, *Econometrica* **3**, 309–326.

Yochmowitz, M. G., and Cornell, R. G. (1978). Stepwise tests for multiplicative components of interaction, *Technometrics* **20**, 79–84.

Note Added in Proof

C. R. Rao (*Biometrika* **35**, 58–79 (1948)) approximated the distribution function of $\Lambda = |S_1(S_1 + S_2)^{-1}|$ with a linear combination of the distribution functions of chi-square variables where S_1 and S_2 are distributed independently as central Wishart matrices with m and n degrees of freedom and $E(S_1/m) = E(S_2/n)$. Also, C. R. Rao (*Bulletin of the International Statistical Institute* **33**, 177–180 (1951)) suggested approximation of the distribution function of $\Lambda^{1/s}$ with a linear combination of the distribution functions of beta variables, of which the first term alone provides a good approximation; the corresponding beta can be transformed into an F for using the F-table for testing significance.

Covariance Analysis of Nonstationary Time Series†

M. M. RAO

DEPARTMENT OF MATHEMATICS
UNIVERSITY OF CALIFORNIA AT RIVERSIDE
RIVERSIDE, CALIFORNIA

† This work was sponsored partly by the Air Force Aerospace Research Laboratories, Air Force Systems Command, USAF, Contract F33615-74-C-4009; and partly by the National Science Foundation under grants MPS 75-13881 and MCS 76-15544. Reproduction in whole or in part is permitted for any purpose of the U.S. government.

ISBN 0-12-4266 01-0

1. INTRODUCTION AND OUTLINE

As the term indicates, nonstationarity is a negative concept. It will be understood here as being not necessarily stationary to give a positive interpretation. A study of such objects essentially involves analysis of certain well-defined classes almost all of which reduce to known facts when the classes are specialized to the stationary ones. Recently there have been many attempts directed at extending the work on stationary time series to various nonstationary types. The fundamental spectral analysis, so useful for the stationary series, does not play such a central role in the general study. Most of the available tools and methods for the analyses of the latter are different and altogether less refined, and still they appear more complicated. However, from the point of view of applications, an analysis of nonstationary time series is perhaps more realistic. Consequently, some results based on certain asymptotic considerations related to a "correlation characteristic" (defined later) are included in this chapter. Moreover, there is also some other work on classes of linear stochastic (or autoregressive) equations both in the discrete as well as continuous time, on certain asymptotic distribution theory of estimators of parameters entering these equations, and on a detailed study of equivalence and likelihood ratios when the time series is Gaussian. An outline of the results is presented here, since it gives the reader an overview of the treatment in this chapter.

Instead of totally abandoning the spectral point of view, in 1959, Kampé de Fériet and Frenkiel, in a remarkable paper, introduced a class of nonstationary time series [called hereafter *class* (KF)] for its covariance analysis (cf. [30]). There they studied in considerable detail a model of the "signal plus noise" form where the signal is a time series with zero mean and a periodic covariance and the noise is a stationary series with zero mean. A detailed numerical study was made to illustrate the usefulness of this class. Also only slightly later, but independently, Parzen [38] has considered such a model briefly and termed it "asymptotically stationary." This class (KF) has been discussed in some detail by Bhagavan [5] in his recent dissertation. In this latter publication, one of the main results asserts that the class of harmonizable time series belongs to class (KF), and then an ergodic theorem is established. In Section 2 the class (KF) has been further analyzed and a more inclusive (new) *almost harmonizable class* is introduced and shown to be of class (KF). This contains the harmonizable case, but the generalization gives a better insight, and shows the essential simplicity of the class (KF). The consistency of an estimator of the averaged mean function and a weak law of large numbers are established. It is of interest to remark that the classical spectral theory of stationary time series again plays a role here, perhaps justifying Parzen's terminology just noted. There are some relations between a class of series introduced by Cramér in 1951, (cf. [10]), to be called *class* (C), and class (KF), although neither one includes the other. A comparison of these approaches is also expounded here, and these relations reappear at many points in this chapter.

Section 3 is devoted to the covariance analysis of time series governed by difference schemes with not necessarily constant coefficients. Such equations are of interest in treating trend and seasonal variations in various situations which are typically (strongly) nonstationary (cf. Hurwicz [24]). Under some conditions on these coefficients, it is shown that some of these generated series belong to class (KF); in general, many do not. A few of the latter cases have been analyzed in detail. In order to understand the dependence, approximate recursion equations for sample covariances, which are the sample analogs of the ancient Yule–Walker equations, are presented when the coefficients have linear "time trend." For a descriptive study, correlograms [the graph of $(k, \rho_t(k))$ where $\rho_t(k)$ is the correlation of X_t with X_{t+k}] and certain approximations are considered. Even here, the computations become involved, but what may be expected in higher order schemes becomes evident. Some other results on estimation and limit behavior of normalized sums are also included.

The continuous time analog of the preceding work refers to the behavior of flows. This is considered in some detail in Section 4. The corresponding schemes are stochastic differential equations whose coefficients are functions of time. The importance of a class of these schemes in industrial applications has been reported by Hartley [22]. If the coefficients are constants, they represent the motion of a simple harmonic oscillator driven by random (i.e., white noise) disturbance; such a model was discussed in 1943 in the important, lengthy article by Chandrasekhar [6] and in earlier classical studies. The latter types of equations have been analyzed by Dym [14], and classified. The general time-dependent case is much more involved, and some properties of such an equation are studied herein. Several results from the theory of ordinary differential equations have special interest in this treatment. The basic sample function continuity of the solutions, the conditions under which they belong to class (KF), and the fact that they always belong to class (C), among others, are established. A specialization of the case in which the coefficients are constants is illustrated, and the correlogram is analyzed for its asymptotic behavior.

Section 5 considers, in detail, the asymptotic distributional problems of the least squares estimators of the structural parameters of a constant coefficient kth-order autoregressive equation. The known results in the general case are summarized. These show the need to distinguish between the stable, unstable, and explosive series accordingly as the roots of the associated characteristic equation lie inside, on, or outside the unit circle in the complex plane. The problem is of interest in dynamic economic models and the like, but is also quite involved. The unstable case has been omitted in the past; several results are known in the explosive case and, in a more or less complete treatment, in the stable case. The latter belongs to class (KF). After outlining this work, the unstable case for the first-order equations is solved herein. This completes the study of the first-order schemes. The present method seems to admit extensions to the higher order equations.

Sections 6 and 7 treat special problems by restricting attention to (nonstationary) Gaussian time series. Most of the work here is stimulated by the Hájek–Feldman dichotomy theorem stating that two Gaussian measures are either equivalent or singular. So it will be of interest to obtain conditions for equivalence and then to calculate the likelihood ratios. There has been a great deal of activity on these problems, each author using a special method in obtaining the corresponding results. Here a general proposition is presented and then it is shown how the work of several other authors may be unified with this proposition. But now more sophisticated tools are necessary to get much of this specialized information, and the techniques use the theory of Aronszajn [3], the efficacy of which has been amply demonstrated by Parzen (e.g., Parzen [40]); its use in covariance analysis was first indicated by Loève ([32], p. 490, Ex. 3). These methods are fully utilized in Sections 6 and 7 in obtaining various conditions for equivalence, simultaneous diagonalization of a pair of covariance kernels, and then computing the likelihood ratio of two arbitrary (equivalent) Gaussian probability measures. Many of the results here are due essentially to Pitcher ([42, 43]), Hájek [20a], Parzen [41], and Root [49] (cf. also Rozanov [50]; Yaglom [56]; Kailath and Weinert [28]).

The results presented here show how new problems must be resolved for a more complete understanding of these general classes of time series. In particular, a classification of solutions of the stochastic equations of Section 4, analogous to [14], will be feasible and very interesting. Some aspects of this problem for vector-valued time series have already been given by Goldstein [17], but the sample function behavior when the disturbance is white noise, as in Dym [14], has not been done. Class (KF) should be analyzed further. The limit distribution theory of Section 5 can now be extended further and perhaps completed, at least in the unstable case. Similarly, constructions of explicit tests for families of Gaussian processes, extending the classical work by Grenander [20] and Kelly *et al.* [31], are desirable. Here the theory of Sections 6 and 7 should be helpful. In the following pages many other connections and scenic byways as well as trappings are noted but not pursued to their logical completion. It is hoped that such work will be considered in the future. The present account is only aimed at describing the state of the subject in the nonstationary case, and is not meant to be the last word. About half the material is a unified account of the works of other authors, usually with different demonstrations, and the rest is due to the author. An extensive (but not exhaustive) set of references, with due credits to the various authors, is included.

It is recognized that, especially in time series, no result can be taken without adequate substantiation (or at least the explanations that can easily be made precise). For this reason essentially all proofs accompany the statements of results if an immediate reference is not at hand. Consequently, readers pressed for time or those primarily interested in the results are advised to skip the proofs and proceed with the statements, discussions, and remarks of this chapter. This perhaps applies particularly to Sections 6 and 7 in which somewhat more advanced techniques unavoidably appear, but which class admits more complete analytical calculations useful for many applications.

2. A GENERAL CLASS OF NONSTATIONARY TIME SERIES

Let $X = \{X_t, t \in \mathbb{R}\}$ be a second-order (real or complex) time series with zero mean and covariance $K(s, t) = E(X_s \bar{X}_t)$ (where the expectation E is on the base probability space, and the overbar denotes the complex conjugate). Suppose that $K(\cdot, \cdot)$ satisfies the following condition:

$$\text{(KF)} \qquad r(h) = \lim_{T\to\infty} \frac{1}{T}\int_{|h|/2}^{T-|h|/2} K\left(s - \frac{h}{2}, s + \frac{h}{2}\right) ds$$

$$\left[= \lim_{T\to\infty} \frac{1}{T}\int_0^{T-|h|} K(s, s + |h|)\, ds = \lim_{T\to\infty} r_T(h)\right], \qquad h \in \mathbb{R}$$

where the limit is assumed to exist. This condition was introduced by Kampé de Fériet and Frenkiel [30]. The class of time series X satisfying this condition is called of *class* (KF), and it is analyzed in detail in what follows. It will become clear that this class (KF) is sufficiently general and is very useful in applications. The interest in this definition stems from the positive definiteness of r_T and r, as stressed by Kampé de Fériet and Frenkiel [30]. (In what follows, "processes" and "series" are synonymous.)

It is clear that if X is (real and) stationary, so that $K(s, t) = K(t - s)$, then $r(h) = K(h) = K(|h|)$. Thus every stationary process is in (KF). If $X = Y + Z$, where $Z = \{Z_t, t \in \mathbb{R}\}$, is stationary (with zero mean) and $Y = \{Y_t, t \in \mathbb{R}\}$ has zero mean and a periodic covariance [i.e., $K_Y(s + h_0, t + h_0) = K_Y(s, t)$ for some "period h_0"], $E(Y_t \bar{Z}_t) = 0, t \in \mathbb{R}$, then also $X \in$ class (KF), but now X is evidently not stationary. In this representation, the output X is composed of the "signal" Y and "noise" Z, which are mutually uncorrelated, and the model describes a communication channel. An elementary example of the Y process is the following:

$$Y_t = \alpha \cos 2\pi t, \qquad t \in \mathbb{R} \tag{2.1}$$

where $E(\alpha) = 0$, $E(\alpha^2) = \sigma^2 > 0$. Then Y is nonstationary and

$$K_Y(s, t) = (\alpha^2/2)[\cos \pi(t + s) - \cos \pi(t - s)]$$

In this example $r_T(h)$ of (KF) is given by $r_T(h) = K_Z(h) + (\alpha^2/2) \cos \pi h + o(T)$ so that $X \in$ class (KF). Also, if $K_X(t, t + h) \to a(h)$ as $t \to \infty$ for each h, where $-\infty < a(h) < \infty$, then (by L'Hôpital's rule) it follows that $X \in$ class (KF). Another example of X in (KF) is the important nonstationary class called the *harmonizable* time series. Namely, if the covariance of X is denoted by K_X, then it is representable as

$$K_X(s, t) = \int_{-\infty}^{\infty} \int_{-\infty}^{\infty} e^{is\lambda - it\mu} \, d^2\gamma(\lambda, \mu) \tag{2.2}$$

where $\gamma(\cdot, \cdot)$ is a covariance function of bounded variation on the complex plane (or on the square $(\pi, \pi] \times (-\pi, \pi]$ if the time series is of discrete parameter). K_X is then called a harmonizable covariance. That this series X is in class (KF) is nontrivial and it is one of the main results of the dissertation presented by Bhagavan [5]. There is a more inclusive class of Cramér, to be called class (C), generalizing (2.2); it is recalled here for comparison with (KF) and for later use.

The time series $X = \{X_t, t \in \mathbb{R}\}$ is said to be of *class* (C) if it has mean zero and covariance $\tilde{K}_X$, representable as

$$\tilde{K}_X(s, t) = \int_{-\infty}^{\infty} \int_{-\infty}^{\infty} g(s, \lambda)\overline{g(t, \mu)} \, d^2\gamma(\lambda, \mu) \tag{2.3}$$

where γ is a covariance function of bounded variation on each finite domain of the complex plane, and $\{g(t, \cdot), t \in \mathbb{R}\}$ is a family such that the integral in (2.3) exists. $\tilde{K}_X$ is then called a covariance of *type* (C). Thus, if γ is of bounded variation, $g(s, \lambda) = e^{is\lambda}$, then (2.3) reduces to (2.2). The problem now is to find conditions on g and γ in order that X of class (C) is in class (KF). The class (C) has been analyzed from the point of integral representation by the author [48], and considerable information is available for this family.

For comparison, it is useful to state the result by Bhagavan [5] in the following form. It is seen to be included in a more general case proved in the next result.

Proposition 2.1. Let K be a continuous harmonizable covariance. Then

$$R(h) = \lim_{T \to \infty} \frac{1}{T} \int_0^{T-|h|} K(t, t + |h|) \, dt, \qquad h \in \mathbb{R} \tag{2.4}$$

exists and $R(\cdot)$ is a stationary covariance so that

$$R(h) = \int_{-\infty}^{\infty} e^{ih\lambda}\, dF(\lambda), \qquad h \in \mathbb{R} \tag{2.5}$$

for a unique "distribution" F $[F(+\infty) - F(-\infty) = R(0)]$. Moreover,

$$F(\lambda) = \gamma(\infty, \infty) - \gamma(-\lambda, -\lambda) \tag{2.6}$$

where γ is the covariance given in (2.2).

In the discrete case there is a corresponding formula [analogous to (2.4)–(2.6)], but the relation (2.6) can be given a more detailed form using the possible discontinuities of γ (this was treated by Bhagavan [5]). The proof of this result uses estimates of trigonometrical sums for integrals on bounded rectangles arising from $\{e^{it\lambda}, t \in \mathbb{R}\}$. In the case of (2.3), it is clear that one has to restrict the family $\{g(t, \cdot), t \in \mathbb{R}\}$ regarding its growth in relation to γ. In general, the limit (2.4) for K of type (C) need not exist. Thus the class (C) is not included in class (KF) nor is the class (KF) contained in the class (C). (See the example in the remark following Theorem 4.2.) Some interesting conditions on g will be obtained so that those time series are in class (KF).

Let g of (2.3) be a bounded (jointly) continuous function. It will now be shown that if $g(\cdot, \lambda)$ is almost periodic for almost all λ [in particular, $g(t, \lambda) = e^{it\lambda}$ is automatically included], then the corresponding class (C) covariance actually is in (KF) so that (2.4) and (2.5) are implied. It is necessary to recall the definition of almost periodicity of g depending on the parameter λ to demonstrate the preceding statement.

Definition. Let $D \subset \mathbb{R}^n$ be an open (or a compact) set. A continuous complex function f on $\mathbb{R} \times D$ is said to be *almost periodic* (a.p.) *on* $\mathbb{R}$ *uniformly relative to* D if for each compact set $S \subset D$, $f(\cdot, x)$ is almost periodic for each x in S, i.e., for any $\varepsilon > 0$, and each compact set $S \subset D$, there is a number $l_0 = l_0(\varepsilon, S) > 0$, such that each interval $I \subset \mathbb{R}$ of length l_0 contains a number $\tau \in I$ for which

$$|f(t+\tau, x) - f(t, x)| \leq \varepsilon, \qquad t \in \mathbb{R}, \quad x \in S \tag{2.7}$$

The τ is called an *ε-translation number* of f.

It can be shown that the set of all a.p. functions depending on a parameter, satisfying (2.7), forms an algebra, and if $|f(t, x)| \geq a_S > 0$ for $x \in S \subset D$, and all $t \in \mathbb{R}$, then $1/f$ is also an a.p. function of the same kind. Thus the set $\{e^{it\lambda}, t \in \mathbb{R}, \lambda \in \mathbb{R}\}$ is included in this class and even properly. For an exposition of this class, the reader may consult Yoshizawa ([57], Chap. I). If $D = \{t\}$ is a single point, then this definition reduces to the classical concept of a.p. functions of Bohr. Also, observe that an a.p. function is only locally (i.e., on bounded intervals) integrable. In fact, it is bounded for each $x \in S \subset D$.

The following is the desired generalization.

Proposition 2.2. Let K be a covariance function of type (C), i.e., satisfies (2.3). Suppose that $g(\cdot, \lambda)$ of the integrand in (2.3) is almost periodic uniformerly relative to $(\lambda \in) D = \mathbb{R}$ and with γ as the covariance of bounded variation. Then the time series $X = \{X_t, t \in \mathbb{R}\}$ with zero mean and covariance K belongs to class (C) $\cap$ class (KF). More precisely,

$$R(h) = \lim_{T\to\infty} \frac{1}{T} \int_0^{T-|h|} K(t, t+|h|)\, dt = \lim_{T\to\infty} R_T(h), \qquad h \in \mathbb{R} \quad (2.8)$$

exists and defines a stationary covariance on $\mathbb{R}$.

Remark. If $g(t, \lambda) = e^{it\lambda}$, then the hypothesis of this proposition is satisfied by that of the preceding one so that the main result (2.4) is a consequence of (2.8). Using the special form of this g, it is possible to obtain (2.6) connecting γ and the spectral distribution F of R. However, in the present generality, this is much more involved.

Proof. By symmetry it suffices to consider $h \geq 0$. Now substituting (2.3) into R_T and interchanging the integrals (this is obviously legal) one gets

$$R_T(h) = \int_{-\infty}^{\infty} \int_{-\infty}^{\infty} \frac{1}{T} \int_0^{T-h} g(t, \lambda)\overline{g(t+h, \lambda')}\, dt\, d^2\gamma(\lambda, \lambda') \qquad (2.9)$$

If $S \subset \mathbb{R}$ is any compact set, and $g(\cdot, \lambda)$, $\lambda \in S$, is a.p., then it follows that $g(\cdot, \lambda)\overline{g(\cdot + h, \lambda')}$, $(\lambda, \lambda') \in S \times S$, is also a.p. So for any fixed but arbitrary h, one has by a classical result (cf. Besicovitch [4], p. 15) that

$$\lim_{T\to\infty} \frac{T-h}{T} \lim_{T\to\infty} \frac{1}{T-h} \int_0^{T-h} g(t, \lambda)\overline{g(t+h, \lambda')}\, dt = a(h; \lambda, \lambda') \quad (2.10)$$

exists uniformly in h. But it is clear that $a(h; \lambda, \lambda')$ is bounded for all $h \geq 0$, $(\lambda, \lambda') \in S \times S$, since each a.p. function is bounded. So from (2.9) and (2.10), together with dominated convergence, one gets

$$R(h) = \lim_{T\to\infty} R_T(h) = \int_{-\infty}^{\infty} \int_{-\infty}^{\infty} a(h; \lambda, \lambda')\, d^2\gamma(\lambda, \lambda') \qquad (2.11)$$

But for each T, $R_T(\cdot)$ is clearly positive definite and hence so is $R(\cdot)$. Thus $R(\cdot)$ is a covariance. However, because of the compactness of S and the uniformity involved in (2.10), it follows that $a(\cdot\,; \cdot, \cdot)$ on $\mathbb{R} \times S \times S$ is a continuous complex function. From this one easily concludes that $R(\cdot)$ is a (continuous) stationary covariance and then the representation (2.5) is just the classical Bochner's theorem. This completes the proof.

Comments. 1. Evidently one may prove an analogous result if the time series is of discrete parameter. It should be observed that the long argument of the harmonizable case of Bhagavan ([5], pp. 72–77) is really a specialized version of the existence of the limit (2.10) and for the special $g(t, \lambda) = e^{it\lambda}$ (the "characters" of $\mathbb{R}$) the mean value is given by the number $a(h; \lambda, \lambda')$ of (2.10) on the diagonal ([4], p. 16, No. 4°) so that the simplifications for (2.6) result. In the general case (2.6) no longer holds. Now X is mean continuous, i.e., $E(X_t - X_s)^2 \to 0$ as $s \to t$, so one only has

$$R(h) = \int_{-\infty}^{\infty} \int_{-\infty}^{\infty} a(h; \lambda, \lambda')\, d^2\gamma(\lambda, \lambda') = \int_{-\infty}^{\infty} e^{ihx}\, dF(x), \qquad h \in \mathbb{R} \quad (2.12)$$

by the Bochner theorem. This F is called the *associated spectral distribution* of X, here and hereafter.

2. The time series $X = \{X_t, t \in \mathbb{R}\}$ whose covariance K is of type (C) (and zero mean) for an a.p. function $g(\cdot, \lambda)$, uniformly relative to $\mathbb{R}$, may and will be called an *almost harmonizable series*. This clearly includes the harmonizable case. Under this generalization, one may profitably consider the more inclusive Besicovitch functions g ($= B^2$ – a.p. of Besicovitch [4]), a.p. uniformly relative to $\mathbb{R}$, since for such functions the desired limit (2.10) again exists. This follows from Besicovitch ([4], p. 93), where one uses the fact that such g functions form an algebra, and then Lemma 4 on page 93 of [4] is applied in a slightly modified form (and consequently the arguments of Besicovitch [4], pp. 14–15, hold). Such an extension is necessary to show, for instance, that the Brownian motion (which is not harmonizable) is covered by the preceding. That the latter is in class (KF) is easy to check directly as observed by Kampé de Fériet and Frenkiel [30]. Similarly, the Uhlenbeck–Ornstein process is in (KF). Both of these are now almost harmonizable (g is not continuous, so B^2-a.p. is needed!). It will be interesting to analyze this set, which is contained in class (KF) $\cap$ class (C). Here one should perhaps also observe that, if g is B^2-a.p. such that $a(h; \lambda, \lambda') \to 0$ as $|\lambda| + |\lambda'| \to \infty$ sufficiently rapidly, then γ need only be of bounded variation on each compact rectangle of the complex plane. Hereafter, "almost harmonizable" refers to the class where g is B^p-a.p. (or B^ϕ-a.p.) uniformly in λ ($p \geq 1$), relative to $\mathbb{R}$ (ϕ being a symmetric convex continuous [= Young's] function, $\phi(0) = 0$).

3. It may also be remarked that g of (2.3) can be more general than that noted previously for the existence of the limit in (2.10). For instance, if g is locally (i.e., on compact sets) square integrable, then $g(\cdot, \lambda)\overline{g(\cdot + h, \lambda)}$ will be $(c, 1)$ summable (i.e., in the sense of first arithmetic mean of Cesàro) and γ is of bounded variation. Many good sufficient conditions are available for it (cf. Hardy, [21], pp. 110–112). This shows that class (KF) $\cap$ class (C) contains even the almost harmonizable family as a proper subset.

One of the key applications of this result is in obtaining conditions for the weak or strong law of large numbers. In a different terminology, this is equivalent to estimating the mean (or the average of the mean function) of the time series X consistently. These problems are natural analogs of the well-known stationary theory (cf. Doob [13], pp. 529–530). The following is such an extension; it is due substantially to Bhagavan [5].

Proposition 2.3. Let $X = \{X_t, t \in \mathbb{R}\}$ be an almost harmonizable time series (which is mean continuous). Suppose that its mean function m has the property that $a_0 = \lim_{T\to\infty} (1/T) \int_0^T m(t)\, dt$ exists where $m(t) = E(X_t)$. If $\hat{m}_T = (1/T) \int_0^T X_t\, dt$ (the sample path integral), then

$$\lim_{T\to\infty} E(\hat{m}_T - a_0)^2 = F(0+) - F(0-)$$

where F is the associated spectral distribution of X [cf. (2.12)]. In particular, if F is continuous at 0, then $\hat{m}_T$ is a strongly consistent estimator of a_0 (or the series X obeys the weak law of large numbers when the limit a_0 exists).

Proof. Let $a_T = (1/T) \int_0^T m(t)\, dt$, so that $a_T \to a_0$ as $T \to \infty$. Then writing $K(s, t) = \operatorname{cov}(X_s, X_t)$, the covariance, and noting $E(\hat{m}_T) = a_T$, one has

$$\begin{aligned} E(\hat{m}_T - a_0)^2 &= E[(\hat{m}_T - a_T) + (a_T - a_0)]^2 \\ &= 2[E(\hat{m}_T - a_T)^2 + (a_T - a_0)^2] \\ &= \frac{2}{T^2} \int_0^T \int_0^T K(s, t)\, dt\, ds + 2(a_T - a_0)^2 \\ &= \frac{2}{T} \int_{-T}^T R_T(h)\, dh + 2(a_T - a_0)^2 \qquad \text{[cf. (2.8)]} \end{aligned} \tag{2.13}$$

The last term of (2.13) tends to zero, and the first term may be simplified as follows. If R_T did not depend on T, then the limit of the first term of (2.13) is the desired result, and it is a classical theorem of Bochner (cf. Cramér [8], p. 25). Since by Proposition 2.2 (X being mean continuous), R_T is a continuous positive definite function converging to a continuous positive definite R which then has the representation (2.12), the following modification of Bochner's proof establishes the present case.

Let F_T and F be the bounded nonnegative nondecreasing functions representing R_T and R, as in (2.12). Then $R_T(h) \to R(h)$ uniformly for h on each bounded closed interval, and these are Fourier transforms of F_T, F which must then be uniformly bounded. Moreover, $F_T \to F$ at each continuity point of F(see, e.g., [8]. Theorem 11). Thus

$$\begin{aligned} \frac{1}{2T} \int_{-T}^T R_T(h)\, dh &= \int_{-\infty}^{\infty} \frac{1}{2T} \int_{-T}^T e^{ith}\, dh\, dF_T(t) \\ &= \int_{-\infty}^{\infty} \frac{\sin Tx}{Tx}\, dF_T(x) \end{aligned} \tag{2.14}$$

Since $(\sin Tx)/Tx \to \delta_{0x}$ as $T \to \infty$ (δ_{0x} is the delta function) and $F_T \to F$ the result follows easily. In fact, for each $\varepsilon > 0$,

$$\int_{-\infty}^{\infty} \frac{\sin Tx}{Tx} dF_T(x) = \left(\int_{-\infty}^{-\varepsilon} + \int_{\varepsilon}^{\infty}\right) \frac{\sin Tx}{Tx} dF_T(x) + \int_{-\varepsilon}^{\varepsilon} \frac{\sin Tx}{Tx} dF_T(x) \tag{2.15}$$

The first two integrals on the right-hand side of this equation go to zero as $T \to \infty$ since $|(\sin Tx)/Tx| \leq 1/\varepsilon T$ and $F_T(x) \leq \sup_T F_T(\infty) < \infty$. Let ε be chosen such that $(F(\varepsilon) - F(-\varepsilon)) - (F(0+) - F(0-)) < \varepsilon/2$ and such that $\pm\varepsilon$ is a continuity point of F. This can be done since the continuity set of F is everywhere dense in $\mathbb{R}$. But $|(\sin Tx)/Tx| \leq 1$, and if T is large, then $F_T(\varepsilon) - F_T(-\varepsilon)$ differs arbitrarily small from $F(\varepsilon) - F(-\varepsilon)$. It follows that the last term of (2.15) differs from $F(0+) - F(0-)$ by less than ε. Since $\varepsilon > 0$ is arbitrary, one concludes that

$$\lim_{T\to\infty} \frac{1}{2T}\int_{-T}^{T} R_T(h)\, dh = F(0+) - F(0-) = \lim_{T\to\infty} \frac{1}{2T}\int_{-T}^{T} R(h)\, dh \tag{2.16}$$

The proposition now follows from (2.13) and (2.16). Note that when F has a jump at 0, the estimator $\hat{m}_T$ (of a_0) is not consistent. This completes the proof.

Remark. Since the limit R and hence F are given only indirectly from R_T, it is desirable to have conditions on $\{R_T, T > 0\}$ that will ensure the continuity of F at 0 (or at any given point). This will be true if $R_T(h) \to 0$ as $h \to \infty$ uniformly in T or if $\int_0^\infty |R_T(h)|\, dh$ is bounded as a function of T. The first condition ensures the statement in parentheses (by the classical Riemann–Lebesgue lemma) and the latter gives the continuity of F at 0 as the last part of (2.16) shows.

In view of the importance of R_T and R in the preceding, an immediate statistical problem is to estimate these functions, with the consistency property at least. (Assume $m = 0$, otherwise one may consider the product moments directly.) Thus the natural estimates are

$$\hat{R}_T(h) = \frac{1}{T}\int_0^{T-h} X_t \bar{X}_{t+h}\, dt = \frac{1}{T}\int_{-\infty}^{\infty} X_t{}^T \bar{X}_{t+h}^T\, dt \tag{2.17}$$

where $X_t{}^T = X_t$ for $0 \leq t \leq T$, $X_t{}^T = 0$ for $t > T$. Under the mean continuity of the X, such that $\int_0^T K^2(t, t)\, dt < \infty$, the estimator $\hat{R}_T$ of (2.17) is well defined, $E(\hat{R}_T(h)) = R_T(h)$—an unbiased estimator of R_T. To fulfill the consistency condition, i.e., for $X \in$ class (KF), $\hat{R}_T(h) \to R(h)$, $h \in \mathbb{R}$, in mean, one has to assume somewhat more on X. The following sufficient conditions were given by Parzen [38] (cf. also [30]). Thus consider the following: (i) for each t, assume $E(|X_t|^4) < \infty$; (ii) if $\rho(t_1, t_2, t_3, t_4) = E(X_{t_1} X_{t_2} X_{t_3} X_{t_4})$, then ρ is Lebesgue integrable on each bounded interval of $\mathbb{R}^4$; and (iii) if $C_t(v) = \int_0^t \operatorname{cov}(X_s X_{s+v}, X_t X_{t+v})\, dt$ [which exists by (ii)], then $\lim_{t\to\infty} C_t(v) = 0$ for each $v \geq 0$.

If (i)–(iii) hold, then $\operatorname{var}(\hat{R}_T(h)) \to 0$ as $T \to \infty$, for each $h \geq 0$, so that $E(\hat{R}_T(h) - R(h))^2 \to 0$. This may be checked by a direct computation (cf. also [38]). Kampé de Fériet and Frenkiel [30] give an interesting example of a periodic covariance of a time series consisting of symmetric bounded random variables which satisfies the above conditions. The accuracy of the approximation of the estimator with R was then illustrated by a numerical example. The reader is referred to the instructive case given by Kampé de Fériet and Frenkiel [30] to gain an insight into the generality of the class (KF) of nonstationary time series.

The following two sections are devoted to another class of nonstationary series (processes) generated by certain stochastic difference and differential equations for which a different set of methods is needed. The latter are related to the "correlogram analysis," and these are discussed. Section 5 contains a treatment of the asymptotic distribution of certain estimators in the autoregressive equation problems, and the final two sections concentrate on the analysis of Gaussian time series.

3. NONSTATIONARY SERIES GENERATED BY DIFFERENCE EQUATIONS

3.1. Motivation

If $X = \{X_t, t \in \mathbb{R}\}$ is a stationary time series, then the correlation function $\rho(\cdot)$ is given by $\rho_t(h) = \rho(h) = \operatorname{cov}(X_t, X_{t+h})/\operatorname{var} X_t$, and is independent of t. As a function of h, i.e., of the *lag*, $\rho(\cdot)$ is taken as an indicator of the dependence of X_t over X_{t+h} for large h. Clearly, $\rho(h) = \overline{\rho(-h)}$, $|\rho(h)| \leq 1$. Suppose that X is nonstationary, but is in class (KF). In this case $\rho_t(\cdot)$ depends on t, but for any $a > 0$ and $h \in \mathbb{R}$,

$$\lim_{T\to\infty} \tilde{\rho}_{T,a}(h) = \lim_{T\to\infty} \frac{(T-a)^{-1} \int_a^{T-h} K(t, t+h)\, dt}{(T-a)^{-1} \int_a^{T-h} K(t, t)\, dt}$$

$$= \frac{R(h)}{R(0)} = \bar{\rho}(h) \qquad \text{(say)} \tag{3.1}$$

Here R is the same as in (2.8) with $K(s, t) = \operatorname{cov}(X_s, X_t)$. Since $R(\cdot)$ is a stationary covariance by virtue of the fact that $X \in$ class (KF), $\bar{\rho}(\cdot)$ does not depend on t or a. Furthermore, several properties of the time series X are reflected in the behavior of R and hence of $\bar{\rho}$. If X is stationary, then $\bar{\rho} = \rho$. On the other hand, if $\lim_{t\to\infty} K(t, t+h) = a(h)$ exists [which is stronger than being in (KF)], then the limit of (3.1) exists and one obtains the same $\bar{\rho}(\cdot)$. This is a consequence of some classical analysis (cf. Hardy [21], p. 50). Motivated by this observation, one may consider the behavior of $\tilde{\rho}$ for a class of nonstationary time series where

$$\tilde{\rho}_t(h) = \frac{\operatorname{cov}(X_t, X_{t+h})}{\operatorname{var} X_t} \tag{3.2}$$

called hereafter the *correlation characteristic* of X. In general, $\tilde{\rho}$ is not a correlation function, but asymptotically its behavior is that of the latter [because of (3.1)] at least for the series of class (KF). In this section the time series governed by certain difference (and in Section 4 differential) equations will be studied. Some of the material in this and the following section appears in a very tentative form in my early study [45].†

3.2. A Setting of the Problem

Suppose that the output of a noisy communication channel follows a linear model of the signal plus noise type as follows. The output X_t at time t depends on the immediate past up to k units linearly, and then a white noise disturbance enters. Thus

$$X_t = \sum_{i=1}^{k} a_i(t) X_{t-i} + \varepsilon_t = S_t + \varepsilon_t, \qquad t \geq 1 \tag{3.3}$$

where the $a_i(\cdot)$ are some (nonstochastic) functions of time specified by the type of channel. This model may also be used for a dynamic economic situation where the X_{t-i} are called "lagged values" of the "endogenous" variable X_t or sometimes "exogenous" variables also. Thus the signal S_t is a linear function of the past k terms and $\{\varepsilon_t, t \geq 1\}$ is assumed to be independent identically distributed variables with zero mean and a finite variance $(= \sigma^2$, say). Now to study the properties of X_t, the difference equation (3.3) may be solved with classical methods (due to D. André, 1878) and it is given by the expression

$$X_t = \sum_{i=0}^{t-1} \varphi(t, i)\varepsilon_{t-i} + \sum_{i=0}^{t+k-1} \varphi(t, i) c_{t-i}, \qquad t \geq 1 \tag{3.4}$$

where $X_i = c_i$, $i = -k + 1, \ldots, 0$, are the constant initial values, and

$$\varphi(t, m) = \sum_{k_1 + \cdots + k_i = m} \prod_{j=1}^{i} a_{k_j}\left(t - \sum_{r=0}^{j-1} k_r\right)$$

$$0 \leq m \leq t, \quad k_0 = 0 \tag{3.5}$$

† The author thanks the syndicate of the University of Madras for granting him a subsistence allowance during the period of that investigation.

the sum ranging over all partitions of m into integers k_i (≥ 0) (cf. Jordan [26], p. 588). For instance, if $c_i = 0$ and $a_i(t) = \alpha_i$, a constant, then the complicated looking expressions (3.4) and (3.5) reduce to familiar forms. To see this, let the characteristic equation of the difference equation (3.3), namely

$$\lambda^k - \alpha_1 \lambda^{k-1} - \cdots - \alpha_k = 0 \tag{3.6}$$

have simple roots $\lambda_1, \ldots, \lambda_k$. Then (3.5) becomes

$$\varphi(t, m) = \varphi(t - m) = \sum_{j=1}^{k} \beta_j \lambda_j^{t-m}, \qquad 1 \leq m \leq t, \quad \sum_{j=1}^{k} \beta_j = 1 \tag{3.7}$$

The β_j further satisfy

$$\sum_{j=1}^{k} \beta_j \lambda_j^{t-1} = 0, \qquad t = 0, -1, \ldots, -k+2 \tag{3.7$'$}$$

For a discussion of this case, see Mann and Wald [33] and also [46b, p. 330]. Let us now specify conditions on the model in order that the time series $X = \{X_t, t \geq 1\}$ is in class (KF), and point out instances when it is not in class (KF) but for which $\tilde{\rho}_t$ of (3.2) has a limit as $t \to \infty$.

Since for an asymptotic study the initial values are not always critical, set $c_i = 0$, $i = 0, -1, \ldots, -k+1$, in (3.4). Suppose further that the $a_i(\cdot)$ satisfy

$$\sum_{m=0}^{t} |\varphi(t, m)|^2 \leq M < \infty, \qquad t \geq 1 \tag{3.8}$$

Then from (3.4), since $E(X_t) = 0$, one finds

$$\begin{aligned} r(s, t) = E(X_s \bar{X}_t) &= \sigma^2 \sum_{m=0}^{s} \varphi(s, m)\overline{\varphi(t, m)}, \qquad 1 \leq s \leq t \\ &= \sigma^2 \sum_{m=0}^{s} \varphi(s, m)\overline{\varphi(s+h, m)}, \qquad h = t - s \end{aligned}$$

It follows from (3.8) that $|r(s, s+h)| \leq M$ for all h and $s \geq 1$ (by the Cauchy inequality). Hence for each $h \geq 0$,

$$\lim_{N \to \infty} \frac{N-h}{N} \cdot \frac{1}{N-h} \sum_{s=1}^{N-1} r(s, s+h) = R(h) \tag{3.9}$$

exists [by the $(c, 1)$-summability method], and by symmetry for all h. This implies the following statement.

Proposition 3.1. If a second-order time series $X = \{X_t, t \geq 1\}$ is generated by Eq. (3.3), and the $a_i(\cdot)$ satisfy (3.8), then the nonstationary series X belongs to class (KF).

It should be observed that condition (3.8) does not involve any mention of the roots of (3.6) even when the a_i are constants. To understand the significance of (3.8), it is useful to specialize. Let the a_i be constants and suppose the roots of (3.6) lie inside the unit circle of the complex plane and are distinct. Let $\delta = \max_j |\lambda_j|$ so that $\delta < 1$. Then (3.8) is automatic since

$$\begin{aligned}\sum_{m=0}^{t} |\varphi(t, m)|^2 &= \sum_{m=0}^{t} |\varphi(t-m)|^2 \\ &= \sum_{m=0}^{t} \left| \left[\sum_{j=1}^{k} \beta_j \lambda_j^{t-m} \right] \right|^2 \qquad \text{[by (3.7)]} \\ &\leq \sum_{j,j'=1}^{k} |\beta_j \beta_{j'}| \left(\frac{1 - (\lambda_j \bar{\lambda}_{j'})^{t+1}}{1 - \lambda_j \bar{\lambda}_{j'}} \right) \\ &\leq \sum_{j,j'=1}^{k} |\beta_j \beta_{j'}| \frac{1 + \delta^{2(t+1)}}{1 - \delta^2} \\ &\leq \frac{2}{1-\delta^2} \sum_{j,j'=1}^{k} |\beta_j \beta_{j'}| = M_0 < \infty\end{aligned}$$

Thus the time series generated by (3.3) with constant coefficients having all the roots of its characteristic equation distinct and lying inside the unit circle belongs to class (KF). On the other hand, if at least one of the roots of (3.6) is on or outside the unit circle (i.e., $\delta \geq 1$), then the resulting time series generated by (3.3) is nonstable or explosive and will not be in class (KF). This is now illustrated by a family of time series, investigated in the literature.

Thus in the constant coefficient case of (3.3), let $\delta > 1$, $\delta = \max_j |\lambda_j|$ of (3.6). Then under the same (remaining) hypothesis as the preceding,

$$E(|X_t|^2) = \sum_{t=0}^{t} |\varphi(t, m)|^2 = O(\delta^{2t}) \tag{3.10}$$

so that (3.8) is violated strongly. This was computed by the author ([47], Lemmas 8, 15). Moreover, by ([47], Lemma 9), one finds

$$\tilde{\rho}_t(h) = \frac{\operatorname{cov}(X_t, X_{t+h})}{\operatorname{var} X_t} \to \delta^{-(h-2)} \quad (\neq 0), \qquad t \to \infty \tag{3.11}$$

A similar conclusion holds if $\delta = 1$, $k = 1$, for (3.10) $[E(|X_t|^2) = O(t^2)]$. These two cases imply that the limit demanded for (KF) of Section 2 cannot exist, and so these time series do not belong to class (KF). This example may be taken as a further justification of Parzen's term "asymptotically stationary" for the series in classes like (KF).

The variable coefficient case of (3.3) was found to be of interest in some meteorological applications (cf. Hurwicz [24]). Consequently, a class of simple variable coefficient schemes is now investigated in some detail. These will throw some light on the behavior of the resulting time series.

3.3. A First-Order Model

The correlation characteristic, given by (3.2), is designed to reflect the dependence of X_t on X_{t+h} for large t and h. It was noted that this behaves asymptotically (as $t \to \infty$) as a correlation function for members of class (KF). How does it behave in the "explosive" cases? Equation (3.11) gives an indication of the constant coefficient case. The variable coefficient model of first order is considered herein, for a striking illustration. This is also the one proposed by Hurwicz [24], describing a time series with linear trend, namely

$$X_t = a_t X_{t-1} + \varepsilon_t, \qquad a_t = a_0 + ta_1, \qquad t \geq 1, \quad a_0 \neq 0 \tag{3.12}$$

The series $\{\varepsilon_t, t \geq 1\}$ consists of uncorrelated random variables with mean zero and variance σ^2, $X_t = 0$, $t \leq 0$ (i.e., the initial values are zero).

The solution of (3.12) may be obtained by iteration. Alternatively, let $p_t = \prod_{i=0}^{t-1} a_i$ so $p_1 = a_0$. Setting $Y_t = X_t/p_t$, (3.12) becomes

$$Y_{t+1} - Y_t = \varepsilon_t/p_{t+1}, \qquad t \geq 0 \tag{3.13}$$

Writing $X_t = Y_t p_t$, the solution of (3.13) [thus of (3.12)] is

$$X_t = \sum_{i=0}^{t} p_t \frac{\varepsilon_i}{p_{i+1}} \tag{3.14}$$

Hence for $h \geq 0$, the covariance is given by

$$\operatorname{cov}(X_t, X_{t+h}) = p_t p_{t+h} \sum_{i=0}^{t} \frac{\sigma^2}{p_{i+1}^2}, \qquad t \geq 1 \tag{3.15}$$

Since var X_t is obtained from (3.15) for $h = 0$, (3.2) becomes

$$\tilde{\rho}_t(h) = \frac{p_{t+h}}{p_t} = \prod_{i=t}^{t+h-1} a_i \tag{3.16}$$

This can be made arbitrarily large for large h, for appropriate a_0, a_1 (e.g., if $|a_0 + a_1 t| > 1$). On the other hand, the actual correlation $\rho_t(h)$ is

$$\rho_t(h) = \left(\sum_{i=0}^{t} p_{i+1}^{-2} \Big/ \sum_{i=0}^{t+h} p_{i+1}^{-2} \right)^{1/2} \tag{3.17}$$

and $|\rho_t(h)| \to \alpha(h)$ ($\neq 0$), as $t \to \infty$. Thus for such time series the correlation does not tend to zero for large t and large h. The correlation characteristic $\tilde{\rho}_t(\cdot)$ magnifies this phenomenon. This computation also implies that the *correlogram* [the graph of $(h, \rho_t(h))$, $h \geq 0$, for any t] does not dampen as the lag h increases if the coefficients contain a linear time trend. Since the same character is maintained in using $\tilde{\rho}_t(h)$ instead of $\rho_t(h)$ and since $\tilde{\rho}_t(h)$ is computationally simpler than $\rho_t(h)$, it will be considered in what follows for a structural study of the "explosive" time series. The resulting graph may be called an *approximate correlogram.*

If the linear trend is replaced by the reciprocal trend, i.e., $a_t = a_0 + a_1/t$, $t \geq 1$, then after a similar but more tedious computation one finds that $\tilde{\rho}_t$ still exhibits nearly the same properties unless more stringent conditions such as (3.8) are imposed. The details are omitted.

3.4. Limiting Behavior of Normalized Sums from Class (KF)

The behavior of series given by (3.3), subject to a condition implying (3.8), is reasonable in that such series obey the central limit theorem for dependent variables. Let $S_n = X_1 + \cdots + X_n$, $X_i = 0$, $i \leq 0$. Suppose that (i) the ε_t are independent identically distributed with means zero and variance one; (ii) the $\varphi(t, m) = \varphi(m)$ is independent of t [which is implied by the case in which the $a_i(\cdot)$ of (3.3) are constants]; and (iii) $\sum_{m=1}^{\infty} |\varphi(m)| < \infty$. Under these conditions the following assertion obtains:

Proposition 3.2. Let the time series $\{X_t, t \geq 1\}$ be generated by (3.3) and let conditions (i)–(iii) hold. Then

$$\lim_{n\to\infty} P[S_n < x(\text{var } S_n)^{1/2}] = \frac{1}{(2\pi)^{1/2}} \int_{-\infty}^{x} e^{u^2/2}\, du \tag{3.18}$$

Thus, S_n obeys the central limit law.

This result follows from some known results of Diananda and Anderson (cf. Anderson [1], Theorem 7.7.8). In fact,

$$X_n = \sum_{m=0}^{n} \varphi(m)\varepsilon_{n-m} = \sum_{m=0}^{\infty} \tilde{\varphi}(m)\varepsilon_{n-m}$$

where $\tilde{\varphi}(m) = \varphi(m)$, $1 \leq m \leq n$, $\tilde{\varphi}(m) = 0$ otherwise. Since the ε_m are independent identically distributed, $(\text{var } S_n)/n \to 1$, as a simple computation shows. This is sufficient to invoke the theorem just cited.

The restrictive condition (ii) may be relaxed in some cases. If $a_i(t) = a_i + \tilde{a}_i t^{-2}$, $t \geq 1$, then

$$\varphi(t, m) = \sum_{k_1 + \cdots + k_i = m} a_{k_1}(t) a_{k_2}(t - k_1) \cdots a_{k_i}\left(t - \sum_{j=1}^{i-1} k_j\right), \qquad k_0 = 0$$

$$= \sum a_{k_1} a_{k_2} \cdots a_{k_i} - O\left(\frac{1}{t^2}\right) = \tilde{\varphi}(m) - O\left(\frac{1}{t^2}\right) \tag{3.19}$$

Hence

$$X_t = \sum_{m=1}^{t} \tilde{\varphi}(m)\varepsilon_{t-m} + O\left(\frac{1}{t^2} \sum_{m=1}^{t} \varepsilon_{t-m}\right) \tag{3.20}$$

One checks that the variance of the last term in (3.20) is $o(1/t^3)$. It has mean zero. So if $S_n = \tilde{S}_n + \tilde{S}_n'$, corresponding to the decomposition of (3.20), then $\tilde{S}_n' \to 0$ in probability and $\{S_n, n \geq 1\}$ satisfies (iii). Hence from a form of the classical Slutzky's theorem S_n and $\tilde{S}_n$ have the same limit distribution. Thus the following result holds:

Proposition 3.3. Let $\{X_t, t \geq 1\}$ be given by the scheme (3.3). Suppose that conditions (3.8) and (i) hold where $a_i(t) = a_i + \tilde{a}_i t^{-2}$. If $\tilde{\varphi}(m)$ of (3.19) satisfies (iii), then the series obeys the central limit theorem; i.e., (3.18) holds for this series.

It is clear that other conditions can be formulated on the coefficients $a_i(t)$, to obtain corresponding results. Assuming that the ε_t have four moments, this result with a sketch of proof was indicated in ref. [45], using Lyapunov's theorem somewhat on the lines of Marsaglia ([34], Theorem 3). However, Parzen ([39], p. 254) has a better result on these problems.

3.5. Approximate Recursion Equations for Sample Covariances

Since the difference equations with coefficients depending linearly on time have been noted to be of interest in describing trend variation (cf. Hurwicz [24]), it is useful to have recursion formulas for computing sample covariances from data, even though the series is not in class (KF). Here an account of this problem is presented for a first-order scheme. Similar results for the higher order schemes are much more involved computationally. The recursion equations are just the sample analogs of the ancient *Yule–Walker equations* of the autoregressive systems (cf. Anderson [1], p. 174). These (sample) equations are useful for a descriptive study of such series.

Consider a time series defined by the equation

$$X_{t+1} = (a_0 + a_1 t)X_t + \varepsilon_{t+1}, \qquad t \geq 0 \tag{3.21}$$

where the ε_t are independent, identically distributed mean zero random variables with finite variance, $X_t = 0$ for $t \leq 0$, so that there are no records of the series for the past and it starts from scratch. Here a_0 and a_1 are unknown parameters and can be estimated by the least squares method. Thus a simple minimization based on N observations gives $\hat{a}_{0N}, \hat{a}_{1N}$ as the estimators of a_0, a_1 by the following equations:

$$\begin{aligned}
\hat{a}_{0N} &= \frac{1}{D_N}\left[\left(\sum_{t=1}^{N} tX_{t+1}X_t\right)\left(\sum_{t=1}^{N} tX_t^2\right) - \left(\sum_{t=1}^{N} t^2X_t^2\right)\left(\sum_{t=1}^{N} X_tX_{t+1}\right)\right] \\
\hat{a}_{1N} &= \frac{1}{D_N}\left[\left(\sum_{t=1}^{N} tX_t^2\right)\left(\sum_{t=1}^{N} X_tX_{t+1}\right) - \left(\sum_{t=1}^{N} X_t^2\right)\left(\sum_{t=1}^{N} tX_tX_{t+1}\right)\right] \\
D_N &= \left(\sum_{t=1}^{N} tX_t^2\right)^2 - \left(\sum_{t=1}^{N} X_t^2\right)\left(\sum_{t=1}^{N} t^2X_t^2\right)
\end{aligned} \tag{3.22}$$

It is desirable to establish the consistency of these estimators, i.e., to show that $\hat{a}_{iN} \to a_i$, $i = 1, 2$, in probability as $N \to \infty$. This is true if $a_1 = 0$. The work from ([46b], Theorem 5; [47]) indicates that the general statement is true under some conditions on the distribution of the ε_t. The actual details are nontrivial and are not considered here. Since the general behavior of the series can be understood to some extent from the properties of the correlation characteristic (or the approximate correlogram of the series) the sample product moment equations will now be derived for this model. This already shows the difficulties involved.

It is convenient to adopt the following notation, from Hurwicz [24], for simplification. Thus set (with $X_t = 0$, $t \le 0$), for $t \ge 0$,

$$Z_{1t} = X_t, \quad Z_{2t} = tX_t, \quad \delta_0 = 1, \quad \delta_1 = -a_0, \quad \delta_2 = -a_1 \tag{3.23}$$

Then Eq. (3.21) becomes

$$\delta_0 Z_{1t+1} + \delta_1 Z_{1t} + \delta_2 Z_{2t} = \varepsilon_{t+1}, \qquad t \ge 0 \tag{3.24}$$

For the product moments, the following additional abbreviations are seen to be useful:

$$\begin{aligned} E_k{}^N &= \sum_{t=1}^{N-k} \varepsilon_t \varepsilon_{t+k}, & S_{ik}^N &= \sum_{t=1}^{N-k} Z_{it} Z_{it+k}, \qquad i = 1, 2 \\ S_{12k}^N &= \sum_{t=1}^{N-k} Z_{1t} Z_{2t+k}, & S_{21k}^N &= \sum_{t=1}^{N-k} Z_{2t} Z_{1t+k} \end{aligned} \tag{3.25}$$

Since $\delta_0 = 1$, using (3.24) and after a brief computation, one obtains

$$\begin{aligned} &\delta_1 S_{1k+1}^N + (1 + \delta_1{}^2) S_{1k}^N + \delta_1 S_{1k-1}^N + \delta_2{}^2 S_{2k}^N \\ &\qquad + \delta_2 (S_{12k-1}^N + S_{21k+1}^N) + \delta_1 \delta_2 (S_{12k}^N + S_{21k}^N) = E_k{}^N \end{aligned} \tag{3.26}$$

But $S_{12k}^N = S_{21k}^N + kS_{1k}^N$. So (3.26) can be simplified as

$$\begin{aligned} &\delta_1 S_{1k+1}^N + S_{1k}^N (1 + \delta_1 \delta_2 k + \delta_1{}^2) + S_{1k-1}^N (\delta_1 + (k-1)\delta_2) \\ &\qquad + \delta_2 [S_{21k+1}^N + 2\delta_1 S_{21k}^N + S_{21k-1}^N + \delta_2 S_{2k}^N] = E_k{}^N \end{aligned} \tag{3.27}$$

This is the sample analog (for $k \ge 1$) of the Yule–Walker equation (cf. Anderson [1]). Note that $\{E_k{}^N, N \ge k\}$ is a series of uncorrelated random variables.

If the "initial point" δ_1 of the linear trend is known *a priori*, then (3.27) can be recast in a better form. For then, setting

$$\begin{aligned} Y_t &= X_t + \delta_1 X_{t-1} \\ \tilde{Y}_t &= X_{t-1} + \delta_1 X_t \\ Q_{1k}^N &= \sum_{t=1}^{N-k} [\varepsilon_{t+k} Y_{t+1} - \varepsilon_{t-1} \tilde{Y}_{t+k}] \end{aligned}$$

(3.27) can be expressed, after a brief computation, as

$$\delta_1 S_{1k+1}^N + S_{1k}^N(1 + \delta_1 \delta_2 k) - S_{1k+1}^N(\delta_1 - (k-1)\delta_2) - S_{1k-2}^N = Q_{1k+1}^N \tag{3.28}$$

Treating the Q_{1k}^N as the correlated disturbance, one notes that the product moment equation of the first-order scheme (3.21) [or (3.24)] is a third-order difference equation whose coefficients depend (linearly) on the lag. If $C_k^N = -(1/(N-k))S_{1k}^N$, $Q_k^N = -(1/(N-k))Q_{1k}^N$, then (3.28) reduces to an equation of sample covariances:

$$\delta_1 C_{k+1}^N + (1 + \delta_1 \delta_2 k)C_k^N - (\delta_1 - (k-1)\delta_2)C_{k-1}^N - C_{k-2}^N = Q_k^N \tag{3.29}$$

Taking expectations of this equation, one gets the Yule–Walker relations. If $\delta_2 = 0$ (so there is no trend), then (3.27) and (3.29) reduce to a corresponding known (standard) case (cf. Anderson [1], p. 124).

Similar considerations with a reciprocal trend [i.e., $a_t = a_0 + a_1/t$ for (3.21)] lead to the product moment equations, corresponding to (3.27) or (3.29), with coefficients depending on the reciprocals of the lag. [This and a second-order scheme were considered in [45], but the computations are too ugly for an inclusion here.] From (3.29) one can easily obtain the correlation characteristic $\tilde{\rho}_N(k) = C_k^N/C_0^N$, and then graph the approximate correlogram. An explicit expression for this can be obtained from (3.29), by the method indicated in (3.4) and (3.5). A formula for C_k^N is given below (from (3.29)) without the intermediate computation. Let $f_0 = \delta_1 = -a_0$, $f_1(k) = 1 + \delta_1 \delta_2 k = 1 + k a_0 a_1$, $f_2(k) = -\delta_1 + \delta_2(k-1) = a_0 - (k-1)a_1$, and $f_3 = 1$. Then

$$C_k^N = \sum_{m=1}^{k} f_0^{-m}\left(\sum f_{t_1}(k_1) \cdots f_{t_i}(k_i)\right) Q_{k-m} \tag{3.30}$$

where (i) $t_1 + \cdots + t_i = m$, (ii) $k_1 = k, k_2 = k - t_1, k_3 = k - t_1 - t_2, \ldots$. The graph $\{(k_1, \tilde{\rho}_N(k)), 1 \le k < N\}$ for large enough N, gives an indication of the dependence behavior of the time series described by (3.21), by the earlier treatment.

4. SERIES GENERATED BY DIFFERENTIAL EQUATIONS: FLOWS

4.1. Introduction

The preceding analysis leads to the continuous time analog (or the stochastic differential equations case) of the problem involving again nonstationary time series. This is useful both for a comparison with the discrete case just given and for an independent study. It also brings up some interesting new problems for future investigations.

Let $\{X_t, t \in T, T \subseteq \mathbb{R}\}$ be a continuous parameter time series [X_t and $X(t)$ are synonymous] governed by the differential equation

$$\frac{d^2X(t)}{dt^2} + a(t)\frac{dX(t)}{dt} + b(t)X(t) = \varepsilon(t) \tag{4.1}$$

where $\{\varepsilon(t), t \in T\}$ is the white noise disturbance and a, b are real functions on T. By definition of white noise, $\varepsilon(t)$ is the (generalized) derivative of Brownian motion $\{B(t), t \in T\}$, and thus (4.1) is a symbolic equation which cannot be interpreted in the classical sense of differential calculus. (Such a problem does not, of course, arise in the discrete case.) However, the classical computations carried out formally can be justified (in the integrated form) with the concept of a stochastic integral replacing $\varepsilon(t)\,dt$ by $dB(t)$, and this is made precise in the following. Potential applications of this model abound. Taking $a = 0$, $b(t) = b' + b''t$, Hartley [22] indicated an industrial application and carried out a correlogram analysis of the $X(t)$ series using classical methods (and formal computations). In fact, assuming that the $\varepsilon(t)$ is integrable (in the calculus sense), he has carried out the analysis using "Airy integrals" and then studied the covariance characteristic. (In a conversation at the IMS meetings in Ames, Iowa in 1957, Dr. Hartley noted that the Weber differential equation method would be better suited for such a problem.) If a, b are constants, then a related problem was considered by Nagabhushanam ([35], p. 482) where $X(t)$ is called a "primary process" obtained by an inversion. Since the Brownian motion is nondifferentiable in the classical sense so that the $\varepsilon(t)$ of (4.1) does not exist, a slightly different route is followed here to validate such an analysis for a solution of (4.1).

4.2. A General Second-Order Problem

The method of attack here is quite simple. First consider the problem with a formal manipulation and express the solution in terms of an integral. Replacing $\varepsilon(t)\,dt$ by $dB(t)$, where $\{B(t), t \in T\}$ is the Brownian motion, and then interpreting the integral as the *stochastic integral* of a nonstochastic (or "sure") function relative to Brownian motion, the solution is rigorously definable. [Of course, the $\varepsilon(t)$ process is not as general as that given in Hartley [22], but the present assumption will be in force throughout this section. For a rigorous treatment, some such restriction is necessary.]

The differential equation (4.1), written symbolically as

$$d\dot{X}(t) + a(t)\dot{X}(t)\,dt + b(t)X(t)\,dt = dB(t) \tag{4.2}$$

is then regarded as the equation leading to a well-defined solution, where

$\dot{X}(t) = dX/dt$. To make this more precise, let $T = [a_0, b_0)$, a bounded interval, and $a(\cdot)$, $b(\cdot)$ be continuous real functions on $\bar{T} = [a_0, b_0]$. Set $Q = \dot{X}$,

$$A(t) = \begin{bmatrix} a(t) & b(t) \\ -1 & 0 \end{bmatrix}, \quad W(t) = \begin{bmatrix} B(t) \\ 0 \end{bmatrix}, \quad \eta(t) = \begin{bmatrix} \varepsilon(t) \\ 0 \end{bmatrix}, \quad Z(t) = \begin{bmatrix} Q(t) \\ X(t) \end{bmatrix}$$

Then (4.1) may be expressed compactly as

$$dZ(t) + A(t)Z(t)\, dt = \eta(t)\, dt = dW(t) \tag{4.3}$$

To solve this (vector) differential equation with a standard method in the classical theory of ordinary differential equations, consider the 2×2 matrix differential equation associated with the homogeneous part of Eq. (4.3), i.e., the (nonstochastic) equation:

$$dY(t) = Y(t)A(t)\, dt, \qquad t \in T, \quad \det(Y(a_0)) \neq 0 \tag{4.4}$$

where det stands for determinant. Then premultiplying (4.3) by $Y(t)$ and using (4.4) one gets

$$\frac{d}{dt}(Y(t)Z(t)) = Y(t)\eta(t)$$

so that formally one has the solution of (4.3) as

$$Z(t) = Y(t)^{-1} \int_{a_0}^{t} Y(u)\eta(u)\, du + Y^{-1}(t)Y(a_0)Z(a_0) \tag{4.5}$$

The fact that $Y(t)$ satisfying Eq. (4.4), if nonsingular for one $t \in T$, has the same property for all $t \in T$ is used here. This is because (4.4) implies $\det(Y(t)) = \det(Y(a_0)) \cdot \exp(\int_{a_0}^{t} \mathrm{tr}(A(u))\, du)$ (cf. Coddington and Levinson [7], p. 28). Thus (4.5) is well defined and $Z(t)$ is obtained as soon as $Y(t)$ is solved from (4.4). Since $Y(t)^{-1}$ is bounded for each t, it can be taken inside the integral also. Thus (4.5) can be expressed as

$$Z(t) = \int_{a_0}^{t} Y(t)^{-1} Y(u)\, dW(u) + Y^{-1}(t)Y(a_0)Z(a_0) \tag{4.6}$$

where the first term on the right-hand side is now rigorously defined as the stochastic integral of the first kind (see Doob [12], p. 352). The uniqueness of the solution is immediate for the given initial condition $Z(a_0)$, because if $\tilde{Z}$ is another solution, then $Z - \tilde{Z}$ will be a solution of the homogeneous equation $dU/dt + A(t)U(t) = 0$, with $U(a_0) = 0$ as the initial condition. This is a nonstochastic equation and the standard theory implies that $U \equiv 0$ is its only solution. Thus $Z = \tilde{Z}$. Hence it remains to find $Y(t)$.

Since (4.4) is a nonstochastic equation, one can apply the classical Picard method of approximation. Writing $\tilde{A} = A^*$, $\tilde{Y} = Y^*$ (where the asterisk means transpose) and integrating (4.4),

$$\tilde{Y}(t) = \tilde{Y}(a_0) + \int_{a_0}^{t} \tilde{A}(u)\tilde{Y}(u)\,du \tag{4.7}$$

Now substituting for $\tilde{Y}$ and iterating,

$$\tilde{Y}(t) = \tilde{Y}(a_0) + \int_{a_0}^{t} \tilde{A}(t_1)\tilde{Y}(a_0)\,dt_1 + \cdots + \int_{a_0}^{t} \tilde{A}(t_1) \times \int_{a_0}^{t_1} \tilde{A}(t_2) \cdots \int_{a_0}^{t_{n-1}} \tilde{A}(t_n)\tilde{Y}(a_0)\,dt_n \cdots dt_1 + R_n \tag{4.8}$$

where

$$R_n = \int_{a_0}^{t} \tilde{A}(t_1) \int_{a_0}^{t_1} \tilde{A}(t_2) \cdots \int_{a_0}^{t_n} \tilde{A}(u)\tilde{Y}(u)\,du\,dt_n \cdots dt_1$$

By hypothesis $a(\cdot)$ and $b(\cdot)$ are continuous on the compact interval $\bar{T}$. So $\|\tilde{A}(t)\| = [\mathrm{tr}(\tilde{A}(t)A(t))]^{1/2} \le M < \infty$, $t \in \bar{T}$, and similarly $\|Y(t)\| \le N < \infty$. [Here N is obtained as follows. Let y_1, y_2 be the linearly independent pair of vector solutions of the homogeneous equation of (4.3). Then these are continuous on $\bar{T}$. If N_1 is the upper bound on the norms of the vectors y_i, $i = 1$, 2, then $Y = (y_1, y_2)$ in (4.4) so that $\|Y\| \le \sqrt{2}\,N_1 = N$.] Consequently,

$$\|R_n\| \le M^n N \frac{(t - a_0)^n}{n!} \le M^n N \frac{(b - a_0)^n}{n!}, \qquad t \in \bar{T} \tag{4.9}$$

Hence $R_n \to 0$ uniformly in t as $n \to \infty$, and the series in (4.8) converges absolutely and uniformly and defines a solution of (4.4). In particular, if $\alpha(t, t_0) = \int_{t_0}^{t} \tilde{A}(u)\,du$, then

$$\begin{aligned}\int_{a_0}^{t} \tilde{A}(t_1) \int_{a_0}^{t_1} \tilde{A}(t_2)\,dt_2\,dt_1 &= \int_{a_0 \le t_2 \le t_1 \le t} \tilde{A}(t_1)\tilde{A}(t_2)\,dt_2\,dt_1 \\ &= \int_{a_0 \le t_2 \le t} \alpha_1(t, t_2)\tilde{A}(t_2)\,dt_2 \\ &= \alpha_2(t, a_0) \qquad \text{(say)}\end{aligned}$$

Similarly,

$$\alpha_3(t, a_0) = \int_{a_0}^{t} \tilde{A}(t_1) \int_{a_0}^{t_1} \tilde{A}(t_2) \int_{a_0}^{t_2} \tilde{A}(t_3)\,dt_3\,dt_2\,dt_1$$

Then (4.8) can be expressed more conveniently as [set $\alpha_0(t, a_0) =$ identity)

$$Y^*(t) = \tilde{Y}(t) = \sum_{n=0}^{\infty} \alpha_n(t, a_0)\tilde{Y}(a_0) \tag{4.10}$$

As yet no special properties of Brownian motion, other than the definition of (4.6), have been utilized. Let $Z(a_0) = C$ be a constant (non-stochastic) initial condition. [The existence and uniqueness of the solution of (4.3) rigorously hold if only $Z(a_0)$ is independent of $B(t) - B(a_0)$. However, this is not sufficient for the following analysis.] Also, the continuity of $a(\cdot)$ and $b(\cdot)$ is not crucial. If $a(\cdot)$, $b(\cdot)$ are integrable on T, then one sees easily that the bound in (4.9) holds (cf. also Coddington and Levinson [7], Problem 1, p. 97), and the rest of the argument is valid. Thus one may state the following simple but important result:

Theorem 4.1. Let $T = [a_0, b_0)$ be a bounded interval and $\{B(t), t \in T\}$ be the Brownian motion. If $\{X(t), t \in T\}$ is a time series generated by (4.2) with $a(\cdot)$, $b(\cdot)$ as the (Lebesgue) integrable real functions on T, then there is one and only one such series for each initial condition $X(a) = c_1$, $\dot{X}(a) = c_2$ where c_1, c_2 are real constants. The solution

$$Z(t) = \begin{bmatrix} \dot{X}(t) \\ X(t) \end{bmatrix}$$

of (4.3) is given by (4.6) where Y is defined by (4.10). Moreover, $\{Z(t), t \in T\}$ is a vector Markov Gaussian time series, almost all of whose trajectories are continuous.

Proof. Because of the preceding discussion and computations, only the proof of the last statement remains. Note that by the classical theory, the function $Y: T \to \mathbb{R}^4 - \{0\}$ is continuous and so also is Y^{-1}. By some well-known properties of the simple stochastic integrals (cf. Dobb [13], Chap. 9) the integral in (4.6) is a continuous function of t with probability 1. It follows that the $Z(t)$ process has almost all continuous trajectories with values in $\mathbb{R}^2$.

By definition, the integral in (4.6) is the mean square limit of approximating sums of the form $\sum_{i=1}^{n} Y(t_i)(W(t_{i+1}) - W(t_i))$, where $a_0 = t_1 < t_2 < \cdots < t_n \leq t$. But since $Y(t_i)$ is nonstochastic and $W(t_{i+1}) - W(t_i)$ are independent Gaussian vectors, this sum is Gaussian distributed. Hence its limit-in-mean (i.e., the integral) defines a Gaussian random vector. Since $Z(a_0) = C$, it follows from (4.6) that the $\{Z(t), t \in T\}$ is a vector Gaussian time series, having almost all continuous sample paths. Finally, (4.6) is true if $a < t$ is replaced by $t, t + h$ $(h > 0)$, a pair of points in T. Thus

$$Z(t) = \int_{t}^{t+h} Y(t+h)^{-1} Y(u)\, dW(u) + Y^{-1}(t+h) Y(t) Z(t) \qquad (4.11)$$

By the preceding definition of the integral, $Z(t)$ is determined by $Y(t)$ and $W(u)$ for $a_0 \leq u \leq t$ only. Since $W(\cdot)$ has independent increments, (4.11) is determined by $W(t_{i+1}) - W(t_i)$ for $t \le t_i < t_{i+1} \le t + h$, and is independent of $Z(u)$ for $a \le u < t$. This means for each interval $A \subset \mathbb{R}^2$, $t \ge t_0$,

$$P[Z_t \in A | Z_s, s \le t_0] = P[Z_t \in A | Z_{t_0}] \tag{4.12}$$

with probability 1. But this implies the Markovian property of the $Z(t)$, and completes the proof.

Remark. In general the Z process need not be a martingale and the X process need not be Markovian. Note that the cases $a(t) = a' + a''t$ and $b(t) = b' + b''t$ (the linear trend, and similarly the reciprocal trend if $T = [a_0, b_0)$, $a_0 > 0$) are included in this treatment. Another model treating seasonality [i.e., $a(t) = a_1 + a_2 \sin t$, $b(t) = b_1 + b_2 \cos t$] of the coefficients considered by Hurwicz ([24], p. 334) is also included. However, the latter case has some special properties worthy of a special treatment, but is not considered in this chapter. It is also clear that Theorem 4.1 is valid without real changes if the order of the equation is $n \ge 2$. In this case the $Y(t)$ and $A(t)$ are $n \times n$ matrices.

From (4.6) it is easy to calculate the mean and covariance of the $Z(t)$ vector series. In fact, if $m(t) = E(Z(t))$, $\sum(t, s) = \operatorname{cov}(Z_s, Z_t)$, then from the condition that $Z(a_0) = C$, a constant, one has

$$m(t) = Y^{-1}(t)Y(a_0)C \tag{4.13}$$

and (see (4.6))

$$\begin{aligned}\sum(s, t) &= E[(Z(t) - m(t))(Z(s) - m(s))^*] \\ &= E\left[\left(\int_{a_0}^{t} Y(t)^{-1}Y(u)\,dW(u)\right)\left(\int_{a_0}^{s} Y(s)^{-1}Y(r)\,dW(r)\right)^*\right] \\ &= \int_{a_0}^{s \wedge t} Y(t)^{-1}Y(u)\begin{bmatrix}1 & 0\\ 0 & 0\end{bmatrix} Y^*(u)^{-1}Y^*(s)^{-1}\,du, \quad s \wedge t = \min(s, t)\end{aligned} \tag{4.14}$$

where an elementary property of the stochastic integral is used in the last line. If $D(t, u)$ is the first column of $Y(t)^{-1}Y(u)$, then (4.14) can be written as

$$(\sigma_{ij}(s, t)) = \sum(s, t) = \int_{a_0}^{s \wedge t} D(t, u)D(s, u)^*\,du \tag{4.15}$$

This is the 2×2 covariance matrix function of the vector $Z(t)$ series, which is (Hermitian) positive definite for all $a_0 \le s, t \le b$. From this it follows immediately that the covariance function of the X series $[r(s, t) = \operatorname{cov}(X_s, X_t) = \sigma_{22}(s, t)]$ is given as

$$r(s, t) = \int_{a_0}^{s \wedge t} (D(t, u)D(s, u)^*)_{22}\,du \tag{4.16}$$

where $(\ )_{22}$ is the second diagonal element of the matrix. If $d(s, u)$ is the second element of the vector $D(s, u)$, then (4.16) is just

$$r(s, t) = \int_{a_0}^{s \wedge t} d(s, u)\, \overline{d(t, u)}\, du \tag{4.17}$$

Thus the time series determined by (4.2) is always of class (C). It is interesting to note that the covariance function r of (4.17) is given by Green's function of the homogeneous ordinary differential equation of (4.1) with zero initial conditions, i.e., $X(t) = \dot{X}(t) = 0$ at $t = a_0$. This useful result may be derived as follows.

Consider the homogeneous equation

$$L(X) = \frac{d^2X(t)}{dt^2} + a(t)\frac{dX(t)}{dt} + b(t)X(t) = 0 \tag{4.18}$$

where $X(a) = c_1$, $\dot{X}(a) = c_2$, as the initial data. If $U_i(a) = X^{(i-1)}(a)$ where $X^{(0)} = X$, $X^{(1)} = \dot{X} = dX/dt$, then by the standard theory (cf. Coddington and Levinson [7]) the system $L(X) = 0$ with $U_1(a_0) = 1$, $U_2(a_0) = 0$, has a unique solution, say V_1. Similarly, $L(X) = 0$ with $U_1(a_0) = 0$, $U_2(a_0) = 1$ has a unique solution V_2. If $G(\cdot, \cdot)$ is the Green function associated with $L(X) = 0$, $X(a_0) = 0 = \dot{X}(a_0)$, then the unique solution of (4.2) with the initial condition $Z(a) = C$ is given by [cf. Ince [25], p. 257, and the discussion on the stochastic analog following (4.2)]

$$X(t) = \int_{a_0}^{t} G(t, u)\, dB(u) + C^* V(t) \tag{4.19}$$

where

$$V = \begin{bmatrix} V_1 \\ V_2 \end{bmatrix} \quad \text{and} \quad C = \begin{bmatrix} C_1 \\ C_2 \end{bmatrix}$$

Here G is continuous on $T \times T$ and differentiable [in general, $(n-1)$ times for the nth-order equation] and $\partial G/\partial t$ is continuous in (t, u) for the range $a_0 \leq t \leq u \leq b_0$. Moreover,

$$\frac{\partial G}{\partial t}(u + 0, u) - \frac{\partial G}{\partial t}(u - 0, u) = 1$$

Thus from (4.19) one has $m_1(t) = E(X(t)) = C^* V(t)$, and

$$r(s, t) = \operatorname{cov}(X(s), X(t)) = \int_{a_0}^{s \wedge t} G(s, u)\overline{G(t, u)}\, du \tag{4.20}$$

A comparison of (4.19)–(4.20) with (4.16)–(4.17) shows that $G(s, u) = d(s, u)$. The form of this equation should also be compared with a result of Cramér ([11], p. 18) in terms of which $X(t)$ has multiplicity 1. [For the definition of the latter, see again Cramér [11] and Section 6.]

The preceding work may be summarized in the following:

Theorem 4.2. The vector Gaussian Markov process $\{Z(t), t \in T\}$ of Theorem 4.1 has mean and covariance given by (4.13) and (4.15). The solution process $\{X(t), t \in T\}$ of (4.2) is given by (4.19) with its covariance function then by (4.20) or (4.17), and it is of type (C). If, moreover, the Green kernel $G(\cdot, \cdot)$ of (4.20) satisfies an appropriate growth condition, in particular if $G(\cdot, u)$ is a.p. uniformly in u on compact subsets of T, then the solution process $\{X(t), t \in T\}$ is in class (C) $\cap$ class (KF).

Remark. If $a(t) = a_1 + a_2 t$, $b(t) = b_1 + b_2 t$, then the resulting solution process in Theorem 4.2, although being in class (C), will not be in class (KF). In any case the solution has "multiplicity" 1 in the sense of Cramér ([11], p. 11). This problem and the work of Section 2 show that the classes (C) and (KF) are not included in each other.

In view of the preceding result, it will be of considerable interest to analyze and extend Dym's [14] detailed treatment from the constant coefficient case to the general case which is a solution process of (4.2). Many of Dym's [14] results have nontrivial extensions to the present one. It needs and deserves a separate treatment and is not considered at present. In the following subsection a special problem is discussed for the properties of its correlogram, for it enables a better appreciation of the time series generated by (4.2).

4.3. A Special Case

Let $a(t) = a$, $b(t) = b$, $t \in T$. Then $Y(\cdot)$ can be calculated explicitly in (4.10). For,

$$A(t) = A = \begin{bmatrix} a & b \\ -1 & 0 \end{bmatrix}$$

so that

$$\alpha(t, t_0) = A^*(t - t_0), \qquad \alpha_2(t, t_0) = A^{*2} \frac{(t - t_0)^2}{2}, \qquad \ldots$$

and hence

$$Y^*(t) = \sum_{n=0}^{\infty} A^{*n} \frac{(t - a_0)^n}{n!} Y^*(a_0) = (\exp(t - a_0)A^*)Y^*(a_0)$$

Consequently (4.6) can be simplified to

$$Z(t) = \int_{a_0}^{t} e^{(t-u)A}\, dW(u) + e^{(a_0 - t)A} C \tag{4.21}$$

where the facts that e^{uA} and e^{tA} commute and $Z(a_0) = C$ were used. From this, if

$$e^{(t-u)A} = \begin{pmatrix} d_1(t-u) & d_2(t-u) \\ d_3(t-u) & d_4(t-u) \end{pmatrix}$$

then (4.21) yields

$$\begin{aligned} \dot{X}(t) &= \int_{a_0}^{t} d_1(t-u)\, dB(u) + d_1(a_0 - t)c_1 + d_2(a_0 - t)c_2 \\ X(t) &= \int_{a_0}^{t} d_3(t-u)\, dB(u) + d_3(a_0 - t)c_1 + d_4(a_0 - t)c_2 \end{aligned} \tag{4.22}$$

Comparing these equations with (4.19) one obtains $G(t, u) = G(t - u) = d_3(t - u)$ and $G'(t - u) = d_1(t - u)$, where G is the Green function associated with (4.2). (In the case of constant coefficients G is only a function of the difference of its arguments!) The covariance is thus simplified to ($a_0 = 0$)

$$r(s, t) = \int_0^{s \wedge t} G(s-u)\overline{G(t-u)}\, du \tag{4.23}$$

Even now r is not a stationary covariance, and the solution process need not be in class (KF) without further conditions. [Here a_0 is taken as 0 for ease.]

In the particular case in which $a = 0$, $b > 0$ in (4.21) with $T = [0, \alpha)$, the covariance (4.23) can be simplified further. Then, $d_3(t) = \sin \sqrt{b}\, t$, and

$$r(s, t) = \int_0^{s \wedge t} \sin \sqrt{b}\,(t-u) \sin \sqrt{b}\,(s-u)\, du \tag{4.24}$$

If $t = s + k$, then the correlation character $\tilde{\rho}_s(k) = r(s, s+k)/r(s, s)$ is reduced to

$$\begin{aligned} \tilde{\rho}_s(k) &= \frac{\int_0^s \sin \sqrt{b}\,(k+s-u) \sin \sqrt{b}\,(s-u)\, du}{\int_0^s \sin^2 \sqrt{b}\,(s-u)\, du}, \qquad u \geq 0 \\ &= \frac{2s\sqrt{b} \cos \sqrt{b}\,k + \sin \sqrt{b}\,k - \sin \sqrt{b}\,(k+2s)}{2s\sqrt{b} - \sin 2\sqrt{b}\,s} \end{aligned} \tag{4.25}$$

For large s, $\tilde{\rho}_s(k)$ therefore behaves as $\cos \sqrt{b}\,k$, and hence the approximate correlogram is essentially a harmonic curve. This computation also shows that for *small* t, $r(t, t) = \int_0^t \sin^2 \sqrt{b}\,u\, du \sim (b/3)t^3$. Similar estimates can be obtained for $\sigma_{ij}(t, t)$. Such calculations were carried out by Dym ([14], p. 135) using different arguments, and both these results agree. The Gaussian character of $Z(t)$, when $a(t) = a$, $b(t) = b > 0$, and various moments were established long ago by Chandrasekhar [6], with classical techniques. It will thus be interesting to carry out the corresponding study for the case of variable $a(\cdot)$, $b(\cdot)$, e.g., if these are periodic or more generally (real) analytic functions.

The result of Theorem 4.1 admits substantial extensions for vector-valued processes and the coefficients can then be matrix-valued integrable functions. Such an extension has been carried out by Goldstein [17] using suitable abstract methods. His results subsume all the previously known existence studies on the problem. The natural question here is to consider (and this appears more difficult) Goldstein's results in classifying the solutions in the sense of Dym's work.

5. ASYMPTOTIC DISTRIBUTIONS OF ESTIMATORS

5.1. Some General Results

The asymptotic distributions of estimators of parameters of a stochastic model generating the time series are rather involved, particularly in the nonstationary cases. To indicate the essential nature of the problems in this area, time series generated by linear difference equations (or autoregressive schemes) are discussed in this section. A relatively complete solution can be presented in the case of first-order equations, and this is given in the following subsection.

Consider the stochastic difference equation with constant coefficients, which may thus be written as

$$X_t + a_1 X_{t-1} + \cdots + a_k X_{t-k} = \varepsilon_t, \qquad t \geq 1 \tag{5.1}$$

where $\{\varepsilon_t, t \geq 1\}$ are independent identically distributed with mean zero and finite variance. If the roots of the characteristic equation

$$\lambda^k + a_1 \lambda^{k-1} + \cdots + a_k = 0 \tag{5.2}$$

lie inside the unit circle in the complex plane, then the least squares estimators $\hat{a}_{iN}$, $i = 1, \ldots, k$, which are easily computed [as noted in (3.22)], are consistent and have a limit distribution as follows. The vector $\{\sqrt{N}(\hat{a}_{iN} - a_i), i = 1, \ldots, k\}$ converges in distribution to a k-variate normal with means zero and covariance Q^{-1} where $Q = p\text{-}\lim_{T \to \infty} (\sum_{t=1}^{T} Y_t Y_t^*/T)$, and where $Y_t^* = (X_t, X_{t-1}, \ldots, X_{t-k})$. This result was proved by Mann and Wald [33] when the ε_t have all moments finite, and by T. W. Anderson [2] in the stated form, i.e., with only two moments assumed finite. In this case the X_t process is of class (KF). If the roots of (5.2) are on or outside the unit circle, then $\{X_t, t \geq 1\}$ is not in class (KF), and the asymptotic distribution theory is much more complicated. Some aspects of this are now discussed.

If all the roots of (5.2) are outside the unit circle, then the least squares estimators $\hat{a}_{iN}$ are consistent. If the roots are distinct, and there is a maximal root λ_0 such that $|\lambda_0| > 1$ and the rest are inside the unit circle, then again the consistency statement on $\hat{a}_{iN}$ is valid. These two results are due to Anderson [2] and the author [47]. Some recent (unpublished) work of others aimed at an extension of these two results apparently implies the consistency of $\hat{a}_{iN}$ without any conditions on the roots of (5.2).† In the case $k = 1$, this is known from the work of Rubin [51], but for the case $k > 1$, the computations get very complicated. The following results are known about the asymptotic distributions of $\hat{a}_{iN}$ in the explosive case.

First observe that (5.1) can be expressed in vector notation as

$$Y_t + \alpha Y_{t-1} = \tilde{\eta}_t, \qquad t \geq 1 \tag{5.3}$$

where $Y_t^* = (X_t, \ldots, X_{t-k+1})$, $\tilde{\eta}_t^* = (\varepsilon_t, 0, \ldots, 0)$, and

$$\alpha = \begin{pmatrix} a_1 & a_2 & \cdots & a_{k-1} & a_k \\ 1 & 0 & \cdots & 0 & 0 \\ \vdots & \vdots & \ddots & \vdots & \vdots \\ 0 & 0 & \cdots & 1 & 0 \end{pmatrix}$$

This is analogous to Eq. (4.3). If $\hat{\alpha}_N$ denotes the estimator of α in which a_i is replaced by $\hat{\alpha}_{iN}$, and ε_t are independent $N(0, 1)$ variables, then the vector $(\hat{\alpha}_N - \alpha)\alpha^{N-2}$ converges in distribution to a random matrix Z, as $N \to \infty$. The elements of this Z are *not* normally distributed. In fact, if the ε_t are not $N(0, 1)$, but satisfy the earlier assumptions, then under some rank conditions, one can show that $(\hat{\alpha}_N - \alpha)\alpha^{N-2}$ still has a limit distribution, but it is different from that of the $N(0, 1)$ case. Thus the limit distribution depends on the distribution of the random disturbances ε_t. These results are direct consequences of Anderson's [2] work. On the other hand, if λ_0 is the maximal root of (5.2), $|\lambda_0| > 1$, and the other roots of (5.2) (are distinct and) lie inside the unit circle, then λ_0 can be estimated consistently. Let $\hat{\lambda}_{0N} = (\sum_{t=1}^N X_t X_{t-1})/(\sum_{t=1}^N X_{t-1}^2)$. Then $\hat{\lambda}_{0N} \to \lambda_0$ in probability, under the assumptions on ε_t given earlier. Thus $\hat{\lambda}_{0N}$ is consistent. Moreover, $|\lambda_0|^N(\hat{\lambda}_{0N} - \lambda_0)(\lambda_0^2 - 1)^{-1}$ has a limit distribution as $N \to \infty$. This again depends on the distribution of the disturbances ε_t and it is Cauchy if the ε_t are $N(0, 1)$. These results were proved by the author [47]. It is to be noted that, in both these cases, the limit distributions of the estimators of interest depend on the distribution of the ε_t in sharp contrast to the stable case. Thus no "invariance principle" can hold for such series. The problem corresponding to the roots on the unit circle is even more involved and less understood. However, a complete treatment of the first-order case [$k = 1$ in (5.1)] is possible, and the rest of this section is devoted to it.

† See Stigum [53a]. However, the result of Theorem 3 is unclear to me since there is no "invariance principle" in the explosive case, and the limit depends on the distribution of ε's (cf. [2]); see Remark 1, following the proof of Theorem 5.1.

5.2. A Complete Description of the First-Order Case

For the first-order scheme, all terms of (5.3) are scalars. Thus (5.1) and (5.3) become

$$X_t = \alpha X_{t-1} + \varepsilon_t, \qquad t \geq 1 \tag{5.3$'$}$$

The least squares estimator of α ($= \lambda_0$ in the preceding discussion) is the same as $\hat{\lambda}_{0N}$ and hence $|\alpha|^N(\hat{\alpha}_N - \alpha)(\alpha^2 - 1)$ has a limit distribution, for all real α, with $|\alpha| \neq 1$. If $|\alpha| > 1$, then this distribution depends on that of the ε_t and if the ε_t are $N(0, 1)$, then the limit distribution is Cauchy. This result was first obtained by White [55]. Of course for the stable case (i.e., $|\alpha| < 1$), it is known that $\sqrt{N}(\hat{\alpha}_N - \alpha)$ has a limit normal distribution whatever the distribution of the ε_t with two finite moments, as noted in the preceding subsection. The case of $|\alpha| = 1$ remained unsolved. As noted earlier, Rubin [51] has shown the consistency of $\hat{\alpha}_N$ even in this case, and White [55] has observed that $N/\sqrt{2}$ is the correct normalization for $(\hat{\alpha}_N - \alpha)$ when $|\alpha| = 1$. This latter author has shown that $(N/\sqrt{2})(\hat{\alpha}_N - \alpha)$ has a limit distribution, if the ε_t are $N(0, 1)$, and obtained its characteristic function. However, he was unable to invert the latter to obtain the limit distribution. A solution is presented here and, not surprisingly, it has a complicated expression.

Since $\hat{\lambda}_{0N}$ is $\hat{\alpha}_N$, comparison with its expression [for (5.3′)] gives

$$\hat{\alpha}_N - \alpha = \left(\sum_{t=1}^{N} \varepsilon_t X_{t-1}\right) \Big/ \left(\sum_{t=1}^{N} X_{t-1}^2\right) = \frac{U_N}{V_N} \quad \text{(say)} \tag{5.4}$$

To appreciate the difficulty of the problem, it is necessary to recall the classical result of Erdös and Kaç [15] where they calculated

$$\lim_{N \to \infty} P[(1/N^2)V_N < x] = F(x), \qquad F(0) = 0 \tag{5.5}$$

and the characteristic function of F was found to be $\xi(u) = [\sec(2iu)^{1/2}]^{1/2}$. This was inverted and the (continuous) distribution given by Erdös and Kaç ([15], p. 293) has a complicated form. Using an "invariance principle" these authors showed that the result holds if the ε_t are independent and have the same distribution with two moments finite ($\alpha = 1$ for them), and this again will be true in the present case if the ε_t have two moments $[E(\varepsilon_t) = 0]$ and are symmetrically distributed. A similar result may be shown to be true for the U_N sequence. Here then is the asserted limit distribution, and this constitutes a solution of the problem left open since 1958 at least (cf. [55]).

Theorem 5.1. Let the first-order equation (5.3′) be given, where the ε_t are independent $N(0, 1)$ and $|\alpha| = 1$. Then

$$\lim_{N \to \infty} P\left[\frac{N}{\sqrt{2}}(\hat{\alpha}_N - \alpha) < x\right] = \int_{-\infty}^{x} h(u)\, du$$

where the density function $h(\cdot)$ is defined by the formula

$$h(x) = \frac{1}{\sqrt{8}\pi} \int_{-\infty}^{\infty} \frac{\rho(x, t)}{r(x, t)^{3/2}} \cos(\delta(x, t) - \tfrac{3}{2}\theta(x, t)) \times [\chi_{\mathbb{R}_+^2}(x, t) + \chi_{\mathbb{R}_+^2}(-x, -t)] \frac{dt}{(tx)^{1/2}} \tag{5.6}$$

Here ρ, r, δ, θ are defined by the following expressions:

$$\begin{aligned} r(x, t)^2 &= \sinh^2(2tx)^{1/2} + \cos^2(2tx)^{1/2} \\ &\quad + \frac{t}{2x}(\sinh^2(2tx)^{1/2} + \sin^2(2tx)^{1/2}) \\ &\quad + \frac{\alpha}{2}\left(\frac{t}{x}\right)^{1/2} (\sin(8tx)^{1/2} - \sinh(8tx)^{1/2}) \end{aligned} \tag{5.7}$$

$$\theta(x, t) = \operatorname{arc\,tan}\left\{\left(\frac{1 - (\alpha/2)(t/x)^{1/2}(\coth(2tx)^{1/2} + \cot(2tx)^{1/2})}{1 - (\alpha/2)(t/x)^{1/2}(\tanh(2tx)^{1/2} - \tan(2tx)^{1/2})}\right) \times \tan(2tx)^{1/2} \tanh(2tx)^{1/2}\right\} \tag{5.8}$$

$$\begin{aligned} \rho(x, t)^2 &= 2\left(1 - \frac{\alpha}{\sqrt{8}x}\right)^2 (\sinh^2(2tx)^{1/2} + \sin^2(2tx)^{1/2}) \\ &\quad + \frac{t}{x}(\sinh^2(2tx)^{1/2} + \cos^2(2tx)^{1/2}) \\ &\quad - \alpha\left(\frac{t}{x}\right)^{1/2}\left(1 - \frac{\alpha}{\sqrt{8}x}\right)(\sin(8tx)^{1/2} + \sinh(8tx)^{1/2}) = C^2 + D^2 \end{aligned} \tag{5.9}$$

and

$$\delta(x, t) = \operatorname{arc\,tan} \frac{C \cos\sqrt{2}\,\alpha t - D \sin\sqrt{2}\,\alpha t}{C \sin\sqrt{2}\,\alpha t + D \cos\sqrt{2}\,\alpha t} \tag{5.10}$$

where

$$\begin{aligned} C &= \left(1 - \frac{\alpha}{\sqrt{8}x}\right)(\sinh(2tx)^{1/2} \cos(2tx)^{1/2} - \cosh(2tx)^{1/2} \sin(2tx)^{1/2}) \\ &\quad + \alpha\left(\frac{t}{x}\right)^{1/2} \sinh(2tx)^{1/2} \sin(2tx)^{1/2} \end{aligned} \tag{5.11}$$

$$\begin{aligned} D &= \left(1 - \frac{\alpha}{\sqrt{8}x}\right)(\sinh(2tx)^{1/2} \cos(2tx)^{1/2} + \cosh(2tx)^{1/2} \sin(2tx)^{1/2}) \\ &\quad - \alpha\left(\frac{t}{x}\right)^{1/2} \cosh(2tx)^{1/2} \cos(2tx)^{1/2} \end{aligned} \tag{5.11'}$$

Proof. The solution is based on White's preliminary computations and an interesting result due to Cramér ([8], Theorem 12; and its extension in Cramér [9], p. 317). The reader must patiently verify the following statements since some of them involve certain trigonometric (and hyperbolic) identities.

Suppose that the ε_t are $N(0, 1)$, and consider the characteristic function:

$$\varphi_N(t, u) = E\left[\exp\left(\frac{\sqrt{2}\,it}{N} U_N + \frac{2iu}{N^2} V_N\right)\right]$$

where U_N and V_N are as in (5.4). It is not hard to evaluate the right-hand side, and in fact White [55] has shown that $\varphi_N(t, u) \to \varphi(t, u)$ as $N \to \infty$ for all (t, u), and that the limit is continuous. Thus $\varphi(\cdot, \cdot)$ is a characteristic function. The explicit expression (shown by White) is

$$\varphi(t, u) = e^{\sqrt{2}\,\alpha it}\left(\cos 2(iu)^{1/2} - \frac{\alpha it}{(2iu)^{1/2}} \sin 2(iu)^{1/2}\right)^{-1/2} \tag{5.12}$$

From (5.12) one finds that $\varphi(0, u) = \xi(2u)$, the characteristic function for (5.5), as it should. Thus $(\sqrt{2}\,U_N/N, 2V_N/N^2)$ converges in distribution to the random vector (U, V) whose characteristic function is given by (5.12), and $P[V > 0] = 1$. Note that $\sqrt{2}\,U_N/N \to U$ in distribution, and the latter has a shifted gamma distribution whose characteristic function is seen as $\varphi(t, 0) = e^{\sqrt{2}\alpha it}(1 - \sqrt{2}\,\alpha it)^{-1/2}$. Further, one finds, after an elementary but somewhat lengthy computation, that

$$|\varphi(t, u)| \to 0 \qquad \text{as} \quad |t| \to \infty \text{ and } |u| \to \infty$$

Thus (by the Riemann–Lebesgue lemma) one may conclude that the distribution of (U, V) is of continuous type and $P(V > 0) = 1$. It is not difficult to check that the distribution of (U, V) has at least two moments [since $\varphi(\cdot, \cdot)$ is differentiable several times at 0]. This information is utilized in finding the asymptotic distribution of $\hat{\alpha}_N$.

Let $g^2(N) = N^2/2$, and $W_N{}^x = U_N/g(N) - x(V_N/g^2(N))$. Consider

$$\begin{aligned}\lim_{N\to\infty} P[g(N)(\hat{\alpha}_N - \alpha) < x] &= \lim_{N\to\infty} P\left[\frac{U_N}{g(N)} < x \frac{V_N}{g^2(N)}\right] \\ &= \lim_{N\to\infty} P[W_N{}^x < 0]\end{aligned} \tag{5.13}$$

This limit exists by (5.11) and (5.12). [In fact by the theory of Erdös and Kaç [15], this limit exists if the ε_t have a common continuous distribution symmetric about the origin when the ε_t are independent $E(\varepsilon_t) = 0$, $E(\varepsilon_t{}^2) = 1$, and then the limit distribution is independent of that of the ε_t; i.e., an invariance principle applies. Hence one can use for convenience, that the ε_t are $N(0, 1)$, and calculate the limit distribution of the right-hand side of (5.13).] Thus

$$E[\exp(i\tau W_N{}^x)] = \varphi_N(\tau, -x\tau) \to \varphi(\tau, -x\tau) \tag{5.14}$$

If W^x is this limit random variable, it is clear that $W^x = U - xV$.

From the preceding established facts, one can apply Cramér's theorem ([8], Theorem 12; and [9], p. 317, Ex. 6) to obtain the desired density:

$$h(x) = \frac{1}{2\pi i} \int_{-\infty}^{\infty} \frac{\partial \varphi}{\partial u}(t, -tx)\, dt \tag{5.15}$$

if the integral converges uniformly in x. Then $h(x) = (d/dx)P[U/V < x]$. The preceding results and computations are used to evaluate the integral in (5.15).

Thus differentiating $\varphi(\cdot, \cdot)$ and simplifying, one gets

$$\frac{\partial \varphi}{\partial u}(t, u) = \frac{e^{\sqrt{2}\alpha i t}[(2)^{1/2} \sin 2(iu)^{1/2}(1 - \alpha t/(8u^2)^{1/2}) + (\alpha i t/(iu)^{1/2}) \cos 2(iu)^{1/2}](i/(8iu)^{1/2})]}{(\cos 2(iu)^{1/2} - (\alpha i t/(2iu)^{1/2}) \sin 2(iu)^{1/2})^{3/2}}$$

Canceling $i = \sqrt{-1}$, and substituting $u = -tx$ one obtains

$$\frac{\partial \varphi}{\partial u}(t, -tx) = \frac{e^{\sqrt{2}\alpha i t}}{(8tx)^{1/2}} \cdot \frac{[(2i)^{1/2} \sinh 2(itx)^{1/2}(1 - \alpha/(8x^2)^{1/2}) - \alpha i (t/x)^{1/2} \cosh 2(itx)^{1/2}]}{(\cosh 2(itx)^{1/2} - \alpha(it/2x)^{1/2} \sinh 2(itx)^{1/2})^{3/2}} \tag{5.16}$$

Observing that $\sqrt{i} = (1 + i)/\sqrt{2}$ and substituting (5.16) into (5.15) yields

$$h(x) = \frac{1}{4\pi i (2x)^{1/2}} \int_{-\infty}^{\infty} \frac{e^{\sqrt{2}\alpha i t}}{\sqrt{t}} \cdot \frac{[(1 + i)(1 - \alpha/(8x^2)^{1/2}) \sinh(1 + i)(2tx)^{1/2} - \alpha i (t/x)^{1/2} \cosh(1 + i)(2tx)^{1/2}]}{(\cosh 2(1 + i)(itx)^{1/2} - (\alpha/2)(t/x)^{1/2}(1 + i) \sinh(1 + i)(2tx)^{1/2})^{3/2}}\, dt \tag{5.17}$$

Set $\beta = (2tx)^{1/2}$ and use the following identities in simplifying the preceding:

$$\begin{aligned} \sinh(1 + i)\beta &= i \cosh \beta \sin \beta + \sinh \beta \cos \beta \\ \cosh(1 + i)\beta &= \cosh \beta \cos \beta + i \sinh \beta \sin \beta \end{aligned} \tag{5.18}$$

After a straightforward, but tedious, computation, one finds

$$\begin{aligned} h(x) &= \frac{1}{4\pi i (2x)^{1/2}} \int_{-\infty}^{\infty} \frac{E + iF}{(A + iB)^{3/2}} \frac{dt}{\sqrt{t}} \\ &= \frac{1}{2^{5/2}\pi} \int_{-\infty}^{\infty} \frac{F - iE}{(A + iB)^{3/2}} \frac{dt}{(tx)^{1/2}} \end{aligned} \tag{5.19}$$

where A, B, E, F are given by

$$A = \cosh(2tx)^{1/2} \cos(2tx)^{1/2} - \frac{\alpha}{2}\left(\frac{t}{x}\right)^{1/2}$$
$$\times (\sinh(2tx)^{1/2} \cos(2tx)^{1/2} - \cosh(2tx)^{1/2} \sin(2tx)^{1/2})$$

$$B = \sin(2tx)^{1/2} \sinh(2tx)^{1/2} - \frac{\alpha}{2}\left(\frac{t}{x}\right)^{1/2}$$
$$\times (\sin(2tx)^{1/2} \cosh(2tx)^{1/2} + \cos(2tx)^{1/2} \sinh(2tx)^{1/2})$$

$$E = C \cos \sqrt{2}\,\alpha t - D \sin \sqrt{2}\,\alpha t$$
$$F = C \sin \sqrt{2}\,\alpha t + D \cos \sqrt{2}\,\alpha t$$

the quantities C and D being those given by (5.11).

It is now necessary to consider the cases (i) $x \geq 0$ and (ii) $x < 0$, by splitting the t integral into $(-\infty, 0)$ and $[0, \infty)$. This is done as follows.

Case i $x \geq 0$. If $t \geq 0$, then $\beta = (2tx)^{1/2} \geq 0$ so that A, B, C, D, E, F are real. Let $F + iE = \rho e^{i\delta}$, $A + iB = re^{i\theta}$ where $\rho^2 = C^2 + D^2 = E^2 + F^2$, $r^2 = A^2 + B^2$ and $\delta = \arctan(E/F)$, $\theta = \arctan(B/A)$, as usual. If $t = -\tau < 0$ $(\tau > 0)$, then one notes that $(A + iB)(-\tau) = \overline{A + iB(\tau)}$ and $C(-\tau) = -iC(\tau)$, $D(-\tau) = iD(\tau)$ so that $(F - iE)(-\tau) = iF(\tau) - E(\tau)$. With these relations (5.19) reduces to

$$h(x) = \frac{1}{2^{5/2}\pi}\left[\int_0^\infty \frac{F - iE}{(A + iB)^{3/2}} \frac{dt}{(tx)^{1/2}} + \int_0^\infty \frac{\overline{F - iE}}{\overline{(A + iB)^{3/2}}} \frac{dt}{(tx)^{1/2}}\right]$$
$$= \frac{1}{\sqrt{8}\,\pi} \int_0^\infty \frac{\rho}{r^{3/2}} \cos\left(\delta - \frac{3\theta}{2}\right) \frac{dt}{(tx)^{1/2}} \tag{5.20}$$

Case ii $x < 0$. If $t < 0$, then $\beta = (2tx)^{1/2} \geq 0$ and A, B, C, D, E, F are again real. If $t \geq 0$, $\tau = -t \leq 0$, then $xt = -x\tau$, and the same relations hold as in the preceding case (i.e., one gets the second integral to be the complex conjugate of the first). Thus (5.20) again holds. The expressions for r, ρ, θ, δ are given by (5.7)–(5.10), where $\alpha^2 = 1$ is used.

Combining these two cases, gives

$$h(x) = \begin{cases} \dfrac{1}{\sqrt{8}\,\pi} \displaystyle\int_0^\infty \frac{\rho(x, t)}{r(x, t)^{3/2}} \cdot \cos(\delta(x, t) - \tfrac{3}{2}\theta(x, t)) \frac{dt}{(tx)^{1/2}}, & x \geq 0 \\ \dfrac{1}{\sqrt{8}\,\pi} \displaystyle\int_{-\infty}^0 \frac{\rho(x, t)}{r(x, t)^{3/2}} \cdot \cos(\delta(x, t) - \tfrac{3}{2}\theta(x, t)) \frac{dt}{(tx)^{1/2}}, & x < 0 \end{cases} \tag{5.21}$$

This is precisely the formula (5.6). Note that ρ, r, δ, θ are *not* symmetric in x

or t. One finds that x is not a singular point in either of the integrals in (5.21) and that the integrals exist uniformly in x. Because of Cramér's theorem noted earlier, this completes the proof.

Remarks. 1. In view of the applicability of the "invariance principle" of Erdös–Kaç [15], one needs the assumption of ε_t being $N(0, 1)$ only for the computations in arriving at (5.12). As seen from the expressions for $X_t = \sum_{i=1}^{t} \alpha^i \varepsilon_i$, it follows that $u_i = \alpha^i \varepsilon_i$ are identically distributed or equivalently that the ε_i are symmetrically distributed about the origin. Thus, it is sufficient for this result to assume that the ε_t are independent, identically, and symmetrically distributed with $E(\varepsilon_t) = 0$, $E(\varepsilon_t^{\,2}) = \sigma^2 < \infty$. For the preceding application of invariance principle, if $\alpha = 1$, the symmetry of the ε_t is unnecessary. In this last case, White has already made a remark ([55], p. 1196) to this effect. The same principle is applicable for the case $|\alpha| < 1$, since then the X_t satisfy the Lindeberg condition and a result of Prokhorov can be applied. (A more general result, of which this is a specialization, appears in [48a].)

2. The preceding analysis shows that the distributional problems of the estimators $\hat{\alpha}_{iN}$, $i = 1, \ldots, k$, in the higher order case, when some roots of (5.2) lie on the unit circle, are essentially more complicated, and need a special treatment. By extending Cramér's theorem to higher dimensions, this work seems capable of a generalization to those cases. However, this aspect of the large sample theory of estimation is at present unexplored.

3. It is of interest to observe that the limit distribution of $\hat{\alpha}_N$ in the stable case ($|\alpha| < 1$) treated by Anderson [2] can be obtained immediately by use of formula (5.15) along with Remark 1. The explicit distribution can be computed with the convenient assumption that the ε_t are $N(0, 1)$. On the other hand, when $|\alpha| > 1$, the invariance principle is not valid as noted by Anderson [2], and this is because the X_t do not satisfy the Lindeberg condition. The latter is necessary for such a principle, as was shown by Prokhorov. In this case, if the ε_t are $N(0, 1)$, using the joint limit characteristic function $\varphi(\cdot, \cdot)$ calculated by White ([55], p. 1193), formula (5.15) yields without difficulty [the right-hand side of (5.15) now can be evaluated easily with a simple contour integral and the Cauchy formula of residues] the fact that $g(n)(\hat{\alpha}_n - \alpha)$ has a limit Cauchy distribution where $g(n) = |\alpha|^n(\alpha^2 - 1)^{-1}$. Thus the present method unifies all the cases, and deserves a further study, as indicated in preceding remarks.

6. EQUIVALENCE OF GAUSSIAN TIME SERIES

6.1. Introduction

The preceding considerations are specialized in this and the final section to the Gaussian case. It is then possible to get more detailed information, and even explicit expressions for the likelihood ratios which are useful in the testing problem. By the important, and by now classical, dichotomy theorem of Hájek and Feldman two arbitrary Gaussian time series (or equivalently two Gaussian measures) are either equivalent or mutually singular. Consequently it will be useful and essential to obtain conditions for equivalence, *preferably on the mean and covariance functions*, and then calculate the likelihood ratios. The former is the topic of this section, and the latter is treated in the following one.

There is considerable literature on the equivalence problem. Specializing one of the two time series to be relatively simple (e.g., Brownian motion, stationary, or Markovian), it is possible to present simplified and useful sufficient conditions. However, one finds it economical as well as revealing to consider the general (not necessarily stationary) time series and reduce the problem to a stage at which other specializations can be immediately made. The use of mathematical tools that are more sophisticated than those of the preceding sections is involved, and this appears unavoidable in all but essentially finite series. The efficacy of the Aronszajn space (i.e., reproducing kernel Hilbert space, or RKHS) techniques in these problems has been demonstrated by Parzen ([40, 41]) and these relatively advanced tools are increasingly accepted in the literature. It is used here to show that the most general covariance can be represented as a *factorable function* on a suitable function space. This does not seem to have been recognized before and it enables an extension of a particular problem considered by Park [37] to the general case. The result is specialized to deduce essentially all the known cases which are often derived by different methods.

6.2. Technical Background: An Outline

In order to consider the general problem just noted, it is necessary to recall a few results, centering around the integral representation of a covariance kernel, from Aronszajn [3], pp. 369–371.

Let $K\colon T \times T \to \mathbb{C}$ be a covariance kernel on a set T so that it is positive definite. Consider the space $\mathscr{H}$ of scalar functions $f\colon T \to \mathbb{C}$ of the form $f = \sum_{i=1}^{n} c_i K(t_i, \cdot)$, where the c_i are scalars and n is an integer. Then $\mathscr{H}$ is a linear space. Define the inner product on $\mathscr{H}$ by the equation

$$\langle f, g\rangle = \sum_{i=1}^{n} \sum_{j=1}^{m} K(s_i, t_j) c_i \bar{d}_j = \sum_{i=1}^{n} c_i \bar{g}(t_i) = \sum_{j=1}^{n} \bar{d}_j f(t_j) \tag{6.1}$$

where f is as just defined and $g = \sum_{j=1}^{m} d_j K(\cdot, t_j) \in \mathscr{H}$. One checks that $(\mathscr{H}, \langle\cdot, \cdot\rangle)$ is an inner product space with the properties: (a) $f \in \mathscr{H} \Rightarrow f(t) = \langle f, K(t, \cdot)\rangle$; (b) $\langle K(s, \cdot), K(t, \cdot)\rangle = K(s, t)$; and (c) $\{K(s, \cdot), s \in T\}$ generates $\mathscr{H}$. If $\mathscr{H}_K$ is the completion of $\mathscr{H}$ under the norm $\|\cdot\|$ derived from $\langle\cdot, \cdot\rangle$, then it is called the *Aronszajn space* (or *RKHS*). The space $\mathscr{H}_K$

is uniquely determined by the kernel K. The general theory of such a space is given by Aronszajn [3], and restated in [40] in the context of time series. The K is called a reproducing kernel because of property (b). For the present purposes it is of great interest to give an integral representation of $K(s, t)$, rather than only (b). This will be realized through the use of the Hellinger–Hahn theorem from the classical theory of Hilbert spaces, adapted to the present case. Such a possibility was first recognized by Hida [23] for a different purpose (cf. also Cramér [11]), and it will be needed here. Let us recall this result.

Suppose from now on that $T \subset \mathbb{R}$ is an interval. Let $\mathscr{H}_t = \overline{sp}\{K(\cdot, s), s \leq t\} \subset \mathscr{H}_K$ be the closed subspace generated by the $K(\cdot, s)$, $s \leq t$, so that for $t_1 < t_2$, $\mathscr{H}_{t_1} \subset \mathscr{H}_{t_2}$. Replacing $\{\mathscr{H}_t, t \in T\}$ by $\{\mathscr{H}_t^* = \bigcup_{s<t} \mathscr{H}_s, t \in T\}$ if necessary, one may assume that for each $t \in T$, $\mathscr{H}_t = \bigcup_{s<t} \mathscr{H}_s$ (denoted $\mathscr{H}_t = \lim_{s\uparrow t} \mathscr{H}_s$) for convenience. Since, if $\mathscr{H}^0 = \bigcap_{t \in T} \mathscr{H}_t$ one may consider $\tilde{\mathscr{H}}_K = \mathscr{H}_K \ominus \mathscr{H}^0$, in what follows one can take $\bigcap_{t \in T} \mathscr{H}_t = \{0\}$ and $\bigcup_{t \in T} \mathscr{H}_t = \mathscr{H}_K$ (by replacing $\mathscr{H}_K$ with $\tilde{\mathscr{H}}_K$). The nontrivial restriction is the assumption: (*) $\mathscr{H}_K$ is *separable*. Relaxing this (as well as taking T to be multidimensional) leads to some real complications in the following analysis. (One has to use the more advanced theory of A. I. Plessner to deal with this case, in lieu of the Hellinger–Hahn theory that follows. This is omitted here.) Let $P(t)$: $\mathscr{H}_K \to \mathscr{H}_t$ be the orthogonal projection on $\mathscr{H}_K$ onto $\mathscr{H}_t$, $t \in T$. Then $\{P(t), t \in T\}$ forms a resolution of the identity in the sense that for any $\cdots < t_1 < t_2 < \cdots < t_n < \cdots$, if $\Delta_k P = P(t_k) - P(t_{k-1})$, then $\sum_k (\Delta_k P)x = x$, $x \in \mathscr{H}_K$ (strong convergence), and the $\Delta_k P$ are mutually orthogonal projections. The Hellinger–Hahn result then is as follows: Given any such resolution of the identity in a separable Hilbert space $\mathscr{H}_K$, there exists a set (nonuniquely but always of the same cardinality) of elements $\{f_n(\lambda), -\infty < \lambda < \infty, -\infty < n < \infty\} \subset \mathscr{H}_K$ such that for each n (i) the "process" $\{f_n(\lambda), \lambda \in \mathbb{R}\}$ has orthogonal increments; (ii) $f_n(\lambda), f_m(\lambda)$ are pairwise orthogonal if $m \neq n$; and (iii) the set of increments $\{f_n(\lambda_2) - f_n(\lambda_1), [\lambda_1, \lambda_2) \subset \Delta = [a, b), n \geq 1\}$ is complete in $\Delta\mathscr{H}_K = \mathscr{H}_b \ominus \mathscr{H}_a$. Then for each n, the function μ_n defined by

$$\mu_n([\lambda_1, \lambda_2)) = \mu_n(\lambda_2) - \mu_n(\lambda_1) = \|f_n(\lambda_2) - f_n(\lambda_1)\|^2 \tag{6.2}$$

can be uniquely extended to the Borel line $(\mathbb{R}, \mathscr{B})$ to give a Lebesgue–Stieltjes measure. Moreover, for each $x \in \mathscr{H}_K$, one can show that

$$\psi_n^x(\lambda) = \frac{d(x, f_n(\cdot))}{d\mu_n}(\lambda), \qquad \text{a.e.} \quad [\mu_n]$$

$$\|x\|^2 = \sum_n \int_{\mathbb{R}} |\psi_n^x(\lambda)|^2 \, d\mu_n(\lambda) \tag{6.3}$$

Now let $x = K(\cdot, t) \in \mathscr{H}_K$ in (6.3), and denote $\psi_n^x(\lambda)$ by $\psi_n(t, \lambda)$, $t \in T$. Let $\mu = \sum_n (1/2^{|n|})\mu_n$. Then μ is a (σ-finite) measure on ($\mathbb{R}$, $\mathscr{B}$) and each μ_n is μ-continuous. If $\tilde{\psi}_n(t, \cdot) = \psi_n(t, \cdot)(d\mu_n/d\mu)^{1/2}$, and $L^2(\mathbb{R}, \mu)$ is the usual L^2-space on ($\mathbb{R}$, $\mathscr{B}$, μ), then $\tilde{\psi}_n(t, \cdot) \in L^2(\mathbb{R}, \mu)$, $t \in T$, and the second part of (6.3) becomes

$$\sum_n \int_{\mathbb{R}} |\psi_n(t, \lambda)|^2 \, d\mu_n = \sum_n \int_{\mathbb{R}} |\tilde{\psi}_n(t, \lambda)|^2 \, d\mu$$

$$= \langle K(\cdot, t), K(\cdot, t) \rangle = K(t, t) \tag{6.4}$$

If $\Psi(t, \lambda) = \{\tilde{\psi}_n(t, \lambda), -\infty < n < \infty\}$ denotes the infinite vector of scalars, if l^2 is the usual Hilbert space of square summable vectors on the integers, and if $L^2(\mathbb{R}, \mu; l^2)$ is the (Hilbert) space of vector- (or l^2-) valued functions on $\mathbb{R}$ which are square integrable [i.e., $f_i = (f_n^i, -\infty < n < \infty)$, $i = 1, 2$, are in this space iff $\int_{\mathbb{R}}((f^i(\lambda), f^i(\lambda))_{l^2})^2 \, d\mu(\lambda) < \infty$, and then $\int_{\mathbb{R}} (f^1(\lambda), f^2(\lambda))_{l^2} \, d\mu$ exists], then (6.4) is equivalent to

$$\int_{\mathbb{R}} (\Psi(t, \lambda), \Psi(s, \lambda))_{l^2} \, d\mu(\lambda) = K(t, s) \tag{6.5}$$

This is the *key* representation needed here. In this form it is useful to describe $\mathscr{H}_K$ more explicitly: Let $\mathscr{F} = \overline{sp}\{\Psi(t, \cdot), t \in T\} \subset L^2(\mathbb{R}, \mu; l^2)$. Let

$$\bar{\mathscr{H}}_K = \{g \in \mathbb{C}^T: g(t) = \int_{\mathbb{R}} (\Psi(t, \lambda), u(\lambda))_{l^2} \, d\mu(\lambda), \text{ for some } u \in \mathscr{F}\} \tag{6.6}$$

If $\|g\|^2 = \int_{\mathbb{R}} (u(\lambda), u(\lambda))_{l^2}^2 \, d\mu(\lambda)$, then one has

(i) $K_t = K(\cdot, t) \in \bar{\mathscr{H}}_K$, by (6.5),
(ii) $\langle g, K_t \rangle = \int_{\mathbb{R}} (\Psi(t, \lambda), u(\lambda))_{l^2} \, d\mu = g(t)$.

Thus $\bar{\mathscr{H}}_K$ is determined by the kernel K and the reproducing property holds. By the uniqueness of these spaces, $\bar{\mathscr{H}}_K = \mathscr{H}_K$, and (6.6) gives the desired representation of $\mathscr{H}_K$. Moreover, for the mapping $\tau: \mathscr{H}_K \to \mathscr{F}$ defined by $\tau: g \mapsto u$ in (6.6), these two spaces are isometrically isomorphic. For a convenient reference, this useful result is stated as follows:

Proposition 6.1. Let K be a covariance function on $T \times T$ such that the associated Aronszajn space $\mathscr{H}_K$ defined by (6.1) is separable. Then there exists a family $\Psi(t, \cdot) = \{\tilde{\psi}_n(t, \cdot), -\infty < n < \infty\}$ and a Lebesgue–Stieltjes measure μ, such that $\Psi(t, \cdot) \in L^2(\mathbb{R}, \mu; l^2)$, $t \in T$, and K is representable by this family as the integral (6.5), and $\mathscr{H}_K$ may be realized by (6.6). The family $\{\Psi(t, \cdot), t \in T\}$ need not be unique, but each such $\{\Psi(t, \cdot), t \in T\}$ has the same cardinality, and each determines K by (6.5) and $\mathscr{H}_K$ by (6.6) uniquely.

Since each $\{\Psi(t, \cdot)\}_{t \in T}$ has the same cardinality, say $n(t)$ $(= n_0)$, $t \in T$, then the unique number n_0, determined by K, is said to be the *multiplicity* of the kernel K $(1 \le n_0 \le \infty)$. Thus if $n_0 = 1$, so that $l^2 = \mathbb{C}$ (or $\mathbb{R}$), the representation of K by (6.5) reduces to what is usually called the *triangular* or *factorable* covariance. In general, the integral may be quite complicated. Thus if $\mathscr{H}_K$ is separable, then K, which always admits the representation (6.5), may be termed a *generalized factorable covariance*. To justify this, let us express (6.5) in the following alternative form.

Let $\mathbb{Z}$ be the integers and $\Omega = \mathbb{R} \times \mathbb{Z}$, $\Sigma = \mathscr{B} \otimes \mathscr{P}$ where $\mathscr{P}$ is the power set of $\mathbb{Z}$, and $\nu = \mu \otimes \alpha$, $\alpha(\cdot)\colon \mathscr{P} \to \mathbb{R}^+$ being the counting measure. Then $\Psi(t, \lambda) = \{\tilde{\psi}_n(t, \lambda), n \in \mathbb{Z}\} = \{\tilde{\Psi}(t, n, \lambda), (n, \lambda) \in \Omega\}$, and

$$\|\Psi(t, \cdot)\|_{2,\nu}^2 = \int_{\mathbb{R}} (\Psi(t, \lambda), \Psi(t, \lambda))_{l^2} \, d\mu(\lambda) = \int_{\Omega} \tilde{\Psi}(t, \omega)\overline{\tilde{\Psi}(t, \omega)} \, d\nu(\omega)$$

Hence (6.5) becomes

$$K(s, t) = \int_{\Omega} \tilde{\Psi}(s, \omega)\overline{\tilde{\Psi}(t, \omega)} \, d\nu(\omega), \qquad t, s \in T \tag{6.7}$$

and if $\mathbb{Z}$ can be replaced by a single point, this K becomes factorable. In this formulation (6.6) reduces to

$$\mathscr{H}_K = \left\{ g \in \mathbb{C}^T\colon g(t) = \int_{\Omega} \tilde{\Psi}(t, \omega)\bar{u}(\omega) \, d\nu(\omega), \, u \in \mathscr{F} \subset L^2(\Omega, \nu) \right\} \tag{6.8}$$

where $\mathscr{F} = \overline{sp}\{\tilde{\Psi}(t, \cdot); t \in T\}$, and $\|g\|^2 = \int_\Omega |u(\omega)|^2 \, d\nu(\omega)$. Moreover, τ: $g \mapsto u(\cdot)$ is an isometry between $\mathscr{H}_K$ and $\mathscr{F}$. This description is used in the following.

6.3. The Equivalence Problem

Let P and Q be a pair of Gaussian measures with means m, n and covariances r, s, respectively. For convenience, denote them as $P = P(m, r)$ and $Q = P(n, s)$. Since a Gaussian process is uniquely determined by its mean and covariance functions, the chain rule for the Radon–Nikodým derivatives implies (by use of the dichotomy theorem of Hájek–Feldman) that $P(m, r) \equiv P(n, s)$ (mutually equivalent) iff $P(m, r) \equiv P(n, r)$ *and* $P(n, r) \equiv P(n, s)$. But $P(n, r) \equiv P(n, s)$ iff $P(0, r) \equiv P(0, s)$ so that $P \equiv Q$ iff $P(m, r) \equiv P(n, r)$ *and* $P(0, r) \equiv P(0, s)$. This fact was first noted and used by Rao and Varadarajan [44]. Hence one can obtain

$$\frac{dQ}{dP} = \frac{dP(n, s)}{dP(m, s)} \cdot \frac{dP(m, s)}{dP(m, r)} \tag{6.9}$$

Also, observe that (by the same chain rule) $P(m, s) \equiv P(n, s)$ iff $P(0, s) \equiv P(\delta, s)$ where $\delta = m - n$. The latter property defines δ as an "admissible translate" of $P(0, s)$ (for a treatment of these translates and related references, see [46c]). Thus $P \equiv Q$ iff (i) $\delta = m - n$ is an admissible translate of $P(0, s)$, and (ii) $P(0, s) \equiv P(0, r)$.

The preceding hypothesis can be translated in terms of conditions on the covariance functions and the Aronszajn spaces $\mathscr{H}_r$, $\mathscr{H}_s$ for the kernels r, s. Also, it may be recalled that the tensor product $\mathscr{H}_r \otimes \mathscr{H}_r$ can be identified with $\mathscr{H}_{r \otimes r}$ where $(r \otimes r)(s_1, s_2; t_1, t_2) = r(s_1, t_1)r(s_2, t_2)$, for $s_i, t_i \in T$, $i = 1, 2$ (cf. Parzen [41]). If $\{X_t, t \in T\}$ is a second-order process with zero mean function and covariance function $r(\cdot, \cdot)$, let $\mathscr{L} = \overline{sp}\{X_t, t \in T\} \subset L^2(P)$ where P is the underlying probability measure. For any $Y \in \mathscr{L}$, let $f(t) = E(Y\bar{X}_t)$, $t \in T$, so that $f = 0$ iff $Y = 0$ a.e. $[P]$. The mapping I: $r(\cdot, t) \mapsto X_t$ from $\mathscr{H}_r$ onto $\mathscr{L}$ is one-to-one and onto. Moreover, it is seen that f is related to Y by this correspondence, i.e., $I(f) = Y$ for $f \in \mathscr{H}_r$. In fact, by definitions of the inner products in $\mathscr{H}_r$ and $\mathscr{L}$,

$$\langle f, r(\cdot, t)\rangle = E(u(f)\overline{u(r(\cdot, t))}) = E(Y\bar{X}_t) = f(t) \tag{6.10}$$

and I: $\mathscr{H}_r \to \mathscr{L}$ defines an isometric isomorphism. Thus $\mathscr{L}$ is separable iff $\mathscr{H}_r$ is. In particular, if the X_t series has the right and left limits in the mean at each t [i.e., $E(|X_t - X_{t+h}|^2)$ has a limit as $h \downarrow 0$ and $h \uparrow 0$], then $\mathscr{L}$ is known to be separable so that $\mathscr{H}_r$ is also. If $r(\cdot, \cdot)$ is continuous on the diagonal of $T \times T$, then $E(|X_t - X_{t+h}|^2) \to 0$ as $h \to 0$ and then these spaces are separable (cf. Cramér [11], p. 6).

The equivalence problem has the following solution (cf. Kallianpur and Oodaira [29], p. 289; Oodaira [36]), in which the preceding facts and notation are utilized.

Theorem 6.1. If $P(m, r)$, $P(n, s)$ are Gaussian measures with mean functions m, n and continuous covariances r, s, then $P(m, r) \equiv P(n, s)$ iff $P(0, r) \equiv P(\delta, s)$ where $\delta = m - n$. Alternatively, the equivalence holds iff (a) $\delta \in \mathscr{H}_r$, (b) $s - r \in \mathscr{H}_r \otimes \mathscr{H}_r \cong \mathscr{H}(r \otimes r)$, and (c) $\mathscr{H}_r = \mathscr{H}_s$, where $\mathscr{H}_r$, $\mathscr{H}_s$ are the Aronszajn spaces, and the equality ($\cong$) in (b) denotes isometric isomorphism.

Using the representation of Proposition 6.1 [or Eq. (6.7)] for r, a more revealing form of this result will be given here, and then various other specializations show its utility. Also it should be observed that $\mathscr{H}_r \otimes \mathscr{H}_r \cong \mathscr{H}_{r \otimes r} \cong \mathscr{L} \otimes \mathscr{L} \subset L^2(P \otimes P) \cong L^2(P) \otimes L^2(P)$, where $L^2(P)$ is the Hilbert space of random variables in which the given process lies (with mean zero and covariance r). Here $I(\mathscr{H}_r) = \mathscr{L}$ in the preceding notation. Also, $\tau(\mathscr{H}_r) = \mathscr{F}$ in (6.8) and so $\mathscr{H}_r \otimes \mathscr{H}_r \cong \mathscr{F} \otimes \mathscr{F}$. Then one has the following result (*hereafter the covariances are always assumed continuous*):

Theorem 6.2. Let $P(m, r)$, $P(n, s)$ be two Gaussian measures on a measurable space. Then they are mutually equivalent iff there exists a measure space (Ω, Σ, ν) and an $R(\cdot, \cdot) \in L^2(\Omega \times \Omega, \Sigma \otimes \Sigma, \nu \otimes \nu)$ such that

(i) $R(\omega, \omega') = \overline{R(\omega', \omega)}$ for almost all (ω, ω'), and for the operator A

on $L^2(\Omega, \Sigma, \nu)$ determined by R [i.e., $Af = \int_\Omega R(\cdot, \omega')f(\omega')\,d\nu(\omega')$ (it is Hilbert–Schmidt)], -1 does not belong to the spectrum $\sigma(A)$ of A;

(ii) $s(u, v) - r(u, v) = \int_\Omega \int_\Omega \tilde{\Psi}(u, \omega)\overline{\tilde{\Psi}(v, \omega')}R(\omega, \omega')\,d\nu(\omega)\,d\nu(\omega')$ where $r(u, v) = \int_\Omega \tilde{\Psi}(u, \omega)\overline{\tilde{\Psi}(v, \omega)}\,d\nu(\omega)$; and

(iii) there exists a $g \in L^2(\Omega, \Sigma, \nu)$ such that

$$m(u) - n(u) = \int_\Omega \tilde{\Psi}(u, \omega)\bar{g}(\omega)\,d\nu(\omega)$$

Here $L^2(\Omega, \Sigma, \nu)$ is taken as the space introduced for (6.7) and in (i) its tensor product is used. [The r here corresponds to the K of (6.7).]

Remark. Specializations and a simple extension are given at the end of the proof. The idea of the proof is to show that these conditions are equivalent to those of the preceding result. Note that means and covariances are prominent now.

Proof. First recall that the abstract theory of tensor products implies that each element V of $\mathscr{F} \otimes \mathscr{F}$ corresponds uniquely to a Hilbert–Schmidt operator on the Hilbert space $\mathscr{F}$ (cf. Schatten [52, pp. 35–36], Theorem 4 and its corollary) and moreover, U is representable by a kernel $K_0(\cdot, \cdot) \in L^2(\mathbb{R} \times \mathbb{R}, \mu \otimes \mu; l^2)$ which belongs to the image of $\mathscr{F} \otimes \mathscr{F}$. Thus if $F \in \mathscr{H}_{r \otimes r}$, then there is a unique K_0 such that

$$F(u, v) = \int_{\mathbb{R}} \int_{\mathbb{R}} (\Psi(u, \lambda), K_0(\lambda, \lambda')\Psi(v, \lambda'))_{l^2}\,d\mu(\lambda)\,d\mu(\lambda') \tag{6.11}$$

and K_0 is Hermitian if F is [i.e., $K_0(\lambda, \lambda')^* = K_0(\lambda', \lambda)$ if $F(u, v)^* = F(v, u)$], where $\{\Psi(u, \cdot), u \in T\}$ is determined by the kernel $\{r(u, t) \colon (u, t) \in T \times T\}$ as in (6.4) and (6.5). [It must be remarked that the actual treatment in the book by Schatten [52] is given if l^2 is replaced by the scalars $\mathbb{C}$. However, the extension needed here presents no problems. If the multiplication of two elements in $\mathbb{C}$ is now interpreted as an appropriate inner product, the proof extends without essential changes. In particular, degenerate kernels will be of the form $\sum_{i=1}^n (a_i f_i(\lambda) \otimes \bar{g}_i(\lambda')$.] Now using (as one may) the measure spaces of (6.7), Eq. (6.11) becomes

$$F(u, v) = \int_\Omega \int_\Omega \tilde{\Psi}(u, \omega)\overline{\tilde{\Psi}(v, \omega')}G(\omega, \omega')\,d\nu(\omega)\,d\nu(\omega') \tag{6.12}$$

for a $G \in L^2(\Omega \times \Omega, \nu \otimes \nu)$, corresponding to K_0, and it is Hermitian symmetric if F is. Let A be the operator on $L^2(\Omega, \nu)$ defined by the kernel R as in the theorem. Then A is Hilbert–Schmidt, by Schatten [52]. (Here R stands for G when $F = s - r$.)

The conditions (i)–(iii) of this theorem may now be shown to be equivalent to those of the preceding result. Conditions (a) and (b) of Theorem 6.1 translate to (ii) and (iii) of the present one. In fact, $\tau(\mathscr{H}_r) = \mathscr{F}$ so that by (6.8) $\tau(\delta) = g \in \mathscr{F}$, and this is simply (iii). Also $s - r \in \mathscr{H}_{r \otimes r}$ is equivalent to saying that $\tau \otimes \tau(s - r) \in \mathscr{F} \otimes \mathscr{F}$ and it is Hermitian. So, as in the representation of (6.12), there is a unique Hermitian $R(\cdot, \cdot) \in L^2(\Omega \times \Omega, \nu \otimes \nu)$ in the range of $\tau \otimes \tau(\mathscr{F} \otimes \mathscr{F})$ such that (iii) holds. Thus only (i) should be shown to be equivalent to (c) of Theorem 6.1, i.e., to $\mathscr{H}_r = \mathscr{H}_s$, or the existence of constants $\alpha, \beta > 0$ such that $\alpha r \ll s \ll \beta r$ by Aronszajn ([3], p. 354), where for a pair of covariances, $K_1 \ll K_2$ means that $K_2 - K_1$ is positive definite.

Since $R \in L^2(\Omega \times \Omega, \nu \otimes \nu)$, the operator $A: L^2(\Omega, \nu) \to L^2(\Omega, \nu)$, defined in (i) with R as its kernel, is Hilbert–Schmidt as noted earlier. Let $\{\lambda_n, n \geq 1\}$ and $\{f_n, n \geq 1\}$ be the eigenvalues and the corresponding eigenfunctions of A. If F_n is defined on T by

$$F_n(t) = \int_\Omega \tilde{\Psi}(t, \omega)\bar{f}_n(\omega)\, d\nu(\omega) \tag{6.13}$$

then by (6.8), $F_n \in \mathscr{H}_r$ and $\tau(F_n) = f_n$, $n \geq 1$. Also $\langle F_n, F_n \rangle_{\mathscr{H}_r} = (f_n, f_n)_{L^2(\nu)} = 1$. Further, since $r(\cdot, t) \in \mathscr{H}_r$, $\langle r(\cdot, t), F_n \rangle = F_n(t)$ and by (ii)

$$\begin{aligned} s(u, v) &= r(u, v) + \int_\Omega \int_\Omega \tilde{\Psi}(u, \omega)\overline{\tilde{\Psi}(v, \omega')}R(\omega, \omega')\, d\nu(\omega)\, d\nu(\omega') \\ &= \int_\Omega \tilde{\Psi}(u, \omega)\left[\overline{\tilde{\Psi}(v, \omega)} + \int_\Omega \overline{\tilde{\Psi}(v, \omega')R(\omega', \omega)}\, d\nu(\omega')\right] d\nu(\omega) \\ &= \int_\Omega \tilde{\Psi}(u, \omega)\overline{g(v, \omega)}\, d\nu(\omega) \end{aligned} \tag{6.14}$$

where g is the function in the brackets of the integrand so that $g \in \mathscr{F} \subset L^2(\Omega, \nu)$. Thus $s(\cdot, v) \in \mathscr{H}_r$ and hence $\mathscr{H}_s \subset \mathscr{H}_r$. By a symmetric argument, $\mathscr{H}_r \subset \mathscr{H}_s$ so that equality holds. Also, $s(\cdot, t)$ and $g(t, \cdot)$ correspond to each other uniquely by (6.8).

To obtain the result that $-1 \notin \sigma(A)$, consider

$$\begin{aligned} \langle s(\cdot, t), F_n \rangle_{\mathscr{H}_r} &= \int_\Omega g(t, \omega)\bar{f}_n(\omega)\, d\nu(\omega) \\ &= F_n(t) + \int_\Omega (A\tilde{\Psi})(t, \omega)\overline{f_n(\omega)}\, d\nu(\omega) \\ &= F_n(t) + \int_\Omega \lambda_n \tilde{\Psi}(t, \omega)\bar{f}_n(\omega)\, d\nu(\omega), \text{ by Fubini} \\ &= F_n(t)(1 + \lambda_n) \end{aligned} \tag{6.15}$$

Thus (c) of Theorem 6.1 is equivalent to the existence of $\alpha, \beta > 0$, such that $\alpha r \ll s \ll \beta r$, and this becomes [by (6.15)]

$$\alpha\langle(r(\cdot, \cdot), F_n), F_n\rangle \le \langle(s(\cdot, \cdot), F_n), F_n\rangle$$
$$\le \beta\langle(r(\cdot, \cdot), F_n), F_n\rangle$$

which may be expressed as

$$\alpha \le 1 + \lambda_n \le \beta, \qquad n \ge 1 \tag{6.16}$$

Since $\alpha, \beta > 0$ neither side can be equal to -1, and $-1 < -1 + \alpha \le \lambda_n \le \beta - 1$ $(\ne -1)$. So $\lambda_n \ne -1$ for all n, and it is not a limit point of λ_n (only zero can be). Thus $-1 \notin \sigma(A)$. Since the reasoning is reversible, this shows $\mathscr{H}_r = \mathscr{H}_s$, and (c) and (i) are equivalent. This completes the proof.

Remark. If the multiplicity of r is 1, then ν and μ are the same measures. So $r(u, v) = \int_{\mathbb{R}} \psi(u, \lambda)\overline{\psi(v, \lambda)}\, d\mu(\lambda)$ in (6.5). Such a function is sometimes called a "factorable covariance." Brownian motion has its covariance, with $\psi(u, \lambda) = \chi_{[0, u)}$, $0 \le u \le 1$, and μ the Lebesgue measure. Note, however, that the multiplicity intervened only in obtaining the general representation of $r(\cdot, \cdot)$ by (6.7) and if (6.7) is given *a priori*, then the rest of the argument of Theorem 6.2 is valid if T is a subset of an arbitrary topological space such that $\mathscr{H}_r$ is separable. In particular, if $\{X_t, t \in T\}$, $T \subset \mathbb{R}^n$, is a Brownian motion with a multiparameter index, then such a representation is valid. For those processes, with the assumption of (6.7), Park [37] has considered this equivalence with essentially the same argument. *The point of the present result is that no such assumption is necessary if* $T \subset \mathbb{R}$ (or for more general T if the corresponding Plessner extended Hellinger–Hahn theory can be used).

The following result is a consequence of Theorem 6.2 if $T = [0, 1]$, $r(u, v) = \min(u, v)$ so that $r(u, v) = \int_T \psi(u, t)\overline{\psi(v, t)}\, dt$ where $\psi(u, t) = \chi_{[0,u)}(t)$. Thus $P(0, r)$ is the Brownian motion (or the Wiener measure).

Corollary 6.1. The Gaussian measures $P(0, r)$ and $P(m, s)$, $r(u, v) = \min(u, v)$, are equivalent iff there exists uniquely an Hermitian $R(\cdot, \cdot) \in L^2(T \times T, dx\, dy)$ such that

(i) $s(u, v) = \min(u, v) + \int_0^u \int_0^v R(\lambda, \lambda')\, d\lambda\, d\lambda'$;
(ii) if A is determined by r on $L^2(T, dt)$, then $-1 \notin \sigma(A)$; and
(iii) there exists a $g \in L^2(T, dt)$ such that $m(t) = \int_0^t g(u)\, du$.

Remark. From (i) and (iii), it follows that $s(\cdot, \cdot)$ is differentiable and $R(\cdot, \cdot)$ and $g(\cdot)$ are then given by $R(u, v) = [\partial^2 s/(\partial u\, \partial v)](u, v)$ and $g(u) = (dm/du)(u)$, a.e. The assertion of this corollary is the basic theorem in the work of Shepp [53] who obtained it by an entirely different method, and with which numerous other results were deduced. For certain related propositions and extensions, see Goloshev [18] and Goloshev and Templ'man [19] as well as Rozanov ([50], p. 59).

In applications certain other forms of Theorem 6.2 are useful. In the following such a result, due to T. S. Pitcher (cf. Root [49], p. 302 and Pitcher [42]), is derived and the present argument is again somewhat different from the original one. For convenience, let $T = [0, 1]$, μ the Lebesgue measure, and the means of both measures zero. [As one sees from the proof, the result holds more generally if T is a topological space, and μ is a finite Stieltjes measure such that $L^2(T, \mu)$ is separable.]

Theorem 6.3. Let $P(0, r)$, $P(0, s)$ with r strictly positive definite and r, $s \in L^2(T \times T, \mu \otimes \mu)$. Let $\hat{R}$, $\hat{S}$ be the operators determined by r and s on $L^2(T, \mu)$. Then $P(0, r) \equiv P(0, s)$ iff $I - \hat{R}^{-1/2}\hat{S}\hat{R}^{-1/2}$ is Hilbert–Schmidt on $L^2(T, \mu)$, or equivalently there is a Hilbert–Schmidt operator J on $L^2(T, \mu)$ such that $\hat{R} - \hat{S} = \hat{R}^{1/2}J\hat{R}^{1/2}$, where r, s are continuous covariances.

Proof. Since $r, s \in L^2(T \times T, \mu \otimes \mu)$, the integral operators $\hat{R}$, $\hat{S}$ defined by the following equations exist:

$$(\hat{R}f)(t) = \int_T r(u, t)f(u)\, d\mu(u)$$
$$(\hat{S}f)(t) = \int_T s(u, t)f(u)\, d\mu(u) \tag{6.17}$$

Moreover $\hat{R}$, $\hat{S}$ are Hilbert–Schmidt on $L^2(T, \mu)$. By a classical theorem of Schmidt (cf. Schatten [52], p. 36), r can be represented as

$$r(u, v) = \sum_{i=1}^{\infty} \alpha_i g_i(u)\overline{g_i(v)}, \qquad g_i \in L^2(T, \mu) \tag{6.18}$$

where $\{\alpha_i > 0\}_1^\infty$ are eigenvalues and $\{g_i\}_1^\infty$ the corresponding eigenfunctions of $\hat{R}$ and $\sum_{i=1}^\infty \alpha_i^2 < \infty$, the series in (6.18) converging in the $L^2(\mu \otimes \mu)$. If $v(n) = \alpha_n$, then $(\mathbb{N}, v)$ is a (σ-finite) measure space and (6.18) may be written compactly as ($g(u, n) = g_n(u)$ here and below):

$$r(u, v) = \int_{\mathbb{N}} g(u, n)\overline{g(v, n)}\, dv(n), \qquad u, v \in T \tag{6.19}$$

so that r is a "factorable" covariance [in the sense of (6.7)]. Hence, as in (6.8), the associated (unique) Aronszajn space $\mathscr{H}_r$ is given by

$$\mathscr{H}_r = \{h\colon h(t) = \int_{\mathbb{N}} g(t, n)\overline{a(n)}\, dv(n), \quad a \in \overline{sp}\{g(t, \cdot), t \in T\} \subset L^2(\mathbb{N}, v)\} \tag{6.20}$$

Further $\|h\|^2 = \int_{\mathbb{N}} |a(n)|^2\, dv(n)$ gives the norm in $\mathscr{H}_r$. To see what this norm really is, consider $\hat{f} = \hat{R}f$ for $f \in \mathscr{L}^2(T, \mu)$. So $\hat{f} \in \mathscr{L}^2(T, \mu)$. It is *claimed* that $\hat{f} \in \mathscr{H}_r$ and $\|\hat{f}\| = \|\hat{R}^{1/2}f\|_2$ where $\hat{R}^{1/2}$ is the positive square root of $\hat{R}$ which exists (and is a bounded but not necessarily a Hilbert–Schmidt operator) so that $\hat{R}^{1/2}(\mathscr{L}^2(T, \mu)) \subset \mathscr{H}_r$. ($\mathscr{L}^2$ becomes L^2 with a.e. identifications.)

Indeed, (6.17)–(6.19) imply

$$\hat{f} = \int_T r(\cdot, t) f(t)\, d\mu(t) = \int_{\mathbb{N}} g(\cdot, n) \int_T \overline{g(t, n)} f(t)\, d\mu(t)\, dv(n)$$

It follows from the Cauchy–Schwarz inequality that $a(\cdot) = \int_T \overline{g(t, \cdot)} f(t)\, d\mu(t) \in L^2(\mathbb{N}, v)$. Hence

$$\begin{aligned}
\|\hat{f}\|^2 &= \int_{\mathbb{N}} |a(n)|^2\, dv(n) \\
&= \sum_{n=1}^{\infty} \alpha_n \int_T \overline{g(t, n)} f(t)\, d\mu(t) \int_T g(u, n) \overline{f(s)}\, d\mu(u) \\
&= \sum_{n=1}^{\infty} \alpha_n \int_T \int_T g(u, n) \overline{g(t, n)} f(t) \overline{f(u)}\, d\mu(u)\, d\mu(t) \\
&= \int_T \int_T r(u, t) f(t) \overline{f(u)}\, d\mu(u)\, d\mu(t) \\
&= \int_T (\hat{R}f)(u) \overline{f(u)}\, d\mu(u) = (\hat{R}f, f) = (\hat{R}^{1/2}f, \hat{R}^{1/2}f)
\end{aligned} \tag{6.21}$$

This shows that $\|\hat{f}\| = \|\hat{R}^{1/2}f\|_2 < \infty$ and $\hat{f} \in \mathscr{H}_r$ as asserted.

Since $\hat{R}$ is strictly positive, $\hat{R}^{-1}$ exists as an unbounded operator ($\hat{R}^{-1/2}$ exists likewise). On the other hand, since $P(0, r)$ is Gaussian, it is known that $\mathscr{H}_r$ is precisely the set of admissible means, i.e., $\mathscr{H}_r = \{m: P(m, r) \equiv P(0, r)\}$ (e.g., see [46c], Prop. 3). It is also known that the set of admissible means of a measure P with mean zero and covariance r (not necessarily Gaussian) is a (generally proper) subset of $\hat{R}^{1/2}(L^2(T, d\mu))$ (cf. e.g., [46d], Theorem 1). It follows, therefore, in the Gaussian case, that $\hat{R}^{1/2}(L^2(T, d\mu)) = \mathscr{H}_r$ and the mapping $\lambda: f \mapsto \hat{f} = \hat{R}f \in \mathscr{H}_r$ satisfies $\|\hat{f}\| = \|\lambda f\| = \|\hat{R}^{1/2}f\|_2 \le \|\hat{R}^{1/2}\|\, \|f\|_2$ so that λ is one-to-one, onto, and continuous. Thus λ^{-1} is also continuous by the closed graph theorem. One should note that $\hat{R}^{1/2}(L^2(T, d\mu))$ is not generally complete in the $L^2(T, \mu)$ norm, but it is so in the new norm $\|\cdot\|$ and is dense in the old norm. With this identification the present result can be deduced from Theorems 6.1 and 6.2.

From the general theory of Aronszajn ([3], p. 372) with each kernel $\Lambda: T \times T \to \mathbb{C}$ such that $\Lambda(\cdot, t) \in \mathscr{H}_r$, $t \in T$, one can uniquely associate a bounded operator A by the equation $(Af)(t) = (f, \Lambda(\cdot, t))$, and A is self-adjoint iff Λ is Hermitian. In particular, the identity operator corresponds to r. Now $P(0, r) \equiv P(0, s)$ iff $r - s \in \mathscr{H}_{r \otimes r}$ and $\mathscr{H}_r = \mathscr{H}_s$ by Theorem 6.1. Hence if S corresponds to $s(\cdot, \cdot)$, then $V = I - S$ acting on $\mathscr{H}_r$ determined by $r - s$ in $\mathscr{H}_{r \otimes r}$ must be Hilbert–Schmidt (cf. Schatten [52], pp. 30, 35). Considering the mappings

$$L^2(T, \mu) \overset{\hat{R}, \hat{S}}{\longrightarrow} L^2(T, d\mu) \overset{\lambda}{\longrightarrow} \mathscr{H}_r \overset{I, S}{\longrightarrow} \mathscr{H}_r \overset{\lambda^{-1}}{\longrightarrow} L^2(T, \mu) \qquad (6.22)$$

one deduces the relations $\hat{R} = \lambda^{-1}\lambda$, $\lambda\lambda^{-1}$ = identity on $\mathscr{H}_r$ (so that $\hat{R}$ corresponds to the identity on $\mathscr{H}_r$), and $\hat{S} = \lambda^{-1} S \lambda$. Thus

$$\hat{R} - \hat{S} = \lambda^{-1}(I - S)\lambda = \lambda^{-1} V \lambda = \hat{R}\lambda^{-1} V \lambda \hat{R} = \hat{R}^{1/2} J \hat{R}^{1/2} \qquad (6.23)$$

where $J = \hat{R}^{1/2}\lambda^{-1} V \lambda \hat{R}^{1/2}$ is bounded and is Hilbert–Schmidt iff $P(0, r) \equiv P(0, s)$. It follows that [since $\hat{R}^{-1}$ exists and using (6.23)]

$$I - B = I - \hat{R}^{-1/2} \hat{S} \hat{R}^{-1/2} = J \qquad (6.24)$$

defines a bounded operator, where B is densely defined and bounded on $L^2(T, \mu)$ and $I - B$ is Hilbert–Schmidt iff the two Gaussian measures are equivalent. This completes the proof.

Remark. An independent proof of this result has been given by Pitcher ([42, 43]) when $T \subset \mathbb{R}^n$ is a compact interval; in this case Theorem 6.1 can therefore be deduced from Theorem 6.3. The fact that the means m, n are here assumed to be zero is not important. [In the contrary case $m - n \in \hat{R}^{1/2}(L^2(T, \mu))$ is also needed.] Since $\hat{R}^{-1}$ is generally unbounded, the consideration of two spaces $L^2(T, \mu)$ and $\mathscr{H}_r$ cannot simply be avoided.

When $P(0, r) \equiv P(0, s)$, the operator $B = \hat{R}^{-1/2} \hat{S} \hat{R}^{-1/2} = I - J$ is densely defined, bounded on $L^2(T, \mu)$, and has a discrete spectrum. This observation is sufficient to present an interesting side result on the simultaneous "diagonalization" of two covariances, without reference to Gaussian measures, in a reasonably general context. Thus the following proposition, based on the properties of B, was established differently by Kadota [27] with further restrictions.

Theorem 6.4. Let T be a set and μ a measure on a σ-algebra of T. Let K_i: $T \times T \to \mathbb{C}$ be Hermitian kernels such that $K_i \in L^2(T \times T, \mu \otimes \mu)$, $i = 1, 2$. If R_i is the integral operator for K_i on $L^2(T, \mu)$, i.e.,

$$(R_i f)(t) = \int_T K_i(s, t) f(s)\, d\mu(s), \qquad f \in L^2(t, \mu) \qquad (6.25)$$

then assume that K_1 is strictly positive definite and K_2 is nonnegative definite such that the operator $B = R_1^{-1/2} R_2 R_1^{-1/2}$ (which is densely defined) has a bounded extension $\hat{B}$ to $L^2(T, \mu)$ and has a discrete spectrum. Let

$\{\alpha_i, i \geq 1\}$ be the eigenvalues and $\{f_n, n \geq 1\}$ the corresponding eigenfunctions of $\hat{B}$ [so that the f_n form a complete orthonormal, or con, set in $L^2(T, \mu)$]. Then

$$K_1(s, t) = \sum_{n=1}^{\infty} (R_1^{1/2} f_n)(s)(R_1^{1/2} f_n)(t) \tag{6.26a}$$

$$K_2(s, t) = \sum_{n=1}^{\infty} \alpha_n (R_1^{1/2} f_n)(s)(R_1^{1/2} f_n)(t) \tag{6.26b}$$

where the series converge in the metric of $L^2(T \times T, \mu \otimes \mu)$.

Proof. By hypothesis $\hat{B} f_n = \alpha_n f_n$, $n \geq 1$. Let $g_n = R_1^{1/2} f_n \in L^2(T, \mu)$ so that $R_2 R_1^{-1} g_n = \alpha_n g_n$, $n \geq 1$. Defining the operator $f_n \otimes \bar{f}_m$ by the equation $(f_n \otimes \bar{f}_m)(h) = (h, f_m) f_n$, for $h \in L^2(T, \mu)$, it follows that $f_n \otimes \bar{f}_n$ is a projection of rank 1, and since $\{f_n, n \geq 1\}$ is *con*, one also has

$$I = \sum_{n=1}^{\infty} f_n \otimes \bar{f}_n \tag{6.27}$$

where I is the identity on $L^2(T, \mu)$. Since $R_1^{-1/2}$ exists, $\{g_n, n \geq 1\}$ are linearly independent and $\sum_{n=1}^{\infty} g_n \otimes \bar{g}_n$ defines a bounded operator. In fact, using $\|\cdot\|_u$ for the operator (or uniform) norm,

$$\begin{aligned}
\sum_{n=1}^{\infty} \|(g_n \otimes g_n)\|_u &= \sum_{n=1}^{\infty} \|R_1^{1/2} f_n \otimes \overline{R_1^{1/2} f_n}\|_u \\
&= \sum_{n=1}^{\infty} \|R_1^{1/2} f_n\|_2^2 \leq \|R_1^{1/2}\|_u \sum_{n=1}^{\infty} \|f_n \otimes \bar{f}_n\|_u \\
&= \|R_1\|_u^{1/2} < \infty
\end{aligned} \tag{6.28}$$

So the following series converge in norm and the operations are valid:

$$\begin{aligned}
\sum_{n=1}^{\infty} g_n \otimes \bar{g}_n &= \sum_{n=1}^{\infty} R_1^{1/2} (f_n \otimes \bar{f}_n) R_1^{1/2} \\
&= R_1^{1/2} \left(\sum_{n=1}^{\infty} f_n \otimes \bar{f}_n \right) R_1^{1/2} = R_1
\end{aligned} \tag{6.29}$$

by (6.27) and the elementary properties of the tensor operation (cf. Schatten [52], p. 7). This gives (6.26a).

For the second series (6.26b), use the spectral theorem for $\hat{B}$. Thus, if $\beta = \|\hat{B}\|_u$, the classical spectral integral gives

$$\hat{B} = \int_{-\beta}^{\beta} \alpha \, dE_\alpha = \sum_{n=1}^{\infty} \alpha_n (f_n \otimes \bar{f}_n) \tag{6.30}$$

where $\{f_n \otimes \bar{f}_n, n \geq 1\}$ are mutually orthogonal one-dimensional projections, and the series on the right converges strongly. So for any $h \in L^2(T, \mu)$,

$$R_1^{-1/2} R_2 R_1^{-1/2} h = \sum_{n=1}^{\infty} \alpha_n (f_n \otimes \bar{f}_n) h \tag{6.31}$$

Replacing h by $R_1^{1/2}\tilde{h}$ in (6.31) and noting that $R_1^{1/2}(L^2(T, \mu))$ is a dense subspace of $L^2(T, \mu)$, one has on multiplying by $R_1^{1/2}$ on the left,

$$R_2 \tilde{h} = \sum_{n=1}^{\infty} \alpha_n R_1^{1/2} (f_n \otimes \bar{f}_n) R_1^{1/2} \tilde{h} \tag{6.32}$$

But $R_1^{1/2}(f_n \otimes \bar{f}_n) R_1^{1/2} = (R_1^{1/2} f_n \otimes \overline{R_1^{1/2} f_n})$. Using this in (6.32) and noting that $\tilde{h}$ in $L^2(T, \mu)$ is arbitrary, it follows that

$$R_2 = \sum_{n=1}^{\infty} \alpha_n R_1^{1/2} f_n \otimes \overline{R_1^{1/2} f_n} \tag{6.33}$$

Because of (6.25), (6.29) and (6.33) are just (6.26a) and (6.26b). This completes the proof.

Remark. If μ concentrates on a finite set of points of T, then $L^2(T, \mu)$ is finite dimensional and K_1, K_2 are matrices. In this case the preceding result reduces to the classical simultaneous diagonalization of two such covariance matrices. Note that (T, μ) is an arbitrary measure space here. Also, on subtracting (6.33) from (6.29) one has

$$R_1 - R_2 = \sum_{n=1}^{\infty} (1 - \alpha_n)(g_n \otimes \bar{g}_n) \tag{6.34}$$

But

$$(g_n, g_n)_{\mathscr{H}_1} = (R_1^{1/2} g_n, R_1^{1/2} g_n)_{L^2} = (f_n, f_m)_{L^2} = \delta_{mn}$$

where $\mathscr{H}_1 = \mathscr{H}_{K_1}$ [cf. (6.21)] so that on $R_1^{1/2}(L^2(T, \mu)) = \mathscr{H}_1$, $\{g_n, n \geq 1\}$ are orthonormal and hence so are $\{g_n \otimes \bar{g}_m\}$ on $\mathscr{H}_{K_1 \otimes K_1}$. Since $R_1 - R_2$ is Hilbert–Schmidt on $\mathscr{H}_1$, it follows from (6.34) that $\sum_{n=1}^{\infty} (1 - \alpha_n)^2 < \infty$. If $T \subset \mathbb{R}$ is a compact interval and K_1, K_2 are continuous, then the preceding statements are related to the classical Mercer expansions, and thus the result includes the main theorem of Kadota [27].

7. LIKELIHOOD RATIOS AND TESTS

In the preceding section various conditions for the equivalence of a pair of Gaussian measures are given. They are of interest when these are translated to deriving the likelihood ratios which then will enable an application to the theory of testing hypotheses. This section is devoted to this problem.

Let $\{X_t, t \in T\}$, $T \subset \mathbb{R}$, be a Gaussian time series with m and r as the mean and covariance functions or with n, s as the alternate functions describing the same situation and such that r is strictly positive definite. Let these underlying probabilities be denoted as $P(m, r)$ and $P(n, s)$, respectively, as in the preceding section. Let μ be the Lebesgue measure on T, and suppose that $P(m, r) \equiv P(n, s)$. Because of (6.9), one may initially assume that $m = n = 0$ (this is removed later). Let $r, s \in L^2((T \times T), \mu \otimes \mu)$ and $\hat{R}$ and $\hat{S}$ be the integral operators determined by r and s on $L^2(T, \mu)$. Then, on $L^2(T, \mu)$, $B = \hat{R}^{-1/2}\hat{S}\hat{R}^{-1/2}$ is densely defined, bounded, and has a pure point spectrum by Theorems 6.3 and 6.4, since $P(0, r) \equiv P(0, s)$. Let $\{\alpha_n, n \geq 1\}$ be the eigenvalues and $\{f_n, n \geq 1\}$ the corresponding eigenfunctions of $\hat{B}$, so that (cf. Theorem 6.4) if $g_n = \hat{R}^{-1/2} f_n [\sum_{n=1}^{\infty} (1 - \alpha_n)^2 < \infty]$,

$$r(u, v) = \sum_{n=1}^{\infty} g_n(u)\bar{g}_n(v), \qquad s(u, v) = \sum_{n=1}^{\infty} \alpha_n g_n(u)\bar{g}_n(v) \tag{7.1}$$

Also the Hermitian function $F = r - s$ $(\in \mathscr{H}_{r \otimes r})$ defines a Hilbert–Schmidt operator A on $L^2(T, \mu)$. Define the functions (r, s being continuous)

$$\xi_n = \int_T X_t f_n(t)\, d\mu(t), \qquad n \geq 1, \text{ (Riemann integral)} \tag{7.2}$$

Then $\xi_n \in \overline{sp}\{X_t, t \in T\} \subset L^2(P(0, r)) \cap L^2(P(0, s))$. In fact the ξ_n are normal, mutually independent, and of means zero. The variance of ξ_n is 1 under $P(0, r)$ and α_n under $P(0, s)$. Moreover,

$$X_t = \sum_{n=1}^{\infty} \xi_n g_n(t), \qquad t \in T \tag{7.3}$$

the series converging in mean [relative to both $P(0, r)$ and $P(0, s)$ measures]. By (7.1), it follows that $E(X_u \bar{X}_v) = r(u, v)$ or $s(u, v)$ accordingly as $P(0, r)$ or $P(0, s)$ is used. This representation is now classical, having been originally obtained by M. Kaç and A. J. F. Siegert in 1947 and effectively used in the time series analysis problems by Grenander [20]. The $\{\xi_n, n \geq 1\}$ are called "observable coordinates," and are useful for most of the computations in deriving the likelihood ratio $dP(0, s)/dP(0, r)$. After obtaining the latter expression, the result will be summarized in Theorem 7.1. Let $\mathscr{T}$ be the Borel σ-field of T.

Let $\mathscr{B}_n = \sigma(\xi_i, 1 \leq i \leq n)$, the σ-field generated by the ξ_i shown. If $\mathscr{B} = \sigma(\bigcup_n \mathscr{B}_n)$ and $\tilde{\mathscr{B}}$ is the completion of $\mathscr{B}$ relative to $P(0, r)$ [and hence also $P(0, s)$], then $\{X_t, t \in T\}$ is measurable for $\tilde{\mathscr{B}} \otimes \mathscr{T}$. If $\mathscr{G} = \sigma(X_t, t \in T)$, then $\mathscr{B} \subset \mathscr{G}$ and one verifies that $P(0, r) \equiv P(0, s)$ on $\mathscr{G}$ iff this holds on $\mathscr{B}$. (This was noted in [46a]; see also Rozanov, [50], p. 53.) Hence one may and does restrict the analysis to $\mathscr{B}$. If $p = dP(0, s)/dP(0, r)$ on $\mathscr{B}$, and $p_n = E^{\mathscr{B}_n}(p)$, the conditional expectation of p relative to $\mathscr{B}_n$, then $p_n \to p$ a.e.

and in $L^1(P(0, r))$ (and then also $1/p_n \to 1/p$ a.e. since $p > 0$ a.e.), by the classical martingale convergence theorem [13]. Such an application was indicated early by Striebel [54]. Thus the problem of obtaining an explicit formula for p will be solved if p_n can be computed on $\mathscr{B}_n$, and this is accomplished as follows. Here and in the following a.e. refers to either (hence both) of the measures.

Let f_n^1 and f_n^2 be the n-dimensional (normal) densities of $(\xi_1, \ldots, \xi_n)$ relative to $P(0, r)$ and $P(0, s)$, respectively. Using the result following (7.2), one obtains

$$p_n(\omega) = \frac{f_n^2}{f_n^1}(\omega) = \left[\prod_{i=1}^{n} \alpha_i\right]^{-1/2} \exp\left[-\tfrac{1}{2}\sum_{i=1}^{n} \xi_i^2(\omega)\left(\frac{1}{\alpha_i} - 1\right)\right], \quad n \geq 1 \tag{7.4}$$

Since $\sum_{n=1}^{\infty} (1 - \alpha_n)^2 < \infty$, by the equivalence, and then $p_n \to p$ a.e., one finds that the limit on the right-hand side of (7.4) exists a.e. But the limit of the product cannot be written as the product of limits. If $\sum_{n=1}^{\infty} |1 - \alpha_n| < \infty$, then $\prod_{n=1}^{\infty} \alpha_n$ exists (called the Fredholm determinant of $\hat{B}$) and so the second limit also must exist. When this stronger condition obtains, $P(0, r)$ and $P(0, s)$ are sometimes called *strongly equivalent*. But this need not be true under the present hypothesis. Nevertheless one can employ a "generalized" (or "regularized") Fredholm determinant theory (cf. Gohberg and Kreĭn [16], p. 166) and simplify the limit of the right-hand side of (7.4). Thus consider

$$p_n(\omega) = \left[\prod_{i=1}^{n} (1 - (1 - \alpha_i))e^{(1-\alpha_i)}\right]^{-1/2}$$
$$\times \exp\left\{-\tfrac{1}{2}\sum_{i=1}^{n}\left[\xi_i^2(\omega)\left(\frac{1 - \alpha_i}{\alpha_i}\right) - (1 - \alpha_i)\right]\right\} \tag{7.5}$$

The fact that $\sum_{i=1}^{\infty} (1 - \alpha_i)^2 < \infty$ implies that

$$\prod_{i=1}^{\infty} (1 - (1 - \alpha_i))e^{(1-\alpha_i)} = D_1^{-1}$$

exists [and $D_1 = \det(\hat{B}\hat{R}^{-1})$ is the regularized determinant]. So the exponential factor in (7.5) must also converge a.e. Thus

$$p(\omega) = D_1^{-1/2} \exp\left\{-\tfrac{1}{2}\sum_{i=1}^{\infty}\left[\xi_i^2(\omega)\left(\frac{1 - \alpha_i}{\alpha_i} - (1 - \alpha_i)\right)\right]\right\}, \quad \text{a.e.} \tag{7.6}$$

To remove the assumption of zero means, one has to use Eq. (6.9). So $P(m, r) \equiv P(n, s)$ iff $P(n - m, s) \equiv P(0, s)$ and $P(0, r) \equiv P(0, s)$, and then

$$\frac{dP(n, s)}{dP(m, r)} = \frac{dP(n - m, s)}{dP(0, s)} \cdot \frac{dP(0, s)}{dP(0, r)}, \quad \text{a.e.} \tag{7.7}$$

But, as is known, this occurs iff $n - m \in \mathscr{H}_s$, $P(0, s) \equiv P(0, r)$, and

$$\frac{dP(n-m, s)}{dP(0, s)} = \exp(Y - \tfrac{1}{2}E(Y^2)) \tag{7.8}$$

for a unique $Y \in \mathscr{L} = \overline{sp}\{X_t, t \in T\} \subset L^2(P(0, s)) \cap L^2(P(0, r))$ (cf. Pitcher [43]). In particular, since T is locally compact, if r and s are continuous and vanish at "∞" of $T \times T$, then (cf. Pitcher [43]; and also [46c] Theorem 1c) there exists a regular bounded Borel measure β on T such that

$$(n - m)(t) = \int_T s(t, u)\, d\beta(u), \qquad Y = \int_T X_t\, d\beta(t) \tag{7.9}$$

where the second symbol denotes the sample function integral. Hence (7.6)–(7.9) imply, on denoting

$$\begin{aligned} D_2 &= \exp(E(Y^2)) \\ &= \exp\left[\int_T \int_T E(X_u \bar{X}_t)\, d\beta(u)\, d\beta(t)\right] = \exp\left[\int_T \int_T s(u, v)\, d\beta(u)\, d\beta(v)\right] \end{aligned}$$

that

$$\begin{aligned} \frac{dP(n, s)}{dP(m, r)} &= (D_1 D_2)^{-1/2} \\ &\quad \times \exp\left\{\int_T X_t\, d\beta(t) - \tfrac{1}{2}\sum_{i=1}^{\infty}\left[\xi_i^2\left(\frac{1-\alpha_i}{\alpha_i}\right) - (1 - \alpha_i)\right]\right\} \end{aligned} \tag{7.10}$$

This may be summarized as follows:

Theorem 7.1. Let $P(m, r) \equiv P(n, s)$ be two equivalent Gaussian probability measures with means m, n and continuous covariances r, s where r is strictly positive definite for the time series $\{X_t, t \in T\}$, and where $T \subset \mathbb{R}$ is an interval. Then the likelihood ratio $dP(n, s)/dP(m, r)$ of this series is given by (7.10) when r, s vanish at "∞" of $T \times T$. The "best" critical region for testing the hypothesis H_0: $P = P(m, r)$ versus H_1: $P = P(n, s)$ is defined by the set A_k where

$$A_k = \left\{\omega: \int_T X_t(\omega)\, d\beta(t) - \tfrac{1}{2}\sum_{i=1}^{\infty}\left[\xi_i^2(\omega)\left(\frac{1-\alpha_i}{\alpha_i}\right) - (1 - \alpha_i)\right] \geq k\right\} \tag{7.11}$$

and k is chosen such that under H_0, $P(A_k) = \delta$, $0 < \delta < 1$, a prescribed "level" of the test.

The last part of the theorem is based on the well-known Grenander extension of the Neyman–Pearson lemma (cf. Grenander [20], p. 210), and the set A_k in (7.11) is immediate from the form of the likelihood ratio (7.10).

These results admit extensions to vector-valued processes, although there is considerable notational complexity (for related work, see also Goloshev and Templ'man [19], Yaglom [56]; Kailath and Weinert [28]). If one of the measures is specialized to a Markov process, for instance, then the expression (7.11) can be simplified as in the works of Goloshev [18], Shepp [53], Hájek [20a], and Rozanov [50], as well as Grenander [20] to begin with. Interested readers will find several examples in these papers which illustrate the foregoing theory.

REFERENCES

1. Anderson, T. W., "The Statistical Analysis of Time Series." Wiley, New York, 1971.
2. Anderson, T. W., On asymptotic distribution of estimates of parameters of stochastic difference equations, *Ann. Math. Statist.* **30** (1959), 676–687.
3. Aronszajn, N., Theory of reproducing kernels, *Trans. Amer. Math. Soc.* **68** (1950), 337–404.
4. Besicovitch, A. S., "Almost Periodic Functions." Cambridge Univ. Press, London 1932.
5. Bhagavan, C. S. K., Nonstationary Processes, Spectra and Some Ergodic Theorems, Thesis, Andhra Univ. Press, Waltair, India (1974).
6. Chandrasekhar, S., Stochastic problems in physics and astronomy, *Rev. Modern Phys.* **15** (1943), 1–89.
7. Coddington, E. A., and Levinson, N., "Theory of Ordinary Differential Equations." McGraw-Hill, New York, 1955.
8. Cramér, H., "Random Variables and Probability Distributions." Cambridge Univ. Press, London and New York, 1937.
9. Cramér, H., "Mathematical Methods of Statistics." Princeton Univ. Press, Princeton, New Jersey, 1946.
10. Cramér, H., A contribution to the theory of stochastic processes, *Proc. Berkeley Symp. Math. Statist. and Probability, 2nd* (J. Neyman, ed.), pp. 329–339. Univ. of California Press, Berkeley, California, (1951).
11. Cramér, H., "Structural and Statistical Problems for a Class of Stochastic Processes." Princeton Univ. Press, Princeton, New Jersey, 1971.
12. Doob, J. L., The Brownian movement and stochastic equations, *Ann. of Math.* **43** (1942), 351–369.
13. Doob, J. L., "Stochastic Processes." Wiley, New York, 1953.
14. Dym, H., Stationary measures for the flow of a linear differential equation driven by white noise, *Trans. Amer. Math. Soc.* **123** (1966), 130–164.
15. Erdös, P., and Kaç, M., On certain limit theorems in the theory of probability, *Bull. Amer. Math. Soc.* **52** (1946), 292–302.
16. Gohberg, I. C., and Kreĭn, M. G., "Introduction to the Theory of Linear Nonselfadjoint Operators." Amer. Math. Soc., Providence, Rhode Island, 1969 (English translation).
17. Goldstein, J. A., An existence theorem for linear stochastic differential equations, *J. Differential Equations* **3** (1967), 78–87.
18. Goloshev, Ju. A., Gaussian measures equivalent to Gaussian Markov measures, *Soviet Math. Dokl.*, **7** (1966), 48–52.
19. Goloshev, Ju. A., and Templ'man, A., On equivalence of measures corresponding to Gaussian vector valued function, *Soviet Math. Dokl.* **10** (1969), 228–232.
20. Grenander, U., Stochastic processes and statistical inference, *Ark. Mat.* **1** (1950), 195–277.

20a. Hájek, J., On linear statistical problems in stochastic processes, *Czeckoslovak Math. J.* **12** (1962), 404–444.
21. Hardy, G. H., "Divergent Series." Oxford Univ. Press, London and New York, 1949.
22. Hartley, H. O., Second order autoregressive schemes with time-trending coefficients, *J. Roy. Statist. Soc. Ser. B* **14** (1952), 229–233.
23. Hida, T., Canonical representations of Gaussian processes and their applications, *Mem. Coll. Sci. Univ. of Kyoto Ser. A* **38** (1960), 109–155.
24. Hurwicz, L., Variable parameters in stochastic processes: trend and seasonality, *in* "Statistical Inference in Dynamic Economic Models" (T. C. Koopmans, ed.), pp. 329–344. Wiley, New York, 1950.
25. Ince, E. L., "Ordinary Differential Equations." Longmans, London, 1926.
26. Jordan, C., "Calculus of Finite Differences." Chelsea, New York, 1950.
27. Kadota, T. T., Simultaneous diagonalization of two covariance kernels and application to second order stochastic processes, *SIAM J. Appl. Math.* **15** (1967), 1470–1480.
28. Kailath, T., and Weinert, H. L., An RKHS approach to detection and estimation problems—Part II: Gaussian signal detection, *IEEE Trans. on Information Theory* **IT-21** (1975), 15–23.
29. Kallianpur, G., and Oodaira, H., The equivalence and singularity of Gaussian measures, *Proc. Symp. Time Series Analysis* (M. Rosenblatt, ed.), pp. 279–291. Wiley, New York, 1963.
30. Kampé de Fériet, J., and Frenkiel, F. N., Correlation and spectra for nonstationary random functions, *Math. Comput.* **16** (1962), 1–21.
31. Kelly, E. J., Reed, I. S., and Root, W. L., The detection of radar echoes in noise, I, II, *SIAM J. Appl. Math.* **8** (1960), 309–341, 481–507.
32. Loève, M., "Probability Theory." Van Nostrand–Reinhold, Princeton, New Jersey, 1954.
33. Mann, H. B., and Wald, A., On the statistical treatment of linear stochastic difference equations, *Econometrica*, **11** (1943), 173–220.
34. Marsaglia, G., Iterated limits and the central limit theorem for dependent random variables, *Proc. Amer. Math. Soc.* **5** (1954), 987–991.
35. Nagabhushanam, K., The primary process of a smoothing relation, *Ark. Mat.* **1** (1951), 421–488.
36. Oodaira, H., The Equivalence of Gaussian Stochastic Processes, Unpublished Ph.D. thesis, Michigan State Univ. (1963). (Referred to in [28] and [37].)
37. Park, W. J., On the equivalence of Gaussian processes with factorable covariance functions, *Proc. Amer. Math. Soc.* **32** (1972), 275–279.
38. Parzen, E., Spectral analysis of asymptotically stationary time series, *Bull. Inst. Internat. Statist.* **39** (livraison 2) (1962), 87–103.
39. Parzen, E., A central limit theorem for multilinear stochastic processes, *Ann. Math. Statist.* **28** (1957), 252–256.
40. Parzen, E., An approach to time series, *Ann. Math. Statist.* **32** (1961), 951–989.
41. Parzen, E., Probability density functionals and reproducing kernel Hilbert spaces, *Proc. Symp. Time Series Analysis* (M. Rosenblatt, ed.), pp. 155–169. Wiley, New York, 1963.
42. Pitcher, T. S., An integral expression for the log likelihood ratio of two Gaussian processes, *SIAM J. Appl. Math.* **14** (1966), 228–233.
43. Pitcher, T. S., Likelihood ratios of Gaussian processes, *Ark. Mat.* **4** (1959), 35–44.
44. Rao, C. R., and Varadarajan, V. S., Discrimination of Gaussian processes, *Sankhyā Ser. A* **25** (1963), 303–330.
45. Rao, M. M., Specification of some stochastic models, Unpublished M.Sc. thesis, Madras Univ., 1955.
46a. Rao, M. M., Inference in stochastic processes—I, *Teor. Veroyatnost.* **8** (1963), 282–298.

46b. Rao, M. M., Inference in stochastic processes—II, *Z. Wahrsch. Verw Geb.* **5** (1966), 317–335.
46c. Rao, M. M., Inference in stochastic processes—V, *Sankhyā, Ser. A* **37** (1975), 538–549.
46d. Rao, M. M., Inference in stochastic processes—VI, *Proc. Symp. Multivariate Anal., 4th* (P. R. Krishnaiah, ed.), pp. 311–324. North-Holland Publ., Amsterdam, 1977.
47. Rao, M. M., Consistency and limit distributions of estimators of parameters in explosive stochastic difference equations, *Ann. Math. Statist.* **32** (1961), 195–218.
48. Rao, M. M., Representation theory of multidimensional generalized random fields, *Proc. Symp. Multivariate Anal. 3rd* (P. R. Krishnaiah, ed.), pp. 411–436. Academic Press, New York, 1969.
48a. Rao, M. M., Asymptotic distribution of an estimator of the boundary parameter of an unstable process, *Ann. Statist.* **6** (1978), 185–190.
49. Root, W. L., Singular Gaussian measures in detection theory, *Proc. Symp. Time Series Anal.* (M. Rosenblatt, ed.), pp. 292–315. Wiley, New York, 1963.
50. Rozanov, Yu. A., "Infinite Dimensional Gaussian Distributions" (English translation). Amer. Math. Soc., Providence, Rhode Island, 1971.
51. Rubin, H., Consistency of maximum likelihood estimates in the explosive case, *in* "Statistical Inference in Dynamic Economic Models" (T. C. Koopmans, ed.), pp. 356–364. Wiley, New York, 1950.
52. Schatten, R., "Norm Ideals of Completely Continuous Operators." Springer-Verlag, Berlin, 1960.
53. Shepp, L. A., Radon-Nikodým derivatives of Gaussian measures, *Ann. Math. Statist.* **37** (1966), 321–354; correction, *Ann. Prob.* **5** (1977), 315–317.
53a. Stigum, B. P., Asymptotic properties of dynamic stochastic parameter estimates (III), *J. Multivariate Anal.* **4** (1974), 351–381.
54. Striebel, C. T., Densities for stochastic processes, *Ann. Math. Statist.* **30** (1959), 559–567.
55. White, J. S., The limiting distribution of the serial correlation coefficient in the explosive case, *Ann. Math. Statist.* **29** (1958), 1188–1197.
56. Yaglom, A. M., On the equivalence and perpendicularity of two Gaussian probability measures in function space, *Proc. Symp. Time Series Anal.* (M. Rosenblatt, ed.), pp. 327–346. Wiley, New York, 1963.
57. Yoshizawa, T., "Stability Theory and the Existence of Periodic Solutions and Almost Periodic Solutions," Appl. Math. Ser. 14. Springer-Verlag, New York, 1975.

Nonparametric Repeated Significance Tests†

PRANAB KUMAR SEN

DEPARTMENT OF BIOSTATISTICS
UNIVERSITY OF NORTH CAROLINA AT CHAPEL HILL
CHAPEL HILL, NORTH CAROLINA

1. INTRODUCTION

Often, the individuals (experimental units) constituting the sample in a study do not enter into the scheme all at the same point of time. For example, in a (comparative) *clinical trial* intended for studying the (relative)

† Work partially supported by U.S. National Heart, Lung, and Blood Institute, Contract NIH-NHLBI-71-2243 from the National Institutes of Health.

ISBN 0-12-4266 01-0

effectiveness of two (or more) competing drugs for a particular treatment, patients undergoing the treatment are alloted (at random) to one of the drugs and statistical conclusions are drawn on the basis of the responses of these drugs. Now, as is the case with many infrequently occurring diseases, there may not be a steady flow of arrivals of patients in the clinic for the particular treatment, so that the recruitment of the desired number of patients in the clinical trial may require a considerable amount of *waiting time*. On the other hand, drawing a reliable and valid statistical inference in such a case, requires setting, in advance, a *target sample size* (usually moderately or genuinely large) on which to base the statistical tests. Thus, a *terminal* (fixed-sample) *test* based on the target sample size may involve a considerable amount of waiting time before it can be administered. From time and cost considerations, under such a sequential entry plan of experimental units in the scheme, it may be advisable to monitor the experiment from the very beginning with the objective of a possible early statistical decision (prior to attaining the target sample size): A *time-sequential* procedure which takes into account the cumulated statistical evidence as more and more units are observable is thought to be ideal. This quasi-sequential setup, through updating of statistical information, allows a possible early termination of experimentation (along with appropriate statistical decision), which results in savings of time and cost of experimentation. On the other hand, such a time-sequential procedure involves *repeated tests of significance* on an increasing dimension of (dependent) observations, and hence, concern for extra manipulations for a valid statistical analysis of the integrated scheme.

Although the possibility of employing a truncated version of the classical sequential probability ratio tests (SPRT) remains open in this setup, its scope of applicability is somewhat limited by the necessity of the formulation of suitable simple null and simple alternative hypotheses in strict parametric models. For composite hypotheses or for nonparametric models, the SPRT encounters considerable difficulties. In this study, we are primarily concerned with a broad class of *nonparametric repeated significance tests* which are valid for a broad class of distributions, adaptive for a wider class of statistical problems, and, at the same time, particularly useful in the setup of the preceding paragraph. In passing, we may briefly refer to the somewhat scattered work done in this area. In the parametric model, Armitage (1975) has considered some repeated significance tests based on the sample proportions and means, and in a rather nonmathematical setup, discussed their role in clinical trials in an elaborate manner. For testing the hypothesis of symmetry of a distribution around a specified point (arising in the so-called one-sample location model), Miller (1970, 1972) has studied a version of a nonparametric test based on a partial sequence of the Wilcoxon signed-rank statistics; similar tests based on a general class of one-sample rank order statistics have been considered by Sen (1974a). For the parametric model, for normally distributed random variables, Samuel-Cahn (1974a,b,c) has studied repeated significance tests for the mean in both the situations in which the variance may or may not be specified. We consider here two broad types of repeated significance tests: first, the genuinely distribution-free tests based on a broad class of nonparametric statistics; and second, the asymptotically distribution-free tests based on a broad class of parametric statistics but having asymptotically nonparametric behavior. Together, these procedures embrace a broad class of problems and provide a unified solution.

Along with the preliminary notions, genuinely distribution-free procedures are introduced in Section 2. Section 3 deals with the asymptotically distribution-free procedures. Section 4 describes the procedures for various problems of real interest. Section 5 is devoted to the asymptotic distribution theory of the allied test statistics under appropriate null hypotheses, and parallel results for (local) alternative hypotheses are considered in Section 6. Section 7 is concerned with the study of the (asymptotic) relative efficiency of the repeated significance tests as well as the conventional terminal (fixed-sample) tests. The concluding section discusses (mostly by way of informative remarks) some related sequential nonparametric tests and presents some observations of general interest.

2. TYPE A REPEATED SIGNIFICANCE TESTS

Keeping the setup of the preceding section in mind, we conceive of a sequence $\mathbf{X} = \{X_i, i \geq 1\}$ of independent random variables (rv), defined on a probability space $(\Omega, \mathscr{A}, P)$, and, for every $n \geq 1$, we denote $\mathbf{X}^{(n)} = (X_1, \ldots, X_n)$ and the corresponding probability measure by $P^{(n)}$. We are concerned with the situation in which the target sample size N (usually large) is fixed in advance. Let $\{T_n = T(\mathbf{X}^{(n)}); n \geq 1\}$ be a sequence of (real-valued) statistics, so that, under our scheme, we are primarily interested in the partial sequence

$$\{T_n: n_0 \leq n \leq N\}, \qquad n_0 \geq 1 \tag{2.1}$$

where n_0 designates the initial sample size with which the statistical screening/monitoring procedure is installed on the experiment. We frame the null hypothesis to be tested as

$$H_0: \quad P \in \mathscr{P}_0 \subset \mathscr{P} \tag{2.2}$$

where $\mathscr{P}$ is a family of probability measures containing P as a member, $\mathscr{P}_0$ is a subset of $\mathscr{P}$, and we are interested in the set of alternative hypotheses

$$H_1: \quad P \in \mathscr{P}_1 \subset \mathscr{P} \backslash \mathscr{P}_0 \tag{2.3}$$

Our problem is to formulate suitable statistical tests for H_0 versus H_1 based on the partial sequence $\{T_n;\ n_0 \le n \le N\}$, in (2.1), in such a way that if at any early stage ($k \le n$), the accumulated statistical evidence evokes a clear decision in favor of H_1, experimentation is curtailed at that stage along with the acceptance of H_1.

In this section, we consider genuinely distribution-free tests. Here we assume that the null hypothesis H_0 induces some sort of invariance of the probability distribution of $\mathbf{X}^{(N)}$ under some groups of transformations (which maps the sample space onto itself) and the partial sequence $\{T_n,\ n \le N\}$ is distribution-free under H_0. Thus, a suitable test based on the set (2.1) will also be nonparametric in nature. With this observation we conceive of a suitable real-valued statistic

$$h_N = h(T_n;\ n_0 \le n \le N) \tag{2.4}$$

to be used as the test statistic. Note that, by definition, for every α: $0 < \alpha < 1$ (the desired level of significance), there exists an $h_{N,\alpha}$ such that

$$(0 \le)\ \alpha_N = P\{h_N > h_{N,\alpha} | H_0\} \le \alpha \le P\{h_N \ge h_{N,\alpha} | H_0\} \tag{2.5}$$

where α_N (specified) may depend on N, and is usually very close to α when N is large.

Now, let us denote by

$$h_N^{(k)} = h_k(T_n;\ n \le k, \text{ target sample size } N), \qquad \forall n_0 \le k \le N \tag{2.6}$$

and assume that for every N,

$$h_N^{(k)} \text{ is nondecreasing in } k\text{:}\quad n_0 \le k \le N; \qquad h_N^{(N)} = h_N \tag{2.7}$$

Then, with the event $[h_N > h_{N,\alpha}]$, we can associate a random variable

$$M_N = \begin{cases} \min\{k\colon h_N^{(k)} > h_{N,\alpha} \text{ and } n_0 \le k \le N\} \\ N \qquad \text{if no such } k \text{ exists} \end{cases} \tag{2.8}$$

We designate M_N as the *stopping number.* Note that, by (2.7) and (2.8), if $M_N < N$, then $h_N^{(k)} > h_{N,\alpha}$, $\forall k \ge M_N$. Hence operationally, we may proceed as follows: As we obtain the partial sequence $\mathbf{X}^{(k)}$, we compute $h_N^{(k)}$, $k \ge n_0$. If, for the first time, for some $k = M_N$ ($n_0 \le M_N \le N$), $h_N^{(M_N)}$ exceeds $h_{N,\alpha}$, experimentation is curtailed following $\mathbf{X}^{(M_N)}$ along with the rejection of H_0. If no such k ($\le N$) exists, sampling is stopped with the target size N (i.e., following $\mathbf{X}^{(N)}$) and H_0 is accepted.

Note that by (2.5), (2.7), and (2.8), the overall level of significance of the repeated significance test is equal to α. Note that in this setup, the procedure allows an early termination of experimentation only when H_0 is judged unacceptable. This feature is quite important in many drug-screening experiments. Suppose that we are testing a newly introduced drug against a standard one. If the new drug performs as well as the standard one, i.e., H_0 holds, continuation of experimentation does not introduce any undesirable element nor affect the medical ethics. On the other hand, if the new drug is judged significantly better (or worse) than the standard one, it is naturally advisable to curtail experimentation as early as possible and to take corrective action for the benefit of the live subjects. Moreover, in most exploratory studies without sufficient background information, we do not desire to accept the response equivalence of the new and the standard drugs, so that the failure to accept the null hypothesis at an early stage is really an advantage of the procedure (in disguise). However, in Section 8 we shall see that some modifications to this feature can be made without much difficulty.

Now, generally, the sequence $\{T_n; n \geq n_0\}$ in (2.1) has some nice behavior (as we shall see in Section 5), and keeping (2.7) in mind, we are inclined to use either of the following two typical statistics. Let us denote by

$$\mu_n{}^0 = E(T_n | H_0) \qquad \text{and} \qquad \sigma_n{}^2 = V(T_n | H_0), \qquad n \geq n_0 \tag{2.9}$$

and assume that both exist, and further

$$\sigma_n{}^2 \text{ is } \nearrow \text{ in } n\ (\geq n_0); \qquad \text{we let } \sigma_k{}^2 = 0, \qquad \forall k \leq n_0 - 1 \tag{2.10}$$

(I) *Kolmogorov–Smirnov type statistics*: We define here

$$h_N^{(k)} = \max_{n_0 \leq n \leq k} (T_n - \mu_n{}^0)/\sigma_N \qquad \text{or} \qquad \max_{n_0 \leq n \leq k} |(T_n - \mu_n{}^0)/\sigma_N| \tag{2.11}$$

for $n_0 \leq k \leq N$. Note that (2.7) holds for the definition in (2.11).

(II) *Cramér–von Mises type statistics*: Let us write

$$\lambda_{Ns} = \sigma_N^{-2}(\sigma_s{}^2 - \sigma_{s-1}^2), \qquad s = n_0, n_0 + 1, \ldots, N \tag{2.12}$$

and define

$$h_N^{(k)} = \sum_{s=n_0}^{k-1} [(T_s - \mu_s{}^0)^2/\sigma_N{}^2]\lambda_{Ns}, \qquad k = n_0, \ldots, N \tag{2.13}$$

Here also (2.7) holds. More intuitive interpretations of (2.11) and (2.13) are provided in Section 5.

3. TYPE B REPEATED SIGNIFICANCE TESTS

In this situation, $\{T_n, n \leq N\}$ in (2.1) is not genuinely distribution-free when H_0 in (2.2) holds, and hence h_N in (2.4) is not distribution-free. Further, in this case, $\sigma_n{}^2$, defined by (2.9), may not be known. Nevertheless, if N is (at least) moderately large, it may be possible to provide a (strongly) consistent estimator of $\sigma_n{}^2$ and proceed on parallel lines.

We assume that

$$\sigma_n{}^2 = V(T_n | H_0) = \sigma^2 \psi(n) \qquad \text{where} \quad \psi(n) \text{ is } \nearrow \text{ in } n \tag{3.1}$$

and $0 < \sigma < \infty$. Further, we assume that there exists a sequence $\{S_n^2 = S_n^2(\mathbf{X}^{(n)});\, n \geq n_0\}$ of estimators of the unknown σ^2, such that

$$S_n^2/\sigma^2 \to 1 \text{ almost surely (a.s.)}, \qquad \text{as} \quad n \to \infty \tag{3.2}$$

while $\psi(n)$ is a known function of n. Let us then define

$$U_n = [T_n - \mu_n^0]/S_n, \qquad n \geq n_0 \tag{3.3}$$

where μ_n^0 is defined by (2.9) and is assumed to be specified, and analogous to (2.11), we let

$$\tilde{h}_N^{(k)} = \max_{n_0 \leq n \leq k} \{U_n/\psi^{1/2}(N)\} \quad \text{or} \quad \max_{n_0 \leq n \leq k} \{|U_n|/\psi^{1/2}(N)\}, \quad \forall n_0 \leq k \leq N \tag{3.4}$$

Note that by (3.4), (2.7) holds for $\{\tilde{h}_N^{(k)}, n_0 \leq k \leq N\}$. To define the Cramér–von Mises type statistics, we let

$$\lambda_{Nk} = [\psi(k) - \psi(k-1)]/\psi(N), \qquad n_0 \leq k \leq N-1; \qquad \lambda_{NN} = 0 \tag{3.5}$$

where we let $\psi(q) = 0$, $\forall q < n_0$, $U_q = 0$, $\forall q < n_0$. Then, parallel to (2.13), we define here

$$\tilde{h}_N^{(k)} = \sum_{s=n_0}^{k-1} [\lambda_{NS} U_s^2/\psi(N)], \qquad n_0 \leq k \leq N, \qquad \tilde{h}_N^{(N)} = \tilde{h}_N \tag{3.6}$$

Here also (2.7) holds for $\{\tilde{h}_N^{(k)};\, n_0 \leq k \leq N\}$.

Having defined $\{h_N^{(k)}\}$ by (3.4) and (3.6) operationally, we may proceed on the same line as in type A tests; the critical value and the stopping number are denoted in this case by $\tilde{h}_{N,\alpha}$ and $\tilde{M}_N$, respectively, so that replacing $h_{N,\alpha}$ and M_N by $\tilde{h}_{N,\alpha}$ and $\tilde{M}_N$, respectively, the procedure remains the same as that following (2.8). The only difference lies in the fact that whereas $h_{N,\alpha}$ is a distribution-free constant, $\tilde{h}_{N,\alpha}$ may depend on the underlying distributions. Nevertheless, it will be shown in Section 5 that under very general conditions, $h_{N,\alpha} - \tilde{h}_{N,\alpha} \to 0$ as $N \to \infty$, so that asymptotically type B tests have the same characteristics as the type A tests.

4. SOME SPECIFIC REPEATED SIGNIFICANCE TESTS

In this section, we consider a variety of nonparametric problems and, in each case, consider suitable repeated significance tests.

4.1. Test for a Population Quantile

We assume that the X_i are independent and identically distributed (i.i.d.) random variables with a continuous distribution function (d.f.) $F(x)$, defined on the real line $(-\infty, \infty)$. The form of F is not specified. For a given p: $0 < p < 1$, the p-quantile of F is defined by

$$\xi_p = \inf\{x: F(x) \geq p\} \tag{4.1}$$

so that $F(\xi_p) \geq p$. In fact, if F is strictly monotonic in some neighborhood of ξ_p, then $F(\xi_p) = p$ and ξ_p can also be uniquely defined as the root of $F(x) = p$. We desire to test

$$H_0: \quad \xi_p = \xi_p{}^0 \qquad \text{versus} \qquad H_1: \quad \xi_p < \xi_p{}^0 \quad \text{or} \quad H_2: \quad \xi_p \neq \xi_p{}^0 \tag{4.2}$$

where $\xi_p{}^0$ is specified. Actually, the most typical value of p is $\frac{1}{2}$, in which case ξ_p is the population median of F. This problem is quite common in clinical, medical, and biological experiments where a dose designed to stimulate a response in a certain percentage of cases is of considerable importance.

Let $c(u) = 1$ or 0 according as $u \geq$ or < 0, and for every $n \geq 1$, we define

$$r_n = \sum_{i=1}^{n} c(\xi_p{}^0 - X_i) = \text{No. of } X_1, \ldots, X_n \text{ less than } \xi_p{}^0 \tag{4.3}$$

Then, under H_0 in (4.2), r_n has the Bernouilli distribution with parameters (n, p). In fact, under H_0, $\{r_n; n \geq 1\}$ is a *random walk* process where at each step it either makes a jump of 1, with probability p, or remains stationary, with probability $1 - p$. The joint distribution of $\{r_n; n \leq N\}$ does not depend on F when H_0 holds, and hence we are in a position to apply type A tests.

Note that by (4.3),

$$\mu_n{}^0 = E[r_n | H_0] = np$$

and

$$\sigma_n{}^2 = v(r_n | H_0) = np(1 - p), \qquad n \geq 1 \tag{4.4}$$

so that (2.11) reduces to

$$h_N^{(k)} = [Np(1 - p)]^{-1/2} \left\{ \max_{1 \leq n \leq k} [r_n - np] \right\} \qquad \text{or}$$

$$[Np(1 - p)]^{-1/2} \left\{ \max_{1 \leq n \leq k} |r_n - np| \right\} \tag{4.5}$$

Also, by (2.12) and (4.4), $\lambda_{Ns} = N^{-1}$, $s = 1, \ldots, N - 1$, so that (2.13) reduces here to

$$h_N^{(k)} = N^{-2}[p(1 - p)]^{-1} \sum_{s=1}^{k-1} (r_s - sp)^2, \qquad k = 2, \ldots, N \tag{4.6}$$

In either case, we may proceed as we did following (2.8). The constant $h_{N,\alpha}$ in

(2.5) may be obtained from the random walk process. For large N, we shall see in Section 5 that $h_{N,\alpha}$ converges to a limit h_α. Other properties of this test are studied in Sections 5 and 6.

4.2. Repeated Rank Order Tests for Location

We assume that X_i, $i \geq 1$, are i.i.d.r.v. with a continuous d.f. $F(x, \theta)$, $-\infty < x < \infty$, where θ is the location parameter of F in the sense that

$$F(x, \theta) = F_0(x - \theta); \qquad F_0(x) + F_0(-x) = 1, \qquad \forall x \tag{4.7}$$

Thus, F_0 is assumed to be symmetric about zero. We want to test

$$H_0\colon\ \theta = \theta_0 \qquad \text{versus} \qquad H_1\colon\ \theta > \theta_0 \quad \text{or} \quad H_2\colon\ \theta \neq \theta_0 \tag{4.8}$$

where θ_0 is specified. In fact, by working with $X_i - \theta_0$, $i \geq 1$, we can always reduce (4.8) to $\theta_0 = 0$. This is the so-called one-sample symmetry problem and a variety of tests is available for (4.8) when one works with the conventional fixed-sample case. We shall consider here type A tests based on a general class of rank order statistics.

For every $n \geq 1$, let $R_{n1}^+, \ldots, R_{nn}^+$ be, respectively, the ranks of $|X_1|, \ldots, |X_n|$ among themselves. Also, let $\phi = \{\phi(u), 0 < u < 1\}$ be a score function, assumed to be square integrable, and for every $n \geq 1$, consider a set of *scores* defined by

$$a_n(i) = E\phi(U_{ni}) \qquad \text{or} \qquad \phi(i/(n+1)), \qquad 1 \leq i \leq n \tag{4.9}$$

where $U_{n1} < \cdots < U_{nn}$ are the ordered random variables of a sample of size n from the rectangular [0, 1] d.f. Then, a typical one-sample signed-rank statistic is defined by

$$S_n = \sum_{i=1}^{n} (\operatorname{Sgn} X_i) a_n(R_{ni}^+), \qquad n \geq 1 \tag{4.10}$$

Now, under (4.7) and $H_0\colon \theta = 0$, the vectors $(\operatorname{Sgn} X_1, \ldots, \operatorname{Sgn} X_n)$ and $(R_{n1}^+, \ldots, R_{nn}^+)$ are mutually independent, the first one has 2^n possible equally likely realizations, and the other assumes all possible $n!$ permutations of $(1, \ldots, n)$ with equal probability $(n!)^{-1}$. Thus, S_n is distribution-free. In fact, $\{S_n, n \leq N\}$ is also a distribution-free sequence (under H_0), its distribution being generated by the $2^N N!$ equally likely realizations of $(\operatorname{Sgn} X_1, \ldots, \operatorname{Sgn} X_N)$ and $(R_{N1}^+, \ldots, R_{NN}^+)$. Thus, type A tests may be applied here.

Note that by definition in (4.10),

$$\mu_n{}^0 = E(S_n | H_0) = 0 \qquad \text{and} \qquad \sigma_n{}^2 = E(S_n{}^2 | H_0) = nA_n{}^2 \tag{4.11}$$

where

$$A_n{}^2 = \frac{1}{n} \sum_{i=1}^{n} a_n{}^2(i), \qquad n \geq 1 \tag{4.12}$$

Thus, in this case (2.11) reduces to

$$h_N^{(k)} = N^{-1/2} A_N^{-1} \left\{ \max_{1 \le n \le k} S_n \right\} \quad \text{or} \quad N^{-1/2} A_N^{-1} \left\{ \max_{1 \le n \le k} |S_n| \right\} \tag{4.13}$$

for $1 \le k \le N$. As we shall see later, under very general conditions, $A_n{}^2 \to A^2 = \int_0^1 \phi^2(u)\,du$ as $n \to \infty$. Hence, we may replace $\sigma_n{}^2$ by nA^2 in (2.12)–(2.13), to obtain a version of $h_N^{(k)}$, parallel to (2.13), as follows:

$$h_N^{(k)} = N^{-2} A^{-2} \sum_{s=1}^{k-1} S_s{}^2, \qquad 2 \le k \le N \tag{4.14}$$

In either case, we may proceed as we did following (2.8).

4.3. Repeated Rank Tests for Bivariate Independence

We assume that $\mathbf{X}_i = (X_{1i}, X_{2i}), i \ge 1$, are i.i.d. r.v. vectors with a continuous bivariate d.f. $F(x_1, x_2)$, $-\infty < x_1, x_2 < \infty$, and we denote the two marginal d.f.'s by $F_1(x) = F(x, \infty)$ and $F_2(x) = F(\infty, x)$, respectively. The null hypothesis of independence is framed as

$$H_0\colon \quad F(x_1, x_2) = F_1(x_1)F_2(x_2), \qquad \forall -\infty < x_1, \quad x_2 < \infty \tag{4.15}$$

We are interested in testing H_0 against suitable alternatives, namely, X_{1i}, X_{2i} have a positive or negative type of association.

Let $R_{ni}^{(j)}$ be the rank of X_{ji} among $X_{j1}, \ldots, X_{jn}$ for $1 \le i \le n; j = 1, 2$, and let $\mathbf{R}_n^{(j)} = (R_{n1}^{(j)}, \ldots, R_{nn}^{(j)})$, $j = 1, 2$; $n \ge 1$. Consider two score functions $\phi_j = \{\phi_j(u);\ 0 < u < 1\}$, $j = 1, 2$, and as in (4.9), let $a_{jn}(i) = E\phi_j(U_{ni})$ or $\phi_j(i/(n+1))$, $1 \le i \le n, j = 1, 2$. Then, a general class of rank order tests for H_0 in (4.15) is based on statistics of the form

$$Q_n = \sum_{i=1}^{n} [a_{1n}(R_{ni}^{(1)}) - \bar{a}_{1n}][a_{2n}(R_{ni}^{(2)}) - \bar{a}_{2n}], \qquad n \ge 1 \tag{4.16}$$

where $\bar{a}_{jn} = n^{-1} \sum_{i=1}^{n} a_{jn}(i), j = 1, 2$. Now, under H_0, $\mathbf{R}_n^{(1)}$, $\mathbf{R}_n^{(2)}$ are stochastically independent, each vector assuming all possible permutations of $(1, \ldots, n)$ with the equal probability $(n!)^{-1}$. Thus, Q_n is distribution-free under H_0. In fact, the same logic applies to the distribution-freeness of $\{Q_n;\ n \le N\}$, under H_0, $\forall N \ge 1$. Hence, here also we may apply a type A test.

Let us define for $n \ge 2$,

$$A_{jn}^2 = (n-1)^{-1} \sum_{i=1}^{n} [a_{jn}(i) - \bar{a}_{jn}]^2, \qquad j = 1, 2; \qquad A_{(n)}^2 = A_{1n}^2 \cdot A_{2n}^2 \tag{4.17}$$

Note that for $n = 1$, $Q_1 = 0$ and we let $A_{(1)}^2 = 0$. Then, by standard steps, we obtain from (4.16) and (4.17) that for every $n \ge 1$,

$$\mu_n{}^0 = E(Q_n | H_0) = 0 \qquad \text{and} \qquad \sigma_n{}^2 = V(Q_n | H_0) = (n-1)A_{(n)}^2 \tag{4.18}$$

Hence, in this case (2.11) reduces to

$$h_N^{(k)} = (N-1)^{-1/2} A_{(N)}^{-1} \left\{ \max_{1 \le n \le k} Q_n \right\} \quad \text{or} \quad (N-1)^{-1/2} A_{(N)}^{-1} \left\{ \max_{1 \le n \le k} |Q_n| \right\} \tag{4.19}$$

and, as in (4.14), we obtain a version of (2.13) as

$$h_N^{(k)} = (N-1)^{-2} A^{-2} \sum_{n=1}^{k-1} Q_n^2, \qquad N \ge 2, \quad k \ge 2 \tag{4.20}$$

where $A^2 = A_1^2 A_2^2$ and $A_j^2 = \int_0^1 \phi_j^2(u)\, du - (\int_0^1 \phi_j(u)\, du)^2$, $j = 1, 2$.

4.4. Repeated Rank Tests for Location/Scale in the Two-Sample Case

Let X_i, $i \ge 1$, be independent random variables with continuous d.f.'s F_i, $i \ge 1$, all defined on the real line. Suppose that

$$F_{2i-1}(x) = F_1(x) \qquad \text{and} \qquad F_{2i}(x) = F_2(x), \qquad \forall i \ge 1 \tag{4.21}$$

so that essentially we reduce it to two sequences $\{X_{2i-1}, i \ge 1\}$ and $\{X_{2i}, i \ge 1\}$ of i.i.d.r.v.'s with continuous d.f.'s F_1 and F_2, respectively. Our null hypothesis of interest is

$$H_0\colon\ F_1 \equiv F_2 \qquad \text{versus} \qquad H_1\colon\ F_1 \text{ and } F_2 \text{ differ in location/scale} \tag{4.22}$$

For example, we may frame $F_2(x) = F_1(x - \Delta)$, where Δ stands for the difference in location of F_2 and F_1, otherwise having the same form, or we may let $F_2(x) = F_1(x/(1 + \Delta))$ $(\Delta > -1)$, where Δ stands for a scale variation.

Let R_{ni} be the rank of X_i among $X_1, \ldots, X_n$ for $i = 1, \ldots, n$ and $\mathbf{R}_n = (R_{n1}, \ldots, R_{nn})$. Define $a_n(i)$ as in (4.5), and let

$$L_n = \sum_{i=1}^{n} (c_i - \bar{c}_n) a_n(R_{ni}), \qquad n \ge 1 \tag{4.23}$$

where $\{c_i, i \ge 1\}$ is a sequence of suitable constants and $\bar{c}_n = n^{-1} \sum_{i=1}^{n} c_i$, $n \ge 1$. In particular, in our problem, we let $c_{2i} = 1$, $c_{2i-1} = -1$, $i \ge 1$. Then a general class of rank tests for H_0 in (4.4) is based on linear rank statistics of the type L_n. Note that under H_0 in (4.21), $\mathbf{R}_n$ assumes possible $n!$ permutations of $(1, \ldots, n)$ with the equal probability $(n!)^{-1}$, so that L_n is a distribution-free statistic; the same argument holds for the entire sequence $\{L_n;\, n \le N\}$ for every $N \ge 2$. Thus, here also we have type A tests.

Defining A_n^2 as in (4.17) with $\phi_1 = \phi$, we have on letting $C_n^2 = \sum_{i=1}^n (c_i - \bar{c}_n)^2$, $n \geq 1$, that

$$\mu_n^0 = E(L_n | H_0) = 0 \quad \text{and} \quad \sigma_n^2 = E(L_n^2 | H_0) = C_n^2 A_n^2 \tag{4.24}$$

so that, in this case (2.11) reduces to

$$h_N^{(k)} = A_N^{-1} C_N^{-1} \left\{ \max_{1 \leq n \leq k} L_n \right\} \quad \text{or} \quad A_N^{-1} C_N^{-1} \left\{ \max_{1 \leq n \leq k} |L_n| \right\}, \qquad N \leq k \leq 2 \tag{4.25}$$

4.5. Repeated Rank Tests for Randomness

The two-sample location or scale models treated earlier are particular cases of more general models. For example, instead of (4.21), we may consider the simple regression model

$$F_i(x) = F(x - \beta_0 - \beta c_i), \qquad i \geq 1 \tag{4.26}$$

where β_0, β are unknown parameters and the c_i are known (regression) constants. Here, the hypothesis of randomness is framed as

$$H_0\colon \ \beta = 0 \qquad \text{versus} \qquad H_1\colon \ \beta > 0 \ \text{ or } \ H_2\colon \ \beta \neq 0 \tag{4.27}$$

where β_0 is treated as a nuisance parameter. For some allied models of randomness, we may refer to Hájek and Šidák (1967, Chap. II). For such problems, the hypothesis of randomness ensures the invariance of the joint distribution of $X_1, \ldots, X_N$ under any permutation of the coordinates and the use of linear rank statistics is quite justified on the ground of local optimality, invariance, and robustness. Thus, we may use the linear rank statistics, defined by (4.23), and proceed as in (4.24)–(4.25).

4.6. Repeated Tests for Regular Functionals of Distribution Functions

Let X_i, $i \geq 1$, be i.i.d.r.v.'s with a d.f. $F(x)$, defined in the p (≥ 1)-dimensional Euclidean space E^p. Consider a functional of F:

$$\begin{aligned} \theta = \theta(F) &= \int_{E^{pm}} \cdots \int \phi(x_1, \ldots, x_m)\, dF(x_1) \cdots dF(x_m) \\ &= E\phi(X_1, \ldots, X_m) \end{aligned} \tag{4.28}$$

where ϕ, symmetric in its arguments, is the *kernel* and m (≥ 1) is the *degree* of θ. Suppose we want to test

$$H_0\colon \ \theta(F) = \theta_0 \qquad \text{versus} \qquad H_1\colon \ \theta(F) > \theta_0 \ \text{ or } \ H_2\colon \ \theta(F) \neq \theta_0 \tag{4.29}$$

where θ_0 is specified.

An unbiased and symmetric estimator of $\theta(F)$, based on a sample of size n ($\geq m$), is given by

$$U_n = U(\mathbf{X}^{(n)}) = \binom{n}{m}^{-1} \sum_{1 \leq i_1 < \cdots < i_m \leq n} \phi(X_{i_1}, \ldots, X_{i_m}) \tag{4.30}$$

Also, note that an unbiased estimator of F is

$$F_n(x) = n^{-1} \sum_{i=1}^{n} I(X_i \leq x), \qquad x \in E^p \tag{4.31}$$

and the von Mises' functional (of F_n) is given by

$$\begin{aligned} \theta(F_n) &= \int_{E^{pm}} \cdots \int \phi(x_1, \ldots, x_m)\, dF_n(x_1) \cdots dF_n(x_m) \\ &= n^{-m} \sum_{i_1=1}^{n} \cdots \sum_{i_m=1}^{n} \phi(X_{i_1}, \ldots, X_{i_m}) \end{aligned} \tag{4.32}$$

$\theta(F_n)$ is a competitor of U_n, although it is not necessarily an unbiased estimator of $\theta(F)$. For $m = 1$, $\theta(F_n) = U_n = n^{-1} \sum_{i=1}^{n} \phi(X_i)$ behaves like the average of n independent r.v.'s, but for $m > 1$, the summands in (4.30) or (4.32) are not all mutually independent. General properties of these estimators have been studied very thoroughly by Hoeffding (1948). It follows from his results that for $n \geq m$,

$$\begin{aligned} \operatorname{var}(U_n) &= \binom{n}{m}^{-1} \sum_{c=1}^{m} \binom{m}{c} \binom{n-m}{m-c} \zeta_c \\ &= m^2 n^{-1} \zeta_1 + O(n^{-2}) \end{aligned} \tag{4.33}$$

where for $0 \leq c \leq m$,

$$\zeta_c = E[\phi(X_1, \ldots, X_m)\phi(X_{m-c+1}, \ldots, X_{2m-c})] - \theta^2(F) \tag{4.34}$$

ζ_m is assumed to exist, and

$$\operatorname{var}(\theta(F_n)) = \operatorname{var}(U_n) + O(n^{-2}) \tag{4.35}$$

Thus, defining T_n by $n(U_n - \theta_0)$ or $n(\theta(F_n) - \theta_0)$, $n \geq m$, we are able to proceed as in Section 3. In general, the joint d.f. of $\{U_n, n \leq N\}$ or $\{\theta(F_n), n \leq N\}$ depends on the underlying d.f. F, and hence type A tests may not be available. However, type B tests can be constructed.

(i) ζ_1 *known*: In this case, we let, for $m \leq k \leq N$,

$$h_N^{(k)} = \left\{ \max_{m \leq n \leq k} n[U_n - \theta_0] \right\} \Big/ mN^{1/2}\zeta_1^{1/2} \quad \text{or}$$

$$\left\{ \max_{m \leq n \leq k} n|U_n - \theta_0| \right\} \Big/ mN^{1/2}\zeta_1^{1/2} \tag{4.36}$$

where we may also replace $\{U_n\}$ by $\{\theta(F_n)\}$; parallel to (2.13), we have

$$h_N^{(k)} = \left(\sum_{n=m}^{k-1} n^2[U_n - \theta_0]^2\right) \Big/ m^2N^2\zeta_1, \qquad m < k \leq N \tag{4.37}$$

where also $\{U_n\}$ may be replaced by $\{\theta(F_n)\}$.

(ii) ζ_1 *unknown*: As is generally the case, ζ_c, $c \geq 1$, are not known and are to be estimated. For this, we proceed as in Sen (1960, 1977a) and consider the following. Let U_{n-1}^i be the U-statistic based on $X_1, \ldots, X_{i-1}, X_{i+1}, \ldots, X_n$, for $1 \leq i \leq n$. Further, let

$$S_n^{\,2} = (n-1) \sum_{i=1}^{n} [U_{n-1}^i - U_n]^2, \qquad n > m \tag{4.38}$$

Then, $S_n^{\,2} \to \zeta_1$ a.s., as $n \to \infty$. Hence, we may proceed as in Section 3, and all we need to do is to replace in (4.36)–(4.37) $\zeta_1^{1/2}$ by S_n, i.e.,

$$\tilde{h}_N^{(k)} = \max_{m \leq n \leq k} (n[U_n - \theta_0]/mN^{1/2}S_n) \quad \text{or}$$

$$\max_{m \leq n \leq k} (n|U_n - \theta_0|/mN^{1/2}S_n), \qquad m < k \leq N \tag{4.39}$$

and for the Cramér–von Mises' type

$$\tilde{h}_N^{(k)} = \left(\sum_{n=m}^{k-1} n^2[U_n - \theta_0]^2/S_n^{\,2}\right) \Big/ m^2N^2 \tag{4.40}$$

where in both (4.39) and (4.40), $\{U_n\}$ may be replaced by $\{\theta(F_n)\}$.

4.7. Repeated Significance Tests Based on Linear Combinations of Order Statistics

Let X_i, $i \geq 1$, be i.i.d.r.v. with a continuous d.f. F, defined on the real line $(-\infty, \infty)$. For every $n \geq 1$, let $X_{n,1} < \cdots < X_{n,n}$ be the ordered random variables (corresponding to $X_1, \ldots, X_n$), and consider the statistics

$$T_n = \sum_{i=1}^{n} c_{n,i} X_{n,i}, \qquad n \geq n_0 \tag{4.41}$$

where the $c_{n,i}$ are suitable constants. For a variety of (parametric as well as nonparametric) problems, T_n is used for the estimation of suitable functionals of F or for testing hypotheses about F. We confine our interest to situations in which for some smooth $\phi = \{\phi(u), 0 < u < 1\}$ and $\phi_n = \{\phi_n(u), 0 < u \leq 1\}$ with $\phi_n(u) \to \phi(u)$ for every $u \in (0, 1)$,

$$c_{n,i} = \phi_n(t) \qquad \text{for} \quad (i-1)/n < t \leq i/n, \quad 1 \leq i \leq n \tag{4.42}$$

Then, T_n may be written as

$$T_n = n \int_{-\infty}^{\infty} x\phi_n(F_n(x))\, dF_n(x) \tag{4.43}$$

where F_n is defined by (4.31). The parameter of interest is

$$\mu = \int_{-\infty}^{\infty} x\phi(F(x))\, dF(x) \tag{4.44}$$

We also define

$$\sigma^2 = 2 \iint_{-\infty<x<y<\infty} F(x)[1 - F(y)]\phi(F(x))\phi(F(y))\, dx\, dy \tag{4.45}$$

Then, under very general conditions $\{\sigma^{-1}(T_n - n\mu), n \geq n_0\}$ has the characteristics of Section 3, suggesting the feasibility of type B tests for testing

$$H_0\colon\ \mu = \mu_0 \qquad \text{versus} \qquad H_1\colon\ \mu > \mu_0 \quad \text{or} \quad H_2\colon\ \mu \neq \mu_0 \tag{4.46}$$

Let us define

$$\hat{\sigma}_{n+1}^2 = \sum_{i=1}^{n} \sum_{j=1}^{n} c_{n,i}c_{n,j}(n+1)^{-2}\{\min(i,j)(n+1) - ij\}$$
$$\times \{X_{n+1,i+1} - X_{n+1,i}\}\{X_{n+1,j+1} - X_{n+1,j}\}, \qquad n \geq n_0 \tag{4.47}$$

Then, $\hat{\sigma}_{n+1}^2$ converges a.s. to σ^2 as $n \to \infty$, under fairly general conditions (see Sen, 1975b). Thus, here we define

$$U_n = (T_n - n\mu)/\hat{\sigma}_n, \qquad n \geq n_0 \tag{4.48}$$

Note that (3.1) holds with $\psi(n) \simeq n$, and hence we may consider $\widetilde{h}_N^{(k)}$ as in (3.4) and (3.6) with $\psi(n)$ and λ_{Nk} being replaced by n and $1/N$, respectively.

4.8. Repeated Significance Tests for the Bundle Strength of Filaments

Here the X_i are nonnegative i.i.d.r.v. with the d.f. F, defined on $[0, \infty)$, the $X_{n,i}$ are defined as preceding (4.41), and we define

$$Z_n = \max_{1 \leq k \leq n} (n - k + 1)X_{n,k}, \qquad n \geq 1 \tag{4.49}$$

Z_n is termed the strength of a bundle of n parallel filaments (see Sen, *et al.*, 1973). It is known that $n^{-1}Z_n$ almost surely converges to

$$\mu = \sup_{x \geq 0} \{x[1 - F(x)]\} \tag{4.50}$$

In this problem also, for testing

$$H_0\colon\ \mu = \mu_0 \qquad \text{versus} \qquad H_1\colon\ \mu > \mu_0 \quad \text{or} \quad H_2\colon\ \mu \neq \mu_0 \tag{4.51}$$

we may use a type B repeated significance test based on

$$U_n = (Z_n - n\mu_0)/S_n, \qquad n \geq n_0 \tag{4.52}$$

where S_n^2 is a suitable estimator of $\sigma_n^2 = n\sigma^2$, $\sigma^2 = \mu_0^2\pi_0/(1 - \pi_0)$, $\pi_0 = F(x_0)$, and x_0 is the (unique) point at which $\mu = x_0[1 - F(x_0)]$. By definition, in (4.49), $Z_n = (n - r_n + 1)X_{n,r_n}$ for some (random) r_n: $1 \leq r_n \leq n$, and it follows from Sen (1973b) that $n^{-1}r_n \to \pi_0$ a.s., as $n \to \infty$. Thus, we may take $S_n^2 = \mu_0^2 r_n/(n - r_n + 1)$. Thus, in this case also $\psi(n) \simeq n$, λ_{Nk} may be replaced by N^{-1}, and we may proceed as in (3.4)–(3.6).

5. ASYMPTOTIC DISTRIBUTION THEORY UNDER NULL HYPOTHESES

The computation of $h_{N,\alpha}$ in (2.5) for a given N and suitable h_N becomes prohibitively laborious as N becomes large. For this reason, in this section we shall consider certain limit theorems pertaining to the various statistics considered in Section 4, and show that in each case $h_{N,\alpha} \to h_\alpha$ $(0 < \alpha < 1)$ as $N \to \infty$.

First, let us consider type A tests. As a basis for this study, we present an invariance principle for $\{T_n, n \leq N\}$, defined by (2.1), which, in turn, provides us with the desired result.

For every N, let us introduce a sequence $\{k_N(t), 0 \leq t \leq 1\}$ of nondecreasing, right-continuous, and nonnegative integers, by letting

$$k_N(t) = \max\{k\colon \sigma_k^2 \leq t\sigma_N^2\}, \qquad 0 \leq t \leq 1 \tag{5.1}$$

where $\{\sigma_n^2, n \geq n_0\}$ is defined by (2.9)–(2.10). We introduce then a stochastic process $W_N = \{W_N(t),\ 0 \leq t \leq 1\}$, by letting

$$W_N(t) = \sigma_N^{-1}\{T_{k_N(t)} - \mu^0_{k_N(t)}\}, \qquad 0 \leq t \leq 1 \tag{5.2}$$

where $\{\mu_n^0,\ n \geq n_0\}$ is defined by (2.9) and for $n < n_0$, we let $T_n - \mu_n^0 = 0$. Then, for every N, W_N belongs to the space $D[0, 1]$ of real-valued functions on the unit interval $[0, 1]$ having no discontinuities of the second kind. We associate with $D[0, 1]$ the Skorokhod J_1-topology specified by the metric

$$\rho_D(x, y) = \inf_{\lambda \in \Lambda} \left[\sup_{0 \leq t \leq 1} \{|x(t) - y(\lambda(t))| + |\lambda(t) - t|\} \right] \tag{5.3}$$

where Λ is the class of strictly increasing, continuous mapping of $[0, 1]$ onto itself. Further, let $W = \{W(t), 0 \leq t \leq 1\}$ be a standard Brownian motion on

[0, 1], so that W belongs to the space $C[0, 1]$ of real continuous functions on [0, 1] (with probability 1), W is Gaussian, and

$$EW(t) = 0, \qquad EW(t)W(t') = \min(t, t'), \qquad \forall 0 \le t, \quad t' \le 1 \tag{5.4}$$

Consider then the functionals

$$\omega^+ = \sup_{0 \le t \le 1} W(t), \qquad \omega = \sup_{0 \le t \le 1} |W(t)|, \qquad \omega^* = \int_0^1 W^2(t)\, dt \tag{5.5}$$

Finally, let us denote the upper $100\alpha\%$ points of the distributions of ω^+, ω, and ω^* by $\omega_\alpha{}^+$, ω_α, and $\omega_\alpha{}^*$, respectively.

Theorem 5.1. Whenever $W_N \xrightarrow{\mathscr{D}} W$ in the J_1-topology on $D[0, 1]$, for the Kolmogorov–Smirnov type statistics in (2.11), for every $0 < \alpha < 1$,

$$h_{N,\alpha} \to \omega_\alpha{}^+ \quad (\text{or } \omega_\alpha) \qquad \text{as} \quad N \to \infty \tag{5.6}$$

and for the Cramér–von Mises type statistics in (2.13)

$$h_{N,\alpha} \to \omega_\alpha{}^* \qquad \text{as} \quad N \to \infty \tag{5.7}$$

Outline of the Proof. By (2.6), (2.7), (2.11), and (5.2), $h_N = h_N^{(N)} = \sup_{0 \le t \le 1} W_N(t)$ (or $\sup_{0 \le t \le 1} |W_N(t)|$), and hence the weak convergence of W_N to W ensures that h_N converges in law to ω^+ (or ω), and this implies (5.6). Similarly, by (2.6), (2.7), (2.12), (2.13), and (5.2) $h_N = \int_0^1 W_N{}^2(t)\, dt$, and hence the weak convergence of W_N to W implies that h_N converges in law to ω^*, which in turn ensures (5.7). Q.E.D.

Now, it is well known that for every $\lambda \ge 0$,

$$\begin{aligned} P\{\omega^+ > \lambda\} &= P\left\{\sup_{0 \le t \le 1} W(t) > \lambda\right\} \\ &= 2P\{W(1) > \lambda\} = \left(\frac{2}{\pi}\right)^{1/2} \int_\lambda^\infty \exp\{-\tfrac{1}{2}t^2\}\, dt \end{aligned} \tag{5.8}$$

$$\begin{aligned} P\{\omega > \lambda\} &= P\left\{\sup_{0 \le t \le 1} |W(t)| > \lambda\right\} \\ &= 1 - \sum_{k=-\infty}^{\infty} (-1)^k [\Phi((2k+1)\lambda) - \Phi((2k-1)\lambda)] \\ &= 2[1 - \Phi(\lambda)] \\ &\quad + 2 \sum_{k=1}^{\infty} (-1)^{k-1} [\Phi((2k+1)\lambda) - \Phi((2k-1)\lambda)] \end{aligned} \tag{5.9}$$

where $\Phi(x)$ is the standard normal d.f. Table 5.1 provides the values of ω_α^+ and ω_α for typical values of α.

TABLE 5.1

Table for the Values of ω_α^+ and ω_α for Some Typical α

α	0.01	0.025	0.05	0.10
ω_α^+	2.576	2.241	1.960	1.645
ω_α	2.807	2.503	2.241	1.960

The case is somewhat different for ω^*. The characteristic function of ω^* is given by

$$\psi(t) = E(e^{it\omega^*}) = \prod_{s=1}^{\infty} (1 - 2tu_s)^{-1/2} \tag{5.10}$$

where

$$u_s = 4\{\pi(2s-1)\}^{-2}, \qquad s \geq 1 \tag{5.11}$$

Thus, if Z_i, $i \geq 1$, are independent standard normal deviates, then

$$\omega^* \stackrel{\mathscr{D}}{=} \sum_{s=1}^{\infty} u_s Z_s^2 \tag{5.12}$$

where $\stackrel{\mathscr{D}}{=}$ stands for the equality of distributions. Unlike (5.8)–(5.9), it is difficult to obtain, in a closed form, the distribution of the right-hand side of (5.12). We quote the following empirical percentile points, obtained by a simulation study (based on 1000 repetitions) by Majumdar and Sen (1978a):

$$\begin{array}{ccccc} \alpha & 0.01 & 0.025 & 0.05 & 0.10 \\ \omega_\alpha^* & 2.87 & 2.08 & 1.67 & 1.20 \end{array} \tag{5.13}$$

Let us now consider the case of type B tests. Recall that here $\sigma_n^2 = \psi(n)\sigma^2$ where $\psi(n) \nearrow \infty$ as $n \to \infty$ and there exists a sequence $\{S_n^2\}$ of strongly consistent estimators of σ^2. Here, (5.1) reduces to

$$k_N(t) = \max\{k: \psi(k) \leq t\psi(N)\}, \qquad 0 \leq t \leq 1 \tag{5.14}$$

where, of course, we let $\sigma_n^2 = 0$, $\forall n < n_0$, i.e., $\psi(n) = 0$, $n < n_0$. We assume that the initial sample size n_0 also increases with N (although at any arbitrary rate), so that

$$n_0 \to \infty \qquad \text{as} \quad N \to \infty \tag{5.15}$$

Also, we introduce the stochastic processes $\tilde{W}_N = \{\tilde{W}_N(t),\ 0 \le t \le 1\}$ for $N \ge N_0$, by letting

$$\tilde{W}_N(t) = [\psi(N)]^{-1/2}[T_{k_N(t)} - \mu^0_{k_N(t)}]/S_{k_N(t)}, \qquad 0 \le t \le 1 \qquad (5.16)$$

Note that by (5.2), (5.15), (5.16), and (3.2),

$$\sup_{0 \le t \le 1} |\tilde{W}_N(t) - W_N(t)| \le \left\{ \sup_{0 \le t \le 1} |W_N(t)| \right\} \left\{ \sup_{k \ge n_0} |1 - \sigma/s_k| \right\} \xrightarrow{p} 0,$$

$$\text{whenever} \quad W_N \xrightarrow{\mathscr{D}} W \qquad (5.17)$$

Hence, from Theorem 5.1, (5.17), (3.4), (3.5), and (3.16), we arrive at the following theorem by a few standard steps.

Theorem 5.2. Under (3.2), (5.15), and the hypothesis of Theorem 5.1, both (5.6) and (5.7) hold for type B tests as well.

In the remainder of this section, we verify (5.6)–(5.7) for all the particular problems sketched in Section 5.

First, we notice that r_n, defined by (4.3), is a special case of U_n and $\theta(F_n)$, defined by (4.30) and (4.32), respectively. Hence, we consider an invariance principle for the latter statistics, which applies equally to (4.3). The following result is due to Miller and Sen (1972):

Let $\{U_n\}$ and $\{\theta(F_n)\}$ be defined by (4.30) and (4.32) and the ζ_c by (4.34). We assume that $\theta(F)$ is stationary of order 0, i.e., $0 < \zeta_1 < (1/m)\zeta_m < \infty$. Let $T_n = n[U_n - \theta(F)]$ or $n[\theta(F_n) - \theta(F)]$, $n \ge m$, and let $k_N(t)$ be defined by (5.14) with $\psi(k) = k$, $k \ge m$. Finally, let $\sigma_N{}^2 = m^2 N \zeta_1$ and as in (5.2), we let $W_N(t) = \sigma_N^{-1}(T_{k_N(t)})$, $0 \le t \le 1$. Then $W_N \to W$, as $N \to \infty$. Further, it follows from Sen (1977a) that $S_n{}^2 \to \zeta_1$ a.s., as $N \to \infty$, where $S_n{}^2$ is defined by (4.38). Thus, for the repeated tests in Sections 4.1 and 4.6, Theorems 5.1 and 5.2 hold, and hence $h_{N,\alpha}$ (or $\tilde{h}_{N,\alpha}$) converges to the limits specified by (5.6)–(5.7).

Consider next the statistics $\{S_n\}$, defined by (4.10), and define $A_n{}^2$ as in (4.12). At this stage assume that the score function ϕ [in (4.9)] is the difference of two nondecreasing and square integrable functions inside (0, 1). Define then $k_N(t)$ as in (5.14) with $\psi(n) = n$, $n \ge 1$, and let $W_N(t) = N^{-1/2} A_N^{-1} S_{k_N(t)}$, $0 \le t \le 1$. Then it follows from Sen (1974a, 1975a) that under (4.8) $W_N \to W$, as $N \to \infty$. Hence, here (5.6) and (5.7) hold.

Let us next consider the statistics $\{Q_n\}$, defined by (4.16). Here we assume that both the score functions ϕ_1 and ϕ_2 are differences of two nondecreasing and square integrable score functions. Also, we define $k_N(t)$ as in (5.14) with $\psi(n) = n$, $n \ge 1$. Finally, we let $W_N(t) = N^{-1/2} A_{(N)}^{-1} Q_{k_N(t)}$, $0 \le t \le 1$, where $A^2_{(N)}$ is defined by (4.17). Then it follows from Sen and Ghosh (1974b) that under (4.15), $W_N \to W$ as $N \to \infty$, and hence, (5.6) and (5.7) hold in this case too.

Consider next the case of linear rank statistics $\{L_n\}$, defined by (4.23). Also, define A_n^2 and C_n^2 as preceding (4.24). Assume that

$$\left[\max_{1 \le k \le n} (c_k - \bar{c}_n)^2\right] \Big/ C_n^2 \to 0 \quad \text{as} \quad n \to \infty \tag{5.18}$$

and the score function ϕ is the difference of two nondecreasing and square integrable functions. In this case, define $k_N(t)$ as in (5.14) with $\psi(n) = C_n^2$, $n \ge 1$, and let $W_N(t) = A_N^{-1} C_N^{-1} L_{k_N(t)}$, $0 \le t \le 1$. Then it follows from Sen and Ghosh (1972) and Sen (1975a) that under (4.22), $W_N \xrightarrow{\mathscr{D}} W$ as $N \to \infty$. Thus, for the repeated significance tests of Sections 4.4 and 4.5, both (5.6) and (5.7) hold.

Let us now consider the case of $\{T_n\}$ defined by (4.43). Here we assume that ϕ has first and second derivatives [inside (0, 1)], $E|X|^r = \int_{-\infty}^{\infty} |x|^r \, dF(x) < \infty$ for some $r > 0$, and for $u \in (0, 1)$,

$$\left| \frac{d^i}{du^i} \phi(u) \right| \le K[u(1-u)]^{-i-1/2+1/r+\delta} \quad \text{for} \quad i = 0, 1 \tag{5.19}$$

Then, defining $k_N(t)$ as in (5.14) with $\psi(n) = n$ and letting $W_N(t) = N^{-1/2}\sigma^{-1}(T_{k_N(t)} - k_N(t)\mu)$, $0 \le t \le 1$, it follows from Sen (1976c) that $W_N \to W$ as $N \to \infty$. Also, it follows from Sen (1975b) that $\hat{\sigma}_n^2$, defined by (4.47), converges almost surely to σ^2 as $N \to \infty$. Hence, here also (5.6) and (5.7) hold.

Let us finally consider the case of $\{Z_n\}$, defined by (4.49). The almost sure convergence of s_n^2 $[= \mu_0^2 r_n/(n - r_n + 1)$ or $n^{-2} Z_n^2 r_n (n - r_n + 1)^{-1}]$ to $\sigma^2 = \mu^2 \pi_0/(1 - \pi_0)$ follows from Sen (1973b) while the weak convergence of $W_N = \{W_N(t),\ 0 \le t \le 1\}$ with $W_N(t) = N^{-1/2}\sigma^{-1}\{Z_{k_N(t)} - k_N(t)\mu\}$, $k_N(t) = \max\{k\colon k/N \le t\}$, $0 \le t \le 1$, follows from Sen (1973c). As such, here also (5.6) and (5.7) hold.

In all these cases, there remains the question of how good the approximation is for moderately large values of N; more work is needed to supplement our current knowledge.

6. ASYMPTOTIC DISTRIBUTION THEORY UNDER LOCAL ALTERNATIVES

Note that the proposed test statistics h_N in (2.11), (2.13), (3.4), and (3.6) are all suitable functionals of some W_N (or $\tilde{W}_N$), defined in Section 5. Hence, one needs to study the distribution theory of W_N (or $\tilde{W}_N$) when the null hypotheses may not hold. For general alternatives, such a study may be quite involved, if not impracticable. Hence, as is the fashion in statistical inference, we take recourse to the large sample case, where the weak convergence of $\{W_N\}$ or $\{\tilde{W}_N\}$ provides us with the necessary tool for studying the power properties of the proposed tests. Again, for fixed alternatives, all these tests can be shown to be consistent under fairly general conditions, and hence, asymptotically, the power will be equal to unity when the null hypothesis is not true. For this reason, we confine our attention to a suitable class of local alternatives for which the asymptotic power of the proposed tests will lie in the open interval (0, 1).

In the setup of Sections 2 and 3, we consider now a statistic h_N as in (2.4), the null hypothesis H_0 as in (2.2), and a sequence $\{H_{(N)}\}$ of alternative hypotheses, such that if $P_N^{(N)}$ and $P_N^{(0)}$ be the probability measures for $\mathbf{X}^{(N)}$ under $H_{(N)}$ and H_0, respectively, then $P_N^{(N)}$ is *contiguous* to $P_N^{(0)}$. The contiguity may be defined either by the convergence (to 0) of the L_1-norm of $P_N^{(N)} - P_N^{(0)}$ or by the less restrictive condition that

$$[P_N^{(0)}(A_N) \to 0] \Rightarrow [P_N^{(N)}(A_N) \to 0], \qquad \forall A_N \in \mathscr{A}_N \tag{6.1}$$

Also, let us define $\{k_N(t), 0 \leq t \leq 1\}$ and $\{W_N\}$ as in (5.1) and (5.2) and assume that under $\{H_{(N)}\}$,

$$W_N \xrightarrow{\mathscr{D}} W + \mu^*, \qquad \text{in the } J_1\text{-topology on } D[0, 1] \tag{6.2}$$

where W is, as before, a standard Brownian motion on [0, 1] and

$$\mu^* = \{\mu^*(t), 0 \leq t \leq 1\} \in C[0, 1] \tag{6.3}$$

Then, for the one-sided Kolmogorov–Smirnov type test, we have the asymptotic power of the repeated test, under $\{H_{(N)}\}$, given by

$$P\{W(t) + \mu^*(t) > \omega_\alpha^+ \text{ for some } t \in [0, 1]\} \tag{6.4}$$

and for the two-sided case, it is

$$P\{|W(t) + \mu^*(t)| > \omega_\alpha \text{ for some } t \in [0, 1]\} \tag{6.5}$$

Similarly, from the Cramér–von Mises type statistics, the asymptotic power is given by

$$P\left\{\int_0^1 [W(t) + \mu^*(t)]^2 \, dt > \omega_\alpha^*\right\} \tag{6.6}$$

For suitable $\mu^*(t)$, such as $\mu^*(t) = t\mu^*, 0 \leq t \leq 1, \mu^* \neq 0$, (6.4) or (6.5) can be expressed in terms of normal distribution and density functions (e.g., see Anderson, 1960). Like the null case [viz., (5.10)–(5.12)], (6.6) poses problems in setting a closed expression. In the remainder of this section, we consider the specific problems of Section 4, and in view of (6.4)–(6.6) try to set some optimality properties of the proposed tests.

For the test for the p-quantile in Section 4.1, H_0 being stated in (4.2), we set $H_{(N)}: \xi_p = \xi_p^{(N)} = \xi_p{}^0 + N^{-1/2}\gamma$ where $\gamma <$ (or $\neq$) 0 and further we assume that if $F_{(N)}(x)$ be the d.f. of X_1 under $H_{(N)}$, then $F_{(N)}(x) = F(x + N^{-1/2}\gamma)$ where F has a continuous and positive density function (f)

in some neighborhood of $\xi_p{}^0$. Then (6.2) holds with $\mu^*(t) = -\gamma t f(\xi_p{}^0)$, $0 \le t \le 1$, so that we have a linear drift in this case.

Consider next the one-sample location problem of Section 4.2. Here, for H_0: $\theta = 0$ in (4.8), we set $H_{(N)}$: $\theta = \theta_N = N^{-1/2}\gamma$, where $\gamma >$ (or $\neq$) 0, and we assume that F_0 possesses an absolutely continuous probability density function (p.d.f.) $f_0(x)$ with a finite Fisher information

$$I(f_0) = \int_{-\infty}^{\infty} \left[\frac{f_0'(x)}{f_0(x)}\right]^2 dF_0(x) < \infty, \qquad f_0'(x) = \frac{d}{dx} f_0(x) \tag{6.7}$$

Let then $\phi_0(u) = -f_0'(F_0^{-1}(u))/f_0(F_0^{-1}(u))$, $0 < u < 1$, and let

$$\rho_1(\phi, \phi_0) = \left(\int_0^1 \phi(u)\phi_0\left(\frac{1+u}{2}\right) du\right) \Big/ \left[I(f_0) \int_0^1 \phi^2(u)\, du\right]^{1/2} \tag{6.8}$$

In this case, it follows from Sen (1975a) that (6.2) holds with

$$\mu^*(t) = t\gamma\rho_1(\phi, \phi_0)[I(f_0)]^{1/2}, \qquad 0 \le t \le 1 \tag{6.9}$$

Note that for (6.4), when $\gamma > 0$, under (6.9), the maximum value is obtained when $\phi(u) = \phi_0((1+u)/2)$, $0 \le u < 1$. Thus, in this problem, the optimal score function is given by $\phi_0((1+u)/2)$, $0 < u < 1$.

Let us next consider the bivariate independence problem. Following Behnen (1971), we consider the following type of alternative hypotheses. Let $\beta_j(t)$, $0 < t < 1$, $j = 1, 2$, be two real-valued functions satisfying (i) $\int_0^1 \beta_j(t)\, dt = 0$ and (ii) $0 < \int_0^1 \beta_j^2(t)\, dt = \sigma_j^2 < \infty$, $j = 1, 2$. Further, let $\gamma_{Nj} = \{\gamma_{Nj}(t), 0 \le t \le 1\}$, $j = 1, 2$, be two functions defined, for each N (≥ 1), such that (i) for every $0 < u, v < 1$, $\int_0^u \gamma_{N1}(t)\, dt \int_0^v \gamma_{N2}(s)\, ds \ge$ (or $\le$) 0 with the strict inequality sign on a set of Lebesgue measure nonzero, and (ii) $\int_0^1 \gamma_{Nj}(t)\, dt = 0$, $\int_0^1 [\gamma_{Nj}(t) - \beta_j(t)]^2\, dt \to 0$, and

$$\sup_{0 \le u, v \le 1} \{N^{-1}[\gamma_{N1}(u)\gamma_{N2}(v)]^4\} \to 0$$

as $N \to \infty$. Finally, let f_1 and f_2 be the p.d.f. of F_1 and F_2, respectively. Then, we consider alternative hypotheses

$$\begin{aligned} H_{(N)} = H_{(N)}^{\Delta}&: f(x, y) \\ &= f_1(x)f_2(y)\,[1 + \Delta N^{-1/2}\gamma_{N1}(F_1(x))\gamma_{N2}(F_2(y))] \end{aligned} \tag{6.10}$$

for $(x, y) \in E^2$, where Δ is a real constant; $\Delta = 0$ corresponds to H_0 in (4.15). In this case, (6.2) holds with

$$\mu^*(t) = \Delta t \rho_1{}^* \rho_2{}^* \sigma_1 \sigma_2, \qquad 0 \le t \le 1 \tag{6.11}$$

where σ_1, σ_2 are as defined earlier and

$$\rho_j{}^* = \left(\int_0^1 \phi_j(t)\beta_j(t)\, dt\right) \Big/ \sigma_j A_j, \qquad j = 1, 2 \tag{6.12}$$

and the A_j are defined following (4.20). In this case, the optimal score functions are given by $\phi_j \equiv \beta_j$, $j = 1, 2$.

We consider now the tests of randomness displayed in Sections 4.4 and 4.5. For brevity, we consider only the case of the H_0 in (4.26) and (4.27) (which includes the two-sample location problem); the case of the scale problem will follow on parallel lines. Here we assume that the d.f. F in (4.26) admits of an absolutely continuous p.d.f. f with a finite Fisher information $I(f)$ [see (6.7)]. We set $H_{(N)}$: $\beta = \beta_N = C_N^{-1}\gamma$ where γ is a real constant and C_N^2 is defined preceding (4.24). Let then $\phi_0(u)$, $0 < u < 1$, be defined as in the text following (6.7) and set

$$\rho_2(\phi, \phi_0) = \left(\int_0^1 \phi(u)\phi_0(u)\, du\right) \Big/ A[I(f)]^{1/2} \tag{6.13}$$

where $A^2 = \int_0^1 \phi^2(u)\, du - (\int_0^1 \phi(u)\, du)^2$. Then, from Sen and Ghosh (1972) and Sen (1975a), it follows that here (6.2) holds with

$$\mu^*(t) = t\gamma\rho_2(\phi, \phi_0)[1(f)]^{1/2}, \qquad 0 \le t \le 1 \tag{6.14}$$

Thus, the optimal score function is $\phi \equiv \phi_0$.

Miller and Sen (1972) and Sen (1974c) have studied the weak convergence of (generalized) U-statistics and von Mises' differentiable statistical functions; the a.s. convergence of S_n^2 to ζ_1 [see (4.34) and (4.38)] follows from Sen (1977a). For testing H_0 in (4.29) versus $H_{(N)}$: $F \equiv F_{(N)}$ for which $\theta(F_{(N)}) \to \theta(F) = \theta_0$ (for some fixed F) in such a way that $N^{1/2}[\theta(F_{(N)}) - \theta_0] \to \gamma$ as $N \to \infty$, where γ is a real constant, it follows that here (6.2) holds (for both ζ_1 known and unknown) with

$$\mu^*(t) = \gamma t/m\zeta_1^{1/2}, \qquad 0 \le t \le 1 \tag{6.15}$$

The case of repeated significance tests based on sample means is a special one where $m = 1$, $\gamma = \lim_{N\to\infty} N^{1/2}[E(X \mid F_{(N)}) - E(X \mid F)]$ and $\zeta_1 = V(X_1)$.

The case of linear combinations of order statistics is somewhat similar to that of U-statistics. Here, under $\{H_{(N)}\}$, we conceive of a sequence $\{F_{(N)}\}$ of d.f., such that for some fixed F [for which μ, defined by (4.44) is equal to μ_0 in (4.46)],

$$\lim_{N\to\infty} N^{1/2} \left[\int_{-\infty}^{\infty} x[\phi(F_{(N)}(x)) - \phi(F(x))]\, dF_{(N)}(x) + \int_{-\infty}^{\infty} x\phi(F(x))\, d[F_{(N)}(x) - F(x)]\right] = \gamma \tag{6.16}$$

exists and $F_{(N)} \to F$ (as $N \to \infty$), at all points of continuity of the latter. Then again (6.2) holds with $\mu^*(t) = \gamma t/\sigma$, $0 \le t \le 1$, where σ^2 is given by (4.45). It follows from the work of Jung and Blom, reported by Sarhan and Greenberg

(1962, Chap. 4), that for a variety of parametric problems, the function ϕ can be so chosen that σ^2 in (4.45) is equal to the reciprocal of the Fisher information of a parameter θ which $n^{-1}T_n$ in (4.43) estimates (asymptotically efficiently), and in such a case, the tests based on $\{T_n\}$ will have the optimality in the class of all linear functions of order statistics.

Finally, let us consider the case of bundle strength of parallel filaments, treated in Section 4.8. From Sen (1973b) and Sen and Bhattacharyya (1976), it follows that (6.2) holds with $\mu^*(t) = t\gamma/\sigma$, $0 \le t \le 1$, where σ^2 is defined following (4.52) and γ is the assumed limit of $N^{1/2}\{\sup_{x \ge 0} x[1 - F_{(N)}(x)] - \sup_{x \ge 0} x[1 - F(x)]\}$, $F_{(N)}$ being the d.f. of X under $H_{(N)}$.

Let us now study the nature of the stopping variable M_N, defined by (2.8). Note that, by definition,

$$N^{-1}EM_N = N^{-1} \sum_{k=n_0}^{n} kP\{M_N = k\}$$

$$= N^{-1}n_0 + N^{-1} \sum_{k=n_0}^{N} P\{M_N > k\} \tag{6.17}$$

Further, note that $\{M_n > k\} \equiv \{h_N^{(l)} \le h_{N,\alpha}, \forall n_0 \le l \le k\}$ (and a similar statement holds for $\bar{h}_N$). Thus, for the one-sided case in (2.11), whenever $N \to \infty$ and $n_0/N \to 0$, under $H_{(N)}$,

$$N^{-1}E[M_N | H_{(N)}] \to \int_0^1 P\{W(s) + \mu^*(s) \le \omega_\alpha^+, \forall 0 \le s \le t\}\, dt \tag{6.18}$$

and similarly for the two-sided Kolmogorov–Smirnov type test, we have

$$N^{-1}E\{M_N | H_{(N)}\} \to \int_0^1 P\{\omega_\alpha - \mu^*(s) \le W(s) \le \omega_\alpha + \mu^*(s), 0 \le s \le t\}\, dt \tag{6.19}$$

For the Cramér–von Mises type test, we obtain

$$N^{-1}E\{M_N | H_{(N)}\} \to \int_0^1 P\left(\int_0^t \{W(s) + \mu^*(s)\}^2\, ds \le \omega_\alpha^*\right) dt \tag{6.20}$$

It may be remarked that if for two drift functions $\mu_j^* = \{\mu_j^*(t), t \in I\}$, $j = 1, 2$, we define an ordered relationship $\mu_1^* \prec \mu_2^*$ by $\mu_1^*(t) \le \mu_2^*(t)$, for every $t \in I$, then, for every $t \in [0, 1]$,

$$\mu_1^* \prec \mu_2^* \Rightarrow P\{W(s) + \mu_1^*(s) \le \omega_\alpha^+, \text{ for every } s \in [0, t]\}$$

$$\ge P\{W(s) + \mu_2^*(s) \le \omega_\alpha^+, \text{ for every } s \in [0, t]\} \tag{6.21}$$

Note that the ordered relationship is only a sufficient condition for the inequality in (6.21); a necessary and sufficient condition for the same is not known. Thus, if we consider two competing tests for the same null hypothesis based on two Kolmogorov–Smirnov type statistics, say $h_{N,1}$ and $h_{N,2}$, and if (6.18) holds for each of them, with the respective drift functions $\mu_1{}^*$ and $\mu_2{}^*$, then

$$\mu_1{}^* \prec \mu_2{}^* \Rightarrow \text{expected stopping time for the first test (under } \{H_{(N)}\}) \text{ is not smaller than the parallel quantity for the second test} \tag{6.22}$$

In this sense, the ordering of the drift function provides us with the relative (asymptotic) performances of competing tests in terms of the smallness of the expected stopping times. The situation becomes more involved in the two-sided case or in the case of the Cramér–von Mises type statistics. There, although a similar statement intuitively seems to be very appealing, it may be considerably difficult to provide an analytical proof. Asymptotic optimality of such tests in terms of stochastic smallness of stopping times remains an open problem.

Note that in the context of rank tests and tests based on linear functions of order statistics, we have considered some optimal score function $\phi(u)$, $0 < u < 1$. In each case, the optimal ϕ leads to the maximization of $\mu^*(t)$ (among the class of applicable ϕ) for each $t \in [0, 1]$. As such, such an optimal ϕ will minimize the right-hand side of (6.18), within the same class. Hence, from the expected stopping number point of view also, the same optimality results hold.

7. EFFICIENCY OF REPEATED SIGNIFICANCE TESTS

In this section, we are primarily concerned with the asymptotic relative efficiency (ARE) of repeated significance tests and the conventional terminal tests based on the target sample size. In view of the fact that the null as well as asymptotic nonnull distributions of the test statistics under a repeated testing scheme and a similar statistic under a terminal testing scheme do not have the same form, it is rather difficult to employ the conventional Pitman ARE measures to study the desired results. We are in a position to employ the Bahadur ARE measure and this will be exploited.

Basically, we are interested in comparing the ARE of h_N, defined by (2.11) and (2.13), and that of the terminal statistic T_N. It follows from (5.2) and Theorem 5.1 that under H_0 and the hypothesis of Theorem 5.1,

$$L((T_N - \mu_N{}^0)/\sigma_N) \to \mathcal{N}(0, 1) \qquad \text{as} \quad N \to \infty \tag{7.1}$$

where $\mu_N{}^0$ and σ_N are defined by (2.9). Similarly, (6.2) ensures that under $\{H_{(N)}\}$,

$$L((T_N - \mu_N{}^0)/\sigma_N) \to \mathcal{N}(\mu^*(1), 1) \tag{7.2}$$

In view of the assumed a.s. convergence in (3.2), (7.1) and (7.2) continue to hold for type B tests as well, where σ_N^2 is replaced by $\psi(N)S_n^2$. Now, looking at (5.8), (5.9), (5.12), (6.4)–(6.6), and (7.1)–(7.2), we gather that for local alternatives (viz., $\{H_{(N)}\}$), although the asymptotic power functions are given, equating them in the Pitman fashion gives us a measure of efficiency which depends on α, the level of significance [through ω_α^+, ω_α, or ω_α^* in (6.4)–(6.6)], as well as the particular form of μ^* [viz., γ in (6.9), (6.14), etc.], besides being very involved in computations. It is therefore difficult to interpret the measure in a meaningful way.

For this reason, we take recourse in the Bahadur ARE (see Puri and Sen, 1971, pp. 122–123), which can be adopted here without much difficulty. Note that for the standard Wiener process $W = \{W(t), 0 \le t \le 1\}$,

$$-(2/\lambda^2) \log P\{W(1) \ge \lambda\} \to 1 \qquad \text{as} \quad \lambda \to \infty \tag{7.3}$$

and hence, by (5.8), we have

$$-(2/\lambda^2) \log P\{\omega^+ \ge \lambda\} \to 1 \qquad \text{as} \quad \lambda \to \infty \tag{7.4}$$

Let us now assume that there exists a nondecreasing sequence $\{Q(n)\}$ of positive numbers such that

$$\text{(i)} \qquad Q(n) \to \infty \qquad \text{as} \quad n \to \infty \tag{7.5}$$

$$\text{(ii)} \qquad \text{for every (fixed) } a\text{: } 0 < a < \infty, \quad \lim_{n\to\infty} Q(an)/Q(a) = s(a) \text{ exists} \tag{7.6}$$

where

$$\text{(iii)} \qquad s(a) \text{ is } \uparrow \text{ in } a\text{: } 0 < a < 1; \qquad s(1) = 1$$

$$\text{(iv)} \qquad (T_n - \mu_n^0)/\sigma_n Q^{1/2}(n) \to h \text{ a.s.,} \qquad \text{as} \quad n \to \infty \tag{7.7}$$

where h depends on the underlying probability measure P in (2.2)–(2.3); under H_0, $h = 0$, but it ceases to be equal to zero when H_1 holds. Then, by virtue of (7.3)–(7.7), providing the (approximate) Bahadur slopes, the Bahadur ARE of the one-sided Kolmogorov–Smirnov type test (based on $\{T_k, n \le N\}$) relative to the terminal test (based on T_N) is given by

$$e_{RT} = s^{-1}\left(\left[\limsup_{N\to\infty} \max_{1\le k\le N} h^2(Q(k)\sigma_k^2/\sigma_N^2 Q(N))\right] \Big/ h^2\right) = s^{-1}(1) = 1 \tag{7.8}$$

so that the repeated significance test and the terminal test are asymptotically equally efficient. On the other hand, by constitution, the terminal test demands waiting until all the N observations are available whereas the repeated significance test usually has an expected stopping time smaller than that of the waiting time to get the target sample size. Hence, from time and cost considerations, the repeated significance test is more desirable and it is no less (asymptotically) efficient than the terminal test.

For the two-sided Kolmogorov–Smirnov versus the terminal tests, we note that (7.3)–(7.4) hold even if we replace $W(1)$ by $|W(1)|$ and ω^+ by ω. The only change occurs in (7.8) where h has to be replaced by $|h|$, but the resulting expression is still equal to unity. Hence, the same conclusion holds.

Let us now consider the case of (2.13), i.e., the Cramér–von Mises statistics. In this case, we note that by (5.11)–(5.12)

$$P\{\omega^* \geq \lambda^2\} = P\left\{\sum_{s=1}^{\infty} u_s Z_s^2 \geq \lambda^2\right\}$$
$$\geq P\{u_1 Z_1^2 \geq \lambda^2\} = P\{|Z_1| \geq \lambda/u_1^{1/2}\} \tag{7.9}$$

where as $\lambda \to \infty$,

$$-(2/\lambda^2)\log P\{|Z_1| \geq \lambda/\sqrt{u_1}\} \to 1/u_1 = \pi^2/4 \tag{7.10}$$

On the other hand, for every $\lambda > 0$, defining $t^* = \frac{1}{2}u_1$,

$$P\{\omega^* \geq \lambda^2\} = P\left\{\sum_{s=1}^{\infty} u_s Z_s^2 \geq \lambda^2\right\}$$
$$\leq \inf_{0<t<t^*}\left\{\exp(-t\lambda^2)E\left[\exp\left(t\sum_{s=1}^{\infty} u_s Z_s^2\right)\right]\right\}$$
$$= \inf_{0<t<t^*}\left\{\exp(-t\lambda^2)\prod_{s=1}^{\infty}(1-2tu_s)^{-1/2}\right\} \tag{7.11}$$

so that for every ε: $0 < \pi^2/8 - 1$, we have from (7.11),

$$P(\omega^* \geq \lambda^2) \leq \exp[-(1+\varepsilon)\lambda^2](1-2(1+\varepsilon)u_1)^{-1/2}\prod_{s=2}^{\infty}(1-2(1+\varepsilon)u_s)^{-1/2}$$
$$\leq C_\varepsilon \exp[-(1+\varepsilon)\lambda^2], \qquad C_\varepsilon < \infty \tag{7.12}$$

where C_ε does not depend on λ. Hence, from (7.9)–(7.12), we obtain on letting $\varepsilon \to \pi^2/8 - 1$ that

$$1+\varepsilon \leq \liminf_{\lambda\to\infty}(-1/\lambda^2)\log P(\omega^* \geq \lambda)$$
$$\leq \limsup_{\lambda\to\infty}(-1/\lambda^2)\log P(\omega^* \geq \lambda) \leq \pi^2/8 \tag{7.13}$$

which ensures that

$$-(1/\lambda^2)\log P(\omega^* \geq \lambda) \to \pi^2/8 \qquad \text{as} \quad \lambda \to \infty \tag{7.14}$$

Let us define λ_{Ns} as in (2.12) and note that defining $Q(n)$ as in (7.5)–(7.7), for h_N defined by (2.13), we have

$$\frac{h_N}{Q(N)} = \sum_{s=n_0}^{N-1} \left\{\frac{(T_s - \mu_s{}^0)^2}{\sigma_s{}^2 Q(s)}\right\} \lambda_{Ns} \left\{\frac{Q(s)\sigma_s{}^2}{Q(N)\sigma_N{}^2}\right\} \tag{7.15}$$

so that in addition to (7.7), if we assume that

$$\lim_{N\to\infty} \sum_{s=n_0}^{N-1} \lambda_{Ns} \left\{\frac{Q(s)\sigma_s{}^2}{Q(N)\sigma_N{}^2}\right\} = q^* \text{ exists} \tag{7.16}$$

then by (7.7), (7.15), and (7.16), we have

$$h_N/Q(N) \to q^*h^2 \text{ a.s.,} \qquad \text{as} \quad N \to \infty \tag{7.17}$$

Thus, from (7.14), (7.17), (7.3), and (7.7), we obtain that for the Cramér–von Mises type test, the Bahadur ARE with respect to the terminal test (based on T_N) is given by

$$e^*_{RT} = (q^*h^2\pi^2/8)/(\tfrac{1}{2}h^2) = q^*\pi^2/4 \tag{7.18}$$

By virtue of (7.8), the Bahadur ARE of the Cramér–von Mises type test with respect to the Kolmogorov–Smirnov type test based on the same (partial) sequence $\{T_n, n \leq N\}$ is given by

$$e_{CK} = e^*_{RT} = q^*\pi^2/4 \tag{7.19}$$

As is usually the case (cf., the examples of Section 4), q^* is typically equal to $\frac{1}{3}$, so that (7.18) or (7.19) reduces to

$$e_{CK} = e^*_{RT} = \pi^2/12 = 0.82225 \tag{7.20}$$

Thus, from the point of view of the Bahadur ARE, the Kolmogorov–Smirnov type test has a superiority over the Cramér–von Mises type test.

For the repeated significance tests based on rank statistics and linear functions of order statistics, it is also possible to compute the Bahadur ARE of one score function relative to another competitor. The results simplify considerably when we consider the limiting Bahadur ARE where we confine ourselves to fixed alternatives "close" to the null hypotheses case. For example, in the case of the one-sample location problem, this limiting ARE can be shown to be equal to

$$e^0_{\phi_1,\phi_2} = s^{-1}(\rho_1{}^2(\phi_1, \phi_0))/s^{-1}(\rho_1{}^2(\phi_2, \phi_0)) \tag{7.21}$$

where $\rho_1(\phi_j, \phi_0)$ is defined by (6.8) (for $\phi_j = \phi$) and $s(\cdot)$ is defined by (7.6). Since $s^{-1}(1) = 1$ and $s(a)$ is $\uparrow$ in a, (7.21) implies that the optimal score function is again $\phi(u) = \phi_0((1+u)/2), 0 < u < 1$. Similar results hold for the bivariate independence, randomness, and other testing problems studied in

Section 4. In passing, we may remark, in particular, for the test for randomness treated in Section 4.5 that for (4.26)–(4.27) with equally spaced c_i (i.e., $c_i = a_0 + a_1 i/N$, $1 \le i \le N$, with a_0 and a_1 being arbitrary but fixed numbers), (7.6) holds with $s(a) = a^3$, so that $s^{-1}(u) = u^{1/3}$. Hence, in this case, the limiting Bahadur ARE of a given score function ϕ with respect to the optimal score function ϕ_0 [as defined following (6.7)] is equal to

$$e^0_{\phi,\phi_0} = s^{-1}(\rho_2{}^2(\phi, \phi_0)) = \rho_2^{2/3}(\phi, \phi_0) \qquad [\ge \rho_2{}^2(\phi, \phi_0)] \tag{7.22}$$

where ρ_2 is defined by (6.13). Thus, the Bahadur ARE in the limiting case here is greater than or equal to the conventional Pitman ARE for the corresponding terminal tests.

Throughout this section, we have explicitly studied the case of type A repeated significance tests. The case with type B tests can be dealt with in a similar fashion; (7.8) and (7.20) hold also for type B tests.

We conclude this section with the remark that by (2.13) and (5.1)–(5.2), the proposed Cramér–von Mises type statistics, expressible as $\int_0^1 W_N{}^2(t)\,dt$, are really the unweighted versions. A more general case may be $\int_0^1 \gamma(t) W_N{}^2(t)\,dt$ where $\gamma(t)$, $0 < t < 1$, is an integrable weight function. For general $\gamma(t)$, one can still derive (5.12), but the coefficients $\{u_s, s \ge 1\}$ will not be specified by (5.11) and they involve more complicated computational schemes. However, in such a general case, one may also proceed along the lines of (7.9)–(7.18) (with different u_s, $s \ge 1$, and q^*), so that an optimal γ can be found for which the corresponding e^*_{RT} will be greater than (or equal to) $\pi^2 q^*/4$. In this way, the Bahadur ARE can be maximized for the Cramér–von Mises type statistics when a weighted version is used. Intuitively, we feel that a smooth and nonincreasing $\gamma(t)$ will also make the expected stopping time smaller. More work in this direction is needed to clarify some of these goals.

8. SOME GENERAL REMARKS

In Section 2, while formulating the repeated significance tests, we adopted the philosophy that the experiment may be terminated prior to achieving the target sample size only if the null hypothesis is not acceptable. There may be some situations where we may like to curtail experimentation at an early stage if there is a clear-cut statistical decision in favor of either the null hypotheses or the alternative. Thus, in this situation, at any intermediate stage, we face a three-decision problem; to accept the alternative hypothesis along with the termination of experimentation, to accept the null hypothesis along with the termination of experimentation, or to continue experimentation. In view of the results of Section 5, it is possible to formulate repeated significance tests for this three-decision problem too. First, we consider the one-sided case. Consider the partial sequence $\{T_n, n \leq N\}$ in (2.1), and keeping in mind the one-sided alternatives of Section 4, we conceive of the following repeated procedure.

Let us assume two sequences $\{L_{Nk}, n_0 \leq k \leq N\}$ and $\{U_{Nk}, n_0 \leq k \leq N\}$ of real numbers satisfying the conditions that

$$L_{Nk} < U_{Nk}, \qquad \forall n_0 \leq k \leq N-1, \qquad \text{and} \qquad L_{NN} = U_{NN} = 0 \tag{8.1}$$

Then, starting with the initial sample of size n_0, we compute for each k, $(T_k - \mu_k^0)/\sigma_N$, where μ_k^0 and σ_N^2 are defined as in (2.9)–(2.10). Continue experimentation so long as

$$L_{Nk} < (T_k - \mu_k^0)/\sigma_N < U_{Nk} \qquad \text{and} \qquad k \leq N \tag{8.2}$$

If, for the first time, for some $k = M_N$ ($\geq n_0$), $(T_k - \mu_k^0)/\sigma_N$ falls below L_{Nk} (or goes above U_{Nk}), we stop experimentation at that stage, along with the acceptance of H_0 (or acceptance of the alternative H_1). Note that for $k = N$, by (8.1), the process terminates whenever $T_N \neq \mu_N^0$. Since T_N need not have a continuous distribution, though $P\{T_N = \mu_N^0\}$ usually goes to 0 as $N \to \infty$, we may use a randomized decision to accept H_0 or H_1 with probability $\frac{1}{2}$ each when $M_N = N$ and $T_N = \mu_N^0$. Thus, in this sense, the procedure is similar to that of a truncated sequential probability ratio test. We need to construct the lower and upper boundaries in (8.2) so that the procedure has bounded type I and II errors, say α and β $(0 < \alpha < \alpha + \beta < 1)$, respectively. In Sections 5 and 6, we have studied the weak convergence of $\{W_N\}$, defined by (5.1)–(5.2), to a (drifted) Brownian motion process when the null or some local alternative hypotheses hold, and the same results can be used now to choose the boundaries in (8.2). For example, if we let $L_{Nk} = L_N < U_N = U_{Nk}$, $\forall k \leq N-1$, then using (6.2) for the null as well as the (local) alternative case, we may make use of the results of Anderson (1960) to determine the asymptotic values of L_N and U_N. Similarly, we can also take triangular boundaries by letting $L_{Nk} = (k/N)L_N$ and $U_{Nk} = (k/N)U_N$ for $k \leq N-1$ and again make use of the boundary crossing probabilities of Wiener processes as given by Anderson (1960).

Let us next consider the case of two-sided alternatives. This is essentially a three-hypothesis situation and a formulation of the Sobel–Wald (1949) approach seems to be a natural choice. Essentially, it amounts to testing simultaneously two one-sided cases: the null versus the lower alternatives and the null versus the upper ones. The procedure sketched in the preceding paragraph can be extended along the lines of Sobel and Wald (1949).

It is worthwhile here to discuss briefly some related repeated significance tests in the nonparametric setup. First, we may consider the analog of the classical sequential probability ratio tests or the sequential likelihood ratio tests based on nonparametric statistics. The principal difference between these sequential tests and the proposed nonparametric tests of this chapter lies in the fact that whereas in the current setup the target sample size is set in advance, in the SPRT type of tests there is no *a priori* bound on it. Thus, whereas the actual sample size can be unbounded in the SPRT case, it has a maximum bound in our case.

There has been a steady flow of work on nonparametric sequential tests. Analogs of SPRT based on ranks have been studied by Savage and Sethuraman (1966, and the references cited therein). Another line of approach is based on the almost sure representation of nonparametric statistics in terms of a (drifted) Wiener process. In this respect, the Skorokhod–Strassen embedding of the Wiener process for martingales has been extensively exploited for various nonparametric statistics; see for example Sen and Ghosh (1971, 1973, 1974a,b), Miller and Sen (1972), Sen (1973a,b,c, 1974a,b), Ghosh and Sen (1972, 1976, 1977), Lai (1975), and others. Sequentialization of weak convergence of nonparametric statistics based on the classical Pyke–Shorack (1968) approach has also been studied by Braun (1976), Wellner (1974), and others. In principle, these developments show that the same procedure employed for the SPRT can be used for nonparametric statistics wherein the sequence of probability ratio (or likelihood ratio) statistics has to be replaced by suitable sequences of nonparametric statistics while the nature of the stopping rule and stopping variable remains the same.

Second, repeated significance tests are very common in life testing or clinical trials problems, where instead of sequential entry of the observations into the scheme we have the following feature. The experiment starts with the entry of n units while the observable random variables are the order statistics. One example of a life testing problem involves the lives of electric lamps of two concerns, where 100 lamps are taken from each and tested; the shortest lifetime among these 200 lamps is observed first, the second shortest second, and so on, until the longest one emerges last. In this case also, from time and cost considerations, it may be necessary to set an upper limit to experimentation, either in terms of a fixed time period or in terms of a preassigned number of failures. Such a terminated scheme is called truncated or censored. In most exploratory studies, early termination may fail to yield a sensitive test whereas prolonging the experiment too much may induce more cost without substantial increase in the precision of the statistical decisions to be made. Thus, a continuous monitoring of the experiment is often advocated. This results in a repeated testing scheme and is termed a progressive censoring scheme. Nonparametric tests for such progressively censored experiments have been proposed by Halperin and Ware (1974), Chatterjee and Sen (1973), Majumdar and Sen (1976, 1977, 1978a,b), and Sen (1976c), among others. Sen (1976b) has also studied progressively censored likelihood ratio tests for some simple models. In this setup, for a given sample size N (usually large), one has a sequence $\{T_{Nk}, k \leq N\}$, where T_{Nk} is based on the smallest k-order statistics of the N units to start with. The test statistics are similar to those in (2.11) or (2.13) with $\{T_n\}$ being replaced by $\{T_{Nn}\}$. However, the main difficulty here lies in the fact that whereas in the setup of our Section 2, $X_1, \ldots, X_N$ are all independent, in the context of progressive censoring the order statistics $X_{N,1} \leq \cdots \leq X_{N,N}$ are, unfortunately, not so. The dependence of successive order statistics introduces extra complications into the statistical analysis of the scheme. Chatterjee and Sen (1973) have exhibited some martingale properties of progressively censored nonparametric statistics and this enables us to use the invariance principles for martingales to obtain results analogous to those in Sections 5 and 6. Similar results for progressively censored likelihood ratio statistics have been studied by Sen (1976b,d) and Gardiner and Sen (1978).

In a variety of clinical trials and life testing problems, a combination of the repeated significance tests of the type considered in Sections 2–4 and of the progressive censoring tests described earlier arises in a very natural way. Consider a life testing model with batch arrivals where all the units do not enter into the schemes at the same point of time; moreover, once a unit enters, it is constantly monitored until its failure occurs or the experiment is terminated, whichever comes first. Suppose that at the start of the experiment (i.e., at $t = 0$), an initial sample of n_0 (≥ 1) units enter into the scheme, and as time progresses on, more and more units are available, until the desired target sample size (N) is achieved at some time point t_0 (> 0). Let n_t be the cumulative sample size at time t ($0 \leq t \leq t_0$) and $n_t = N$ for $t \geq t_0$. Typically, the scheme may be represented as in Fig. 1.

Note that n_t is nondecreasing in t (≥ 0) and for every $0 < a < b$, out of n_b entries prior to time point b, the n_a have been monitored for a period $b - a$ or more, and some failures may occur during the same time. Thus, in the progressive censoring scheme (allowing a maximum experimentation time

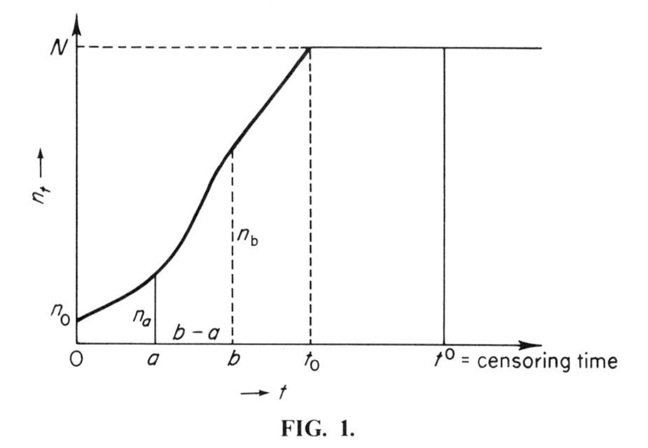

FIG. 1.

t^0), at any particular time t (> 0) of the n_t entries, the effective monitoring periods (prior to t) are the actual differences of their entry points from t. For units surviving at t, these observations are censored, and for failures occurring prior to t, the actual failure times are known. Hence, if the ith unit enters into the scheme at time t_i, $i = 1, \ldots, N$ (where $0 \le t_1 \le \cdots \le t_N \le t_0$), denoting their actual failure times by X_i (random variables), $i = 1, \ldots, N$, at time t for the n_t units for which $t_i \le t$, we actually observe

$$Y_i(t) = \min(X_i, t - t_i), \qquad \delta_{it} = \begin{cases} 1 & \text{if } X_i \le t - t_i \\ 0 & \text{otherwise}, \quad 1 \le i \le n_t \end{cases} \tag{8.3}$$

For the set of random vectors $\{Y_i(t), \delta_{it};\ 1 \le i \le n_t\}$, we construct a suitable test statistic which we denote by

$$T_N(t) \qquad \text{for} \quad 0 \le t \le t^0 \tag{8.4}$$

Then, in the same fashion as in (2.11)–(2.13), we are interested in using as a test statistic

$$h_N = \sup_{0 \le t \le t^0} T_N(t) \quad \text{or} \quad \sup_{0 \le t \le t^0} |T_N(t)| \quad \text{or} \quad \int_0^{t_0} T_N{}^2(t)\, dt \tag{8.5}$$

where, of course, in (8.5), to make the expressions simple, we have assumed that under the null hypothesis under consideration, for every t ($\in [0, t^0]$), $T_N(t)$ has a location 0 [in the sense that the mean, or some other measure of central tendency, of the distribution of $T_N(t)$ is equal to 0].

In most of the cases of real interest, distribution theory for h_N in (8.5) is highly complicated even in the situation in which the null hypotheses are true. However, in many cases, it is possible to consider some closely related statistics for which we have more leverage on the distribution theory.

For every n: $1 \le n \le N$ corresponding to $X_1, \ldots, X_n$, the order statistics are denoted by $X_{n,1} \le \cdots \le X_{n,n}$, so that we really have a triangular array of order statistics:

$$\{X_{n,1} \le \cdots \le X_{n,n}\colon 1 \le n \le N\} \tag{8.6}$$

where the elements in the same or different rows are not mutually stochastically independent. Let T^*_{nm} be a suitable statistic based on the partial sequence $X_{n,1}, \ldots, X_{n,m}$ (for $1 \le m \le n \le N$) designed to test an appropriate null hypothesis against suitable alternatives when $X_{n,1}, \ldots, X_{n,m}$ are observable. Then, as we progress, we have the double sequence

$$\{T^*_{nm},\ 1 \le m \le n;\ n = 1, \ldots, N\} \tag{8.7}$$

In a variety of situations, it may be possible to define a sequence $\{k_n;\ 1 \le n \le N\}$ of nonnegative integers (depending on the t_i, $i \le N$, t_0 and t^0)

and a sequence of integer-valued random variables $\{k_n^*: 1 \leq n \leq N\}$ such that

$$\sup_{0 \leq t \leq t^0} T_N(t) = \max_{1 \leq n \leq N} \left\{ \max_{k \leq k_n^*} T_{nk}^* \right\} \tag{8.8}$$

where with probability arbitrarily close to 1 (as $N \to \infty$),

$$\max_{1 \leq n \leq N} \left\{ \max_{k \leq k_n^*} T_{nk}^* \right\} \leq \max_{1 \leq n \leq N} \left\{ \max_{k \leq k_n} T_{nk}^* \right\} \tag{8.9}$$

(i.e., $k_n^* \leq k_n$, $n \leq N$), and similar inequalities hold for the two other forms in (8.5). Note that (8.9) holds trivially when $k_n = n$, $n \leq N$. In the latter case, however, the difference of the two sides of (8.9) may not be small, and thus employing the distribution of the right-hand side may result in a conservativeness of the test based on the left-hand side. Thus, our goal will be to choose $\{k_n\}$ in such a way that the percentile points of the distributions of the statistics on both sides of (8.9) are close to each other. This is often possible by virtue of invariance principles for the double sequence in (8.7) which along with the stochastic convergence of $n^{-1}k_n^*$ to some positive number bounded from above by $n^{-1}k_n$, $n \leq N$, ensures some smooth distribution, at least in the asymptotic case. For example, let us consider the two-sample location/scale problem treated in Sections 4.4 and 4.5, where we conceive of a batch arrival model, so that under a progressive censoring scheme, at the time point t, we have a bunch of censored linear rank statistics T_{nk}^* (for $n \leq n_t$ and $k \leq k_n^*$ where k_n^* is the number of failures among the n units exposed from their entries to the time point t). For fixed n, behavior of the sequence $\{T_{nk}^*, k \leq k_n^*\}$ has been thoroughly studied by Chatterjee and Sen (1973). On the other hand, in our case here, we allow n to vary between n_0 and n_t, so that with the variation of t ($0 \leq t \leq t^0$), we allow n to vary between n_0 and $N \wedge n_{t^0}$. This requires the formulation of a two-dimensional time parameter stochastic process based on $\{T_{nk}^*, 1 \leq k \leq n \leq N\}$, which has been actually considered by Sen (1976a) where under very general regularity conditions, it is shown that such processes converge weakly (in the $D[0, 1]^2$ space) to a two-parameter Gaussian process termed the *Brownian sheet*. Actually, for most of the cases treated in Section 4, similar invariance principles hold for (8.7). As a result, if $W = \{W(\mathbf{t}); \mathbf{0} \leq \mathbf{t} \leq \mathbf{1}\}$ is a Brownian sheet on $[0, 1]^2$ (so that $EW = 0$ and $EW(\mathbf{s})W(\mathbf{t}) = \mathbf{s} \wedge \mathbf{t}$, $\mathbf{s}, \mathbf{t} \in [0, 1]^2$), and if A is a closed subspace of $[0, 1]^2$, we confront the problem of finding out the distributions of functionals like

$$\sup_{\mathbf{t} \in A} W(\mathbf{t}), \qquad \sup_{\mathbf{t} \in A} |W(\mathbf{t})|, \qquad \text{or} \qquad \int_A W^2(\mathbf{t})\, d\mathbf{t} \tag{8.10}$$

For the particular case of $A = [0, 1]^2$, Dugue (1969) has obtained the characteristic functions of some of these statistics. However, the form does not yield the corresponding distribution functions in very tractable forms. Usually A is a proper subset of $[0, 1]^2$, including $\mathbf{t} = \mathbf{0}$ as its lower vertex and parts of both lines $t_1 = 0$, $t_2 = 0$ as its edges. For such a subset of $[0, 1]^2$ (or even for $A = [0, 1]^2$), unfortunately, the current status of knowledge on the distribution theory of functionals of Brownian sheets does not provide us with the desired algebraic forms for the distributions of the functions in (8.10). Nevertheless, the prospect for simulation studies of these distributions (via the invariance principles for suitable double arrays of independent random variables) appears to be bright. Wellner (1975) has considered some simulations for W itself, while more extensive simulation studies of the distributions of the random functions in (8.10) are underway. Pursuit of analytical techniques to derive such distributions remains open.

We conclude this section with a brief discussion on some further extensions of repeated significance tests, particularly relating to the situation underlying (8.3)–(8.7). In the context of time-sequential studies, a common problem arises from the (possible) withdrawals of subjects from the experimental scheme, either due to some chance causes or to other accountable ones. Thus, in a batch arrival model allowing different points of entries and accommodating withdrawals, one needs to consider a picture deeper than in Fig. 1. Here, at any time point t (> 0), one has the accounting of (i) the number n_t of entries prior to t, (ii) the failure and withdrawal times of these entries occurring prior to t, and (iii) the surviving individuals. Since the entry points are not all equal and also withdrawals occur at possibly different points of time, more complications arise in the construction of suitable statistics $\{T_N(t), t > 0\}$ in (8.4). Among other possibilities, it is often assumed that withdrawal times are random variables whose distribution satisfies suitable side conditions enabling one to simplify the model considerably. For example, in the two-sample problem of Section 4 [i.e., (4.21)–(4.22)], the actual observable random variable may be

$$X_i^0 = \min\{X_i, W_i\}, \qquad i = 1, \ldots, N \tag{8.11}$$

where W_i stands for the withdrawal time (possibly $+\infty$) for the ith individual, $i = 1, \ldots, N$. Now, if the ith individual enters into the experimental scheme at time point t_i (≥ 0), then at a time point $t \geq t_i$, one really observes

$$Y_i^0(t) = \min\{X_i^0, (t - t_i)\} \tag{8.12}$$

Thus, in such a case, our repeated significance test is actually based on the collection of random functions $\{Y_i^0(t), t \geq t_i, i = 1, \ldots, N\}$. If we assume that X_i and W_i are independent and further that the distribution of W_i does not depend on i, then under H_0 in (4.22), X_{2i}^0 and X_{2i-1}^0, $i \geq 1$, all have the same distribution; hence we can proceed as in (8.3)–(8.4) where we replace the

$Y_i(t)$ by the corresponding $Y_i^0(t)$. Here the assumption of identity of the d.f.'s of the W_i is crucial, for without this assumption, even under H_0 in (4.22), X_{2i}^0 and X_{2i-1}^0 may not have a common distribution, so that the construction of a suitable (genuinely or asymptotically) distribution-free test based on the $Y_i^0(t)$ may be difficult. Although this assumption does not appear to be very unrealistic, there are situations in which it may not be tenable. For example, in clinical trials, the pattern (distribution) of withdrawals may be related to the effectiveness of the treatments—under steady improvement (or consistent deterioration) quitting the treatment (or changing over to a different one) at an intermediate stage is not unlikely. Elimination of the effects of heterogeneity of withdrawal distributions on the testing procedures poses a complicated problem requiring detailed (and somewhat involved) statistical analysis of the whole complex [see Majumdar and Sen (1978b)].

Another important situation arises as a result of the presence of concomitant variables in the experimental setup. In the conventional terminal testing procedures, the analysis of covariance technique is usually employed for the testing problem. In the framework of Sections 2–4, it is not difficult to make adjustments for concomitant variates. For nonparametric testing problems, analysis of covariance tests are available in the literature; some of these are discussed by Puri and Sen (1971, Chaps. 5, 7). In fact, all we need is to work with such covariate-adjusted statistics in (2.1), (2.4), (2.6), (2.11), and (2.13). In the setup of Section 5, these adjustments call for the development of suitable invariance principles for such statistics in the general multivariate case which may be utilized to obtain results parallel to the ones in our Theorems 5.1 and 5.2. Some of these developments are in progress now. The situation becomes somewhat more complicated when one considers the batch arrival–progressive censoring model underlying (8.3)–(8.7) and there are concomitant variates in the experimental setup. The basic problem is that the estimation of the nuisance parameters in general linear models (in an analysis of covariance setup) based on partial (censored) data (as needed for eliminating the effects of the covariates in the testing problem) introduces complicated types of dependence patterns in the partial array of (covariate adjusted) $\{T_{nm}^*\}$, defined as in (8.7). This, in turn, makes it difficult to study the necessary distribution theory in simple forms. The usual canonical reduction technique used for the (multivariate) analysis of variance problems also faces some difficulties here because of the basic fact that as one obtains more and more observations, not only is an updating of the canonical form needed, but also the complete picture of the whole set of observations at an intermediate stage is not available. More work in this potential area is desired. In the context of nonparametric testing for some multiple regression models under the progressive censoring setup, Majumdar and Sen (1978a) have studied a general class of rank order tests involving Kolmogorov–Smirnov and Cramér–von Mises type statistics. A natural extension of their procedures in the setup of batch arrival models involves a repeated testing situation as in our (8.3)–(8.7). However, it is not difficult to extend their results in this direction, and these are underway.

We conclude this section with a discussion of some related repeated significance tests (in the parametric case) treated in Armitage (1975) and elsewhere. Armitage, for example, has presented a general discussion of some repeated significance tests for the mean of a normal distribution (or the true probability of a binomial distribution) based on the partial sequence $\{T_k = k^{-1/2}S_k/\sigma, 1 \leq k \leq N\}$, where $S_k = X_1 + \cdots + X_k, k \geq 1, \sigma^2 = \text{var}(X_i)$, $i \geq 1$, and for simplicity we take $EX_i = 0$ for every $i \geq 1$. Notice that for testing $H_0 : EX_i = 0$ versus $EX_i \neq 0$, $i \geq 1$, σ known, T_k is the classical parametric test statistic based on the sample of size k. Thus, these repeated significance tests are based on the partial sequence $\{T_k\}$ of normalized test statistics for the successive sample sizes. If σ^2 is not known, we take $T_k = k^{-1/2}S_k/s_k$, where $s_k^2 = (k-1)^{-1}\sum_{i=1}^{k}(X_i - \bar{X}_k)^2$ and $\bar{X}_k = k^{-1}\sum_{i=1}^{k} X_i$; we need to start, then with $k \geq 2$. On the other hand, in our Section 4.6, we considered a class of repeated significance tests based on U-statistics and von Mises' differentiable statistical functions, which for the particular case of a kernel of degree 1, involves partial sequences of the form $\{N^{-1/2}S_k/\sigma, 1 \leq k \leq N\}$ (or the corresponding sequence of studentized partial sums). Although related, the test statistics for the two procedures have different distributions. For the test based on $\{T_k, 1 \leq k \leq N\}$, one needs to know the boundary crossing probabilities for $\{t^{-1/2}W(t), t \in [0, 1]\}$, where $W = \{W(t), t \in [0, 1]\}$ is a standard Wiener process. Since $t^{-1/2}W(t)$ is not well behaved as $t \to 0$ and, moreover, the weak convergence of stochastic processes constructed from $\{T_k\}$ may not hold generally if one includes the lower end-point of the unit interval, one really needs to base a repeated significance test on the partial sequence $\{T_k : k_N \leq k \leq N\}$, where $k_N \to \infty$ but $N^{-1}k_N \to 0$ as $N \to \infty$ (otherwise the distribution theory becomes quite intractable). Even for some $\varepsilon : 0 < \varepsilon < 1$, analytical forms for the distributions of $\sup\{W(t)/t^{1/2} : \varepsilon \leq t \leq 1\}$ or $\sup\{|W(t)|/t^{1/2} : \varepsilon \leq t \leq 1\}$ are not precisely known. Nevertheless, the simulation study made by Majumdar and Sen (1977) provides us with the approximate percentile points of these distributions for some typical values of ε (the entries are sensitive to very small values of ε, but are fairly stable when ε is greater than 0.05). These critical values can be incorporated in the construction of nonparametric repeated significance tests based on $\{k^{-1/2}S_k/\sigma, k \leq N\}$ or on general $h_N^{(k)}$ in (2.11), where in the denominator of the statistics we replace σ_N by σ_n. In that case, we need to start testing when a sample of size k_N has been reached, where $k_N \to \infty$ but $N^{-1}k_N \to 0$ as $N \to \infty$.

REFERENCES

Anderson, T. W. (1960). A modification of the sequential probability ratio tests to reduce the sample size, *Ann. Math. Statist.* **31**, 165–197.

Armitage, P. (1975). "Sequential Medical Trials" (2nd ed.). Wiley, New York.

Behnen, K. (1971). Asymptotic optimality and ARE of certain rank order tests under contiguity, *Ann. Math. Statist.* **42**, 325–329.

Braun, H. I. (1976). Weak convergence of sequential linear rank statistics, *Ann. Statist.* **4**, 559–575.

Chatterjee, S. K., and Sen, P. K. (1973). Nonparametric testing under progressive censoring, *Calcutta Statist. Assoc. Bull.* **22**, 13–50.

Dugue, D. (1969). Characteristic functions of random variables connected with Brownian motion and of the von Mises' multidimensional ω_n^2." *Multivariate Analysis*—II (P. R. Krishnaiah, ed.), pp. 289–301. Academic Press, New York.

Gardiner, J. C., and Sen, P. K. (1978). Asymptotic normality of a class of time-sequential statistics and applications. *Comm. Statist.* **7** (in press).

Ghosh, M., and Sen, P. K. (1972). On bounded length confidence intervals for the regression coefficient based on a class of rank statistics, *Sankhyā Ser. A* **34**, 33–52.

Ghosh, M., and Sen, P. K. (1976). Asymptotic theory of sequential tests based on linear functions of order statistics. "Essays in Probability & Statistics" (Ogawa Volume: (S. Ikeda, ed.), pp. 480–499. Shinko Tsusho, Tokyo.

Ghosh, M., and Sen, P. K. (1977). Sequential rank tests for regression, *Sankhyā Ser. A* **39** 45–62.

Hájek, J., and Šidák, Z. (1967). "Theory of Rank Tests." Academic Press, New York.

Halperin, M., and Ware, J. (1974). Early decision in a censored Wilcoxon two-sample test for accumulating survival data, *J. Amer. Statist. Assoc.* **69**, 414–422.

Hoeffding, W. (1948). A class of statistics with asymptotically normal distribution, *Ann. Math. Statist.* **19**, 293–325.

Lai, T. L. (1975). On Chernoff-Savage theorem and sequential rank tests, *Ann. Statist.* **3**, 825–845.

Majumdar, H., and Sen, P. K. (1976). Chi-square tests for general models under progressive censoring with batch arrivals. Inst. of Statist., Univ. North Carolina, Mimeo Ser. Rep. No. 1058.

Majumdar, H., and Sen, P. K. (1977). Rank order tests for grouped data under progressive censoring, *Comm. Statist.* **6**, 507–524.

Majumdar, H., and Sen, P. K. (1978a). Nonparametric tests for multiple regression under progressive censoring *J. Multivar. Anal.* **8**, in press.

Majumdar, H., and Sen, P. K. (1978b). Nonparametric testing for simple regression under progressive censoring with staggering entry and random withdrawal. *Comm. Statist.* **7**, in press.

Miller, R. G. Jr. (1970). A sequential signed rank test, *J. Amer. Statist. Assoc.* **65**, 1554–1561.

Miller, R. G. Jr. (1972). Sequential rank tests—one sample case, *Proc. Berkeley Symp. Math. Statist. Probability, 6th* **1**, 97–108.

Miller, R. G. Jr., and Sen, P. K. (1972). Weak convergence of U-statistics and von Mises' differentiable statistical functions, *Ann. Math. Statist.* **43**, 31–41.

Puri, M. L., and Sen, P. K. (1971). "Nonparametric Methods in Multivariate Analysis." Wiley, New York.

Pyke, R., and Shorack, G. (1968). Weak convergence of a two sample empirical process and a new approach to Chernoff-Savage theorems, *Ann. Math. Statist.* **39**, 755–771.

Samuel-Cahn, E. (1974a). Two kinds of repeated significance tests and their applications for the uniform distribution. *Comm. Statist.* **3**, 419–432.

Samuel-Cahn, E. (1974b). Repeated significance tests II for hypotheses about the normal distribution, *Comm. Statist.* **3**, 711–734.

Samuel-Cahn, E. (1974c). Repeated significance tests I and II. Generalizations, *Comm. Statist.* **3**, 735–744.

Sarhan, A. E., and Greenberg, B. G. (1962). "Contributions to Order Statistics." Wiley, New York.

Savage, I. R., and Sethuraman, J. (1966). Stopping time of a rank order sequential probability ratio test based on Lehmann alternatives, *Ann. Math. Statist.* **37**, 1154–1160.

Sen, P. K. (1960). On some convergence properties of U-statistics, *Calcutta Statist. Assoc. Bull.* **10**, 1–18.

Sen, P. K. (1973a). Asymptotic sequential tests for regular functionals of distribution functions, *Theor. Probability Appl.* **18**, 235–249.

Sen, P. K. (1973b). On fixed size confidence bond for the bundle strength of filaments, *Ann. Statist.* **1**, 526–537.

Sen, P. K. (1973c). An asymptotically optimal test for the bundle strength of filaments, *J. Appl. Probability* **10**, 586–596.

Sen, P. K. (1974a). The invariance principle for one sample rank order statistics, *Ann. Statist.* **2**, 49–62.

Sen, P. K. (1974b). Almost sure behavior of U-statistics and von Mises' differentiable statistical functions, *Ann. Statist.* **2**, 387–395.

Sen, P. K. (1974c). Weak convergence of generalized U-statistics, *Ann. Probability* **2**, 90–102.

Sen, P. K. (1975a). Rank statistics, martingales and limit theorems, "Statistical Inference of Related Topics" (M. L. Puri, ed.), pp. 129–158. Academic Press, New York.

Sen, P. K. (1975b). An invariance principle for linear combinations of order statistics, Inst. of Statist., Univ. North Carolina, Mimeo Ser. Rep. No. 1047; to appear in *Z. Wahrsch. Verw Geb.* (1978).

Sen, P. K. (1976a). A two-dimensional functional central limit theorem for linear rank statistics, *Ann. Probability* **4**, 13–27.

Sen, P. K. (1976b). Weak convergence of progressively censored likelihood ratio statistics and its role in asymptotic theory of life testing, *Ann. Statist.* **4**, 1247–1257.

Sen, P. K. (1976c). On Wiener process embedding for linear combinations of order statistics, *Sankhyā Ser. A.* **38**, 190–193.

Sen, P. K. (1976d). Weak convergence of some quantile processes arising in progressively censored test, Inst. of Statist., Univ. North Carolina, Mimeo Ser. Rep. No. 1092; to appear in *Ann. Statist.* (1979).

Sen, P. K. (1976e). Asymptotically optimal rank order tests for progressive censoring, *Calcutta Statist. Assoc. Bull.* **25**, 65–78.

Sen, P. K. (1977a). Some invariance principles relating to jack-knifing and their role in sequential analysis, *Ann. Statist.* **5**, 316–329.

Sen, P. K. (1977b). Tied-down Wiener process approximations for aligned rank order statistics and some applications, *Ann. Statist.* **5**, 1107–1123.

Sen, P. K., and Bhattacharyya, B. B. (1976). Asymptotic normality of the extrema of certain sample functions, *Z. Wahrsch. Verw Geb.* **34**, 113–118.

Sen, P. K., and Ghosh, M. (1971). On bounded length sequential confidence intervals based on one-sample rank order statistics, *Ann. Math. Statist.* **42**, 189–203.

Sen, P. K., and Ghosh, M. (1972). On strong convergence of regression rank statistics, *Sankhyā Ser. A.* **34**, 335–348.

Sen, P. K., and Ghosh, M. (1973a). A Chernoff-Savage representation of rank order statistics for stationary ϕ-mixing processes, *Sankhyā, Ser. A.* **35**, 153–172.

Sen, P. K., and Ghosh, M. (1973b). A law of iterated logarithm for one sample rank order statistics and an application, *Ann. Statist.* **1**, 568–576.

Sen, P. K., and Ghosh, M. (1974a). Sequential rank tests for location, *Ann. Statist.* **2**, 540–552.

Sen, P. K., and Ghosh, M. (1974b). Some invariance principles for rank statistics for testing independence, *Z. Wahrsch. Verw Geb.* **29**, 93–108.

Sen, P. K., Bhattacharyya, B. B., and Suh, M. W. (1973). Limiting behavior of the extrema of certain sample functions, *Ann. Statist.* **1**, 297–311.

Sobel, M., and Wald, A. (1949). A sequential decision procedure for choosing one of three hypotheses concerning the unknown mean of a normal distribution, *Ann. Math. Statist.* **20**, 502–522.

Wellner, J. A. (1974). Convergence of the sequential uniform empirical process with bounds for centered Beta r.v.'s and a log-log law, Tech. Report 31, Dept. of Mathematics, Univ. of Washington, Seattle.

Wellner, J. A. (1975). Monte-Carlo of two-dimensional Brownian sheets, "Statistical Inference and Fielded Topics" (M. L. Puri, ed.), pp. 59–76. Academic Press, New York.

A Review of Some Recent Work on Discrete Optimal Factorial Designs for Statisticians and Experimenters†

J. N. SRIVASTAVA

DEPARTMENT OF MATHEMATICS
COLORADO STATE UNIVERSITY
FORT COLLINS, COLORADO

1. INTRODUCTION

In this chapter we discuss classical factorial experiments in which each factor has two levels. We consider the problem of obtaining designs for every value of N within a certain practical range, such that these designs maximize

† This work was supported by NSF Grant No. MCS76-23282.

ISBN 0-12-4266 01-0

the amount of information with respect to certain widely used optimality criteria. The optimality is within the class of balanced designs explained more elaborately in Section 2. Although the optimality is yet provable only in a restricted class, it appears that in the majority of cases the designs are either optimal or near optimal in the class of all designs whether balanced or unbalanced. Furthermore, the property of balance gives rise to ease in the analysis and interpretation of the results. This chapter has been written in a readable form both for the experimenters and the users of designs on the one hand, and for mathematical statisticians and researchers on the other. Sections 2 and 13 are written especially for the user. In other sections, we consider the structure of factorial designs, the counting operator, orthogonal fractional factorial designs, balanced designs and balanced arrays, existence of balanced designs for the 2^m factorial for different values of m, computation of optimal designs, discussion of optimality criteria, and some comments on search designs, which constitute the most recent development in the field. For lack of space, we restrict to designs of resolution $2l+1$, for $l = 2, 3$, which allow the estimation of general mean, the main effects, and all interactions involving l or lesser number of factors. Besides the review of earlier work, some new results are also presented.

Those readers who are mainly interested in using the designs presented in this chapter should proceed to Sections 2 and 13, although Section 2 can be omitted by those readers who are familiar with the basic definitions and concepts. These sections have been written in a nonmathematical language, and should be accessible to persons who are not primarily statisticians. There is a slight amount of repetition between the material in these two sections and that in other sections, which is necessary to ensure uniformity of presentation and readability. It is hoped, therefore, that this will not be minded.

The subject of statistics has two branches, statistical planning and statistical inference. These branches are deeply interrelated, and should be regarded as the two faces of the same coin. The subject of statistical planning deals with how to collect data intelligently and appropriately, whether they are collected from experiments or investigations and surveys, are univariate or multivariate, or are collected in one or more than one stage. The data may deal with nonstochastic problems or they may involve observations on complex stochastic processes. The subject of statistical inference, on the other hand, is concerned with how to draw inferences from the data once they have been collected. In other words, it deals with how to analyze data and summarize them appropriately for various uses and purposes. Naturally, the nature of statistical analysis and inference will depend on the structure of the data or the way the data were collected. On the other hand, how to collect data properly is a question which cannot be answered well unless we know how the collected data will be analyzed. Thus, this deep interrelationship between the two branches is inherent in all statistical problems and activities.

This chapter is concerned with a subbranch of statistical planning, namely design of experiments. In this subbranch, the available theory is concerned mostly with those statistical problems which can be described relatively satisfactorily by a linear model, i.e., where the expected values of observations are known linear functions of certain (possibly) unknown parameters. Usually, the observations have to be independent or at least have known correlations. Thus, the situation is essentially describable as follows. We have a vector of N observations $\mathbf{y}(N \times 1)$, which satisfies the conditions

$$\mathrm{Exp}(\mathbf{y}) = A\xi, \qquad \mathrm{var}(\mathbf{y}) = \sigma^2 I_N \tag{1.1}$$

For estimation purposes, the major part of the theory does not need to assume any particular distribution for the observations. For confidence intervals and testing of hypotheses, usually normality is assumed. There are, however, important procedures available for the distribution-free case, based usually on rank order statistics (e.g., see Puri and Sen, 1971). Since this chapter is concerned mostly with the design problem, we shall assume that we are mostly concerned with the problem of estimation of the unknown parameters $\xi(v \times 1)$, where σ^2 is unknown and $A(N \times v)$ is a known matrix. The design problem is this. There are potentially many different sets of N observations $\mathbf{y}$ that one could collect. We may denote the ith set by $\mathbf{y}^{(i)}(N \times 1)$, $i \in I$, where I is a set of indices which could be finite or infinite. It is assumed that for each i the expected value of $\mathbf{y}^{(i)}$ is $A^{(i)}\xi$. Thus, the same ξ is involved for each i, and the different matrices $A^{(i)}$ are known. A question as to which $\mathbf{y}^{(i)}$ should be selected for estimating ξ is then the main design question or the question of statistical planning. Now, if we restrict ourself to obtaining an unbiased estimate of ξ, which has minimum variance, then it is well known that under (1.1), the estimate of ξ which has minimum "variance" is given by the solutions of the *normal equations*:

$$A'A\hat{\xi} = A'\mathbf{y}, \qquad \hat{\xi} = (A'A)^{\mathrm{C}}A'\mathbf{y} \tag{1.2}$$

also, then we have

$$\mathrm{var}(\hat{\xi}) = \sigma^2(A'A)(A'A)(A'A)^{\mathrm{C}} \tag{1.3}$$

where $(A'A)^{\mathrm{C}}$ is any conditional inverse of $(A'A)$. It is well known that the estimate $\hat{\xi}$ is in general nonunique and depends on the chosen conditional inverse, as in the second condition of (1.2). However, if $\mathbf{l}'\xi$ is any known linear function of the parameter ξ [in other words, if $\mathbf{l}'\xi$ is such that there exists a known linear function of the observations $\mathbf{b}'\mathbf{y}$ such that $\text{Exp}(\mathbf{b}'\mathbf{y}) = \mathbf{l}'\xi$, for all ξ], then $\mathbf{l}'\hat{\xi}$ and its variance are unique and we have

$$\text{var}(\mathbf{l}'\hat{\xi}) = \sigma^2 \mathbf{l}'(A'A)^{\text{C}}\mathbf{l} \qquad \text{if} \quad \mathbf{l}'\xi \text{ is estimable} \tag{1.4}$$

Equation (1.4) shows that the variance of the estimate of estimable linear functions of ξ depends on the matrix $(A'A)$; the "larger" $(A'A)$, the "smaller" is the variance. For this reason, the matrix $(A'A)$ is usually called the information matrix. Thus, one way to select a good design (in other words, a good set of observations $\mathbf{y}^{(i)}$) for estimating ξ (or its linear functions) is to choose i in such a way that the corresponding information matrix $A^{(i)\prime}A^{(i)}$ is "large" in some sense.

How to measure the "largeness" of $A'A$? This is commonly done in terms of characteristic roots of $A'A$. For simplicity, we shall consider the case when $A'A$ is nonsingular so that $v = \text{rank } A = \text{rank } A'A \leq N$. Let $\chi_1 \geq \chi_2 \geq \cdots \geq \chi_v$ be the characteristic roots of $A'A$. The three popular criteria are as follows:

(i) *A-Optimality: The trace criterion:* This says that we should maximize the average of $\chi_1^{-1}, \ldots, \chi_v^{-1}$; i.e., maximize $(\sum_{j=1}^{v} \chi_j^{-1})^{-1}$.
(ii) *D-Optimality: The determinant criterion:* Maximize $|A'A|$, the determinant of $A'A$ which also equals $\prod_{j=1}^{v} \chi_j$.
(iii) *E-Optimality: The largest root criterion:* Maximize ψ_v, an extreme root.

These criteria have the following broad interpretation. Minimizing the trace criterion is equivalent to minimizing the average variance of the estimates of all normalized linear functions of the parameters. Maximizing the determinant of $A'A$ amounts to maximizing the volume of the ellipsoid of concentration. The determinant of $A'A$ is also sometimes called the generalized variance. The largest root criterion is a minimax criterion and corresponds to minimizing the largest variance that the estimate of a normalized linear combination of parameters could have. Among these, for use in discrete factorial designs, the author favors the trace criterion. Some remarks in this connection are made in Section 14.

The (linear model-based) theory of optimal designs is concerned with the problem of choosing the observations $\mathbf{y}$ (or, equivalently, the matrix A) such that one of these criteria, or some other criterion (based on A), is optimized. There exists a large body of work in this direction. Some of the more important papers of the last two decades are listed in the references at the end of this chapter. Here, mention must be made of the pioneering work of Kiefer. Other important contributors include Wolfowitz, Karlin, Studden, and Sacks. Among the more recent contributors are Atwood, Herzberg, and Wynn. The work of Wahba and Ylvisaker is in the direction of stochastic processes. A very important book on optimal designs which also contains references to contributors of authors from Eastern Europe and the Soviet Union is the book by Fedorov. Some of the work on optimal design theory is among the most brilliant pieces of work in all of statistics.

The authors mentioned in the preceding paragraph have approached the optimal design problem mostly by measure-theoretic methods. Thus, they deal with factorial designs, but largely the case in which the factors are continuous. A large body of work has also been done on discrete optimal designs, where the factors have a finite number of levels which are discrete. In this regard, mention must be made of the work of Kiefer, Hedayat, and Seiden. Some of the latest work concerns the construction of optimum generalized Youden squares (GYD). Other work in the area of discrete designs concerns the classical factorial experiment. Some of the main work in this area is due to the author and his associates. Lately, this work has been generalized by Yamamoto and his associates, particularly Shirakura and Kuwada.

The work on discrete optimal designs has not been summarized in any book. Thus, this chapter should be found useful. In this chapter, we shall restrict ourselves to factorial designs of the type 2^m. Very little work has been done on optimal designs of classical factorial types which do not belong to the 2^m series. For the 2^m case, work has been done on the designs of both odd and even resolutions. A design is said to be of resolution $2l + 1$ if the general mean, the main effects, and all interactions involving l or lesser number of factors can be estimated, assuming that higher order interactions are zero. Also, a design is said to be of resolution $2l + 2$ if all of the interactions and other factorial effects can be estimated assuming that interactions involving $2l + 2$ or more factors are negligible, although interactions involving $2l + 1$ factors may not be negligible. In this chapter, we shall restrict our attention to designs of odd resolution, particularly those with $l = 2$ and 3. Balanced resolution IV designs, optimal in a certain important subclass, are discussed in the paper by Srivastava and Anderson (1970).

There are many, many important papers of the aforementioned workers and of others, which have not been mentioned in this chapter for lack of space. The selection of papers included in the reference list, which deals with subjects not concerned with the main body of this chapter, is not extensive, and is meant for illustration.

We conclude this section with some comments on open problems and further lines of work in the field of this chapter. The reader will find that in many cases (particularly for 2^m factorials with $m \geq 8$, and the smaller values of N for these cases), the absolute efficiency of the optimal balanced designs is rather low. We do not know how close the optimal balanced designs are (in terms of absolute efficiency) to designs which are optimal in the class of *all* designs whether or not balanced; the balanced designs may or may not be sufficiently close. Thus, research in optimal designs, without the condition of balance, is warranted. Another useful direction of work would be the combinatorial problems dealt with in Sections 10–12. A look at the original paper dealing with this method would show that in most cases, the proofs are long and tedious. Simplified proofs would be useful not only here but also in other areas of designs and combinatorics. Finally, a very significant area of work is the development of optimal search designs, or designs optimal under bias-free optimality criteria.

2. PRELIMINARIES FOR USERS OF DESIGNS

This chapter is concerned with the presentation of certain factorial designs which are good in a certain sense, and also the theory leading to the development of these. Almost all the work presented here is a summarization from earlier papers. However, the summary has been made in such a way as to make the chapter readable and also to give a flavor of the proofs of many results, so as to create insight into the subject.

In each experiment or investigation, there are variables which are called factors and responses. A response variable is a variable in an experiment in which the effect of something is studied. A factor is a variable whose effect is studied on a response variable. For example, in an experiment comparing several varieties of wheat, a response variable could be the yield of wheat, while a factor would be the different varieties of wheat. The various varieties of wheat would constitute levels of the factor. In another experiment, one may have a factor such as nitrogen fertilizer. This fertilizer can be applied at any desired rate to the different plots. The various rates at which the fertilizer can be applied can then be thought of as the levels of this particular factor. An experiment is conducted on a set of experimental units. On each unit of an experiment, various combinations of levels of factors are applied, and the response or characteristic under study is measured for each unit. A more detailed introduction to the concept of factors, responses, etc., would be found in the book by Roy *et al.* (1970, Chap. II).

It is clear that some factors, like nitrogen fertilizer, are basically continuous factors in the sense that the rate at which they can be applied is a continuous variable. However, certain other factors are discrete, for example various methods of cultivation of a particular crop. Although in many experiments certain factors are continuous, for various reasons they are sometimes studied in a discrete manner. Thus, for example, although nitrogen fertilizer can be applied at any desired rate, usually in agricultural experiments a few discrete levels are taken, for example 0, 20, or 40 lb/acre. In this case, a factor which is basically continuous is studied in a discrete fashion. This particular chapter deals with experiments in which there are many factors, and each factor is either discrete or is treated as discrete. Such experiments are called classical or discrete factorial experiments.

In most experiments, there occur factors which may be termed nuisance factors. Usually, there is one such factor, and the set of levels of this factor is called a set of blocks. A good experimenter tries his best to classify the experimental units with respect to the distinct levels of this factor, so that experimental units having the same level of the factor (in other words, experimental units which fall in the same block) are as homogeneous as possible, while units falling in different blocks could be as different as possible with respect to the response under study. In this chapter, almost all of the discussion corresponds to the case in which there is no such nuisance factor present. In other words, in the usual terminology, we shall assume that there are no block effects.

Consider an experiment with m factors. Suppose that the ith factor ($i = 1, \ldots, m$) has s_i levels. If the s_i are all equal, then the experiment is said to be a symmetrical factorial experiment; otherwise, the experiment is said to be an asymmetrical factorial experiment. In this chapter, we shall restrict ourselves to symmetrical factorial experiments of the 2^m type, which means that we have m factors each at two levels. Now consider a 2^m factorial experiment. Level combinations (also called treatment combinations, treatments, assemblies, or runs) are denoted by m-tuples $(j_1, j_2, \ldots, j_m)$, where j_r takes the value 0 or 1. For example, in a 2^3 experiment the treatments are (000), (001), (010), (011), (100), (101), (110), and (111). Here, the level combination $(j_1, j_2, \ldots, j_m)$ denotes that combination of the levels of the m factors in which the rth factor occurs at level j_r, for $r = 1, \ldots, m$. Thus, for example, the level combination (101) in a 2^3 factorial experiment denotes that level combination in which factors 1 and 3 occur at level 1, and factor number 2 occurs at 0.

For our purposes, a design (denoted usually by T) is defined merely as a collection of treatment combinations or assemblies, such that any particular assembly may not occur in T, or may occur one or more times. For example, the following is a design belonging to the 2^4 factorial experiment series:

$$T = \begin{bmatrix} 1111 \\ 1001 \\ 0101 \\ 1001 \\ 1001 \\ 1011 \\ 0100 \end{bmatrix} \tag{2.1}$$

Notice that the assembly (1111) occurs once in T, the assembly (0000) does not occur at all, while (1001) occurs three times, etc.

Corresponding to a 2^m factorial experiment, there exists the same number 2^m of factorial effects. These are generally denoted μ (the general mean); $F_1, \ldots, F_m$ (the main effects); $F_{12}, F_{13}, \ldots, F_{m-1,m}$ (the two-factor interactions); $F_{123}, F_{124}, \ldots, F_{m-2,m-1,m}$ (the three-factor interactions); ...; $F_{12\cdots m}$ (the m-factor interaction). The general mean μ denotes the average of the effects of all treatment combinations. The main effect F_1 denotes the difference between the effects of the two levels of the first factor, when we average out over the levels of the remaining $(m-1)$ factors. The main effects $F_2, F_3, \ldots$ are similarly defined. The interaction F_{12} is the two-factor interaction between the factors number 1 and number 2; it denotes the change in the difference between the effects of the two levels of the first factor when we change the level of the second factor from 0 to 1, and when we average out over the two levels of the remaining $(m-2)$ factors. Other two-factor interactions are similarly defined. Next, consider the three-factor interaction F_{123}. One important interpretation of this factorial effect is as follows. Consider a 2^{m-1} factorial experiment in which the level of the first factor is kept constant and the levels of the other $(m-1)$ factors are varied. In this experiment, we could calculate the interaction between the factors 2 and 3 (say F_{23}^*). Now, consider the other 2^{m-1} factorial experiment which is the same as this last one, except for the fact that the level of the first factor has been changed to 1. Compute the interaction between the second and third factor again, and call it $F_{23\cdot 1}^1$. Then the difference $F_{23\cdot 1}^1 - F_{23\cdot 1}^0$ denotes the change in the interaction between the second and third factors as we change the level of the first factor, while averaging out the levels of all the other factors; this difference is the interaction F_{123}. It can be shown that we shall get the same value of the interaction F_{123} irrespective of which factor we hold constant for the 2^{m-1} experiment. Higher order interactions are similarly defined. For a detailed introduction, the reader is referred to texts such as those of Cochran and Cox (1970) and Kempthorne (1971).

For any treatment combination $(j_1, \ldots, j_m)$ let $\tau(j_1, \ldots, j_m)$ denote its "true effect." The true effect of any treatment is the value of the yield or response that we would observe in the ideal case in which we have a set of units which are completely homogeneous and in which there are no random fluctuations. Thus, in a 2^m factorial experiment there exists the same number 2^m of true effects of level combinations. There is a relationship between the 2^m factorial effects and 2^m true effects of treatments. An easy to remember rule is as follows. The parameter μ equals the sum of all the 2^m true effects of the various treatments. Also, any main effect or interaction equals $\tau_1 - \tau_0$, where τ_1 denotes the sum of the true effects of a certain set of 2^{m-1} distinct treatment combinations, and τ_0 represents the same sum for the remaining 2^{m-1} treatments, where τ_1 and τ_0 are defined as follows. It is clear that $\tau_1 - \tau_0 = 2\tau_1 - \mu$, so that it is sufficient to define τ_1. Take any interaction (including any main effect) involving an odd (even) number of factors. Then corresponding to this interaction, τ_1 is the sum of the true effects of all those treatments each of which have an odd (even) number of factors occurring at level 1 in common with the factors occurring in this particular interaction. Thus, for the main effect F_2, the value of τ_1 is the sum of the true effects of all treatments in which the second factor occurs at level 1. For the interaction F_{13}, τ_1 is the sum of the true effects of all treatments in which factors 1 and 3 are both at level 1, or are both at level 0. Thus, for example, if we are considering a 2^5 factorial, then the true effect $\tau(1, 0, 1, 0, 1)$ is included in τ_1 corresponding to F_{13} because the treatment (10101) has an even number of factors at level 1 (namely, the factors 1 and 3) in common with the two factors (namely, factors 1 and 3) which occur in the interaction F_{13}. Similarly, in this case, a true effect $\tau(0, 0, 0, 1, 1)$ is also included in τ_1 because the treatment (00011) has an even number (namely, zero) of factors at level 1 which also occur in the interaction F_{13}. As a last example, consider the interaction F_{124} for a 2^7 factorial. It is clear that true effects of treatments like (1000000), (1101001), (0001010) are included in τ_1 since each of these treatments has an odd number of factors at level 1 out of the set of three factors, namely factors 1, 2, and 4.

We shall illustrate from a 2^4 factorial. To save space, we shall use matrix notation wherever possible which we now explain for the benefit of readers not familiar with it. A matrix of size $(m \times n)$ is an array of numbers with m rows and n columns. Let A be a matrix of size $(m \times n)$ and let B be a matrix of size $(p \times q)$. Then (i) A and B can be added if and only if they have the same number of rows and the same number of columns, i.e., $m = p$, $n = q$; and (ii) A and B can have two products, namely AB and BA. The product AB is definable if and only if the number of columns in A equals the number of rows in B, i.e., $n = p$. Similarly, the product BA is definable if and only if $q = m$. Let A, B, C be matrices of sizes $(m \times n)$, $(m \times n)$, and $(n \times p)$, respectively, with the following elements:

$$A = \begin{bmatrix} a_{11} & a_{12} & \cdots & a_{1n} \\ a_{21} & a_{22} & \cdots & a_{2n} \\ \vdots & \vdots & & \vdots \\ a_{m1} & a_{m2} & \cdots & a_{mn} \end{bmatrix}, \quad B = \begin{bmatrix} b_{11} & \cdots & b_{1n} \\ \vdots & & \vdots \\ b_{m1} & \cdots & b_{mn} \end{bmatrix}, \quad C = \begin{bmatrix} c_{11} & \cdots & c_{1p} \\ c_{21} & \cdots & c_{2p} \\ \vdots & & \vdots \\ c_{n1} & \cdots & c_{np} \end{bmatrix} \tag{2.2}$$

Then, we define

$$A + B = \begin{bmatrix} a_{11} + b_{11} & a_{12} + b_{12} & \cdots & a_{1n} + b_{1n} \\ \vdots & \vdots & & \vdots \\ a_{m1} + b_{m1} & a_{m2} + b_{m2} & \cdots & a_{mn} + b_{mn} \end{bmatrix} \tag{2.3}$$

$$AC = \begin{bmatrix} d_{11} & d_{12} & \cdots & d_{1p} \\ d_{21} & d_{22} & \cdots & d_{2p} \\ \vdots & \vdots & & \vdots \\ d_{m1} & d_{m2} & \cdots & d_{mp} \end{bmatrix}$$

where, for $i = 1, \ldots, m$ and $j = 1, \ldots, p$, we have

$$d_{ij} = a_{i1}c_{1j} + a_{i2}c_{2j} + a_{i3}c_{3j} + \cdots + a_{in}c_{nj} \tag{2.4}$$

Notice that if X and Y are two matrices, one or both of the products XY and YX may or may not be definable. Also, even if the matrices XY and YX are both definable and are of the same sizes, they may not be equal. This fact is expressed by saying that matrix multiplication does not follow the commutative law.

A matrix with one row is called a row vector, and a matrix with only one column is called a column vector. The zero matrix is a matrix all of whose elements equal zero. If A is a matrix, say the one in (2.2), and u is any number, then we can define the product of the number u and the matrix A, written uA, which is a matrix whose (i, j)th element is ua_{ij}. [The (i, j)th element of a matrix is the element standing in its ith row and jth column. Thus, the (2, 3) element of the matrix B in (2.2) is b_{23}.]

A matrix in which the number of rows and columns is the same is called a square matrix. A square matrix of size $(n \times n)$, which has the number 1 along its diagonal and 0 elsewhere, is called the identity matrix of size $(n \times n)$ and is denoted by I_n. Thus, if the matrix C in (2.2) is an identity matrix, then we must have $n = p$, $c_{11} = c_{22} = \cdots = c_{nn} = 1$, and $c_{ij} = 0$, for all $i \neq j$. Let $X(n \times n)$ be a square matrix. Then, X is said to be nonsingular if there exists an $(n \times n)$ matrix Z such that $XZ = I_n$. In this case, we also have $ZX = I_n$, and we write $Z = X^{-1}$. Again, if A is a matrix of size $(m \times n)$, then we denote by A' the $(n \times m)$ matrix whose (j, i)th element is the same as the (i, j)th element of A for $i = 1, \ldots, m$ and $j = 1, \ldots, n$; the matrix A' is said to be the transpose of A. If A is a square matrix and $A = A'$, then A is said to be *symmetric*. It can be easily checked that multiplication of any matrix by an appropriate sized identity matrix leaves that matrix unchanged. In other words, if $A(m \times n)$ is any matrix, then $A = I_m A = AI_n$. Let $\mathbf{a}' = (a_1, \ldots, a_n)$ and $\mathbf{b}' = (b_1, \ldots, b_n)$ be two row vectors of size $(1 \times m)$ each. Then $\mathbf{a}$ and $\mathbf{b}$ (which are, respectively, the transposes of $\mathbf{a}'$ and $\mathbf{b}'$) are said to be orthogonal to each other, if and only if $\mathbf{a}'\mathbf{b} \equiv \mathbf{b}'\mathbf{a} \equiv a_1 b_1 + a_2 b_2 + \cdots + a_n b_n = 0$. Finally, the Schur product of $\mathbf{a}'$ and $\mathbf{b}'$ is the row vector $(a_1 b_1, a_2 b_2, \ldots, a_n b_n)$. Also, if A is any matrix, then a submatrix of A is any matrix obtainable from A by deleting some rows and/or some columns of A.

We now proceed with the illustration of the definition of the factorial effects by considering a 2^4 experiment. It can be easily checked that in view of the foregoing discussion, these effects are given by (2.5) (on page 277) where the symbols $(+)$ and $(-)$ stand for the numbers $(+1)$ and (-1), respectively. These equations can be written in a compact form as

$$\mathbf{F} = D\mathbf{f} \tag{2.6}$$

$$
\begin{bmatrix} \mu \\ F_1 \\ F_2 \\ F_3 \\ F_4 \\ F_{12} \\ F_{13} \\ F_{14} \\ F_{23} \\ F_{24} \\ F_{34} \\ F_{123} \\ F_{124} \\ F_{134} \\ F_{234} \\ F_{1234} \end{bmatrix}
=
\begin{bmatrix}
+ & + & + & + & + & + & + & + & + & + & + & + & + & + & + & + \\
+ & - & + & + & + & - & - & - & + & + & + & - & - & - & + & - \\
+ & + & - & + & + & - & + & + & - & - & + & - & - & + & - & - \\
+ & + & + & - & + & + & - & + & - & + & - & - & + & - & - & - \\
+ & + & + & + & - & + & + & - & + & - & - & + & - & - & - & - \\
+ & - & - & + & + & + & - & - & - & - & + & + & + & - & - & + \\
+ & - & + & - & + & - & + & - & - & + & - & + & - & + & - & + \\
+ & - & + & + & - & - & - & + & + & - & - & - & + & + & - & + \\
+ & + & - & - & + & - & - & + & + & - & - & + & - & - & + & + \\
+ & + & - & + & - & - & + & - & - & + & - & - & + & - & + & + \\
+ & + & + & - & - & + & - & - & - & - & + & - & - & + & + & + \\
+ & - & - & - & + & + & + & - & + & - & - & - & + & + & + & - \\
+ & - & - & + & - & + & - & + & - & + & - & + & - & + & + & - \\
+ & - & + & - & - & - & + & + & - & - & + & + & + & - & + & - \\
+ & + & - & - & - & - & - & - & + & + & + & + & + & + & - & - \\
+ & - & - & - & - & + & + & + & + & + & + & - & - & - & - & +
\end{bmatrix}
\begin{bmatrix} \tau(1111) \\ \tau(0111) \\ \tau(1011) \\ \tau(1101) \\ \tau(1110) \\ \tau(0011) \\ \tau(0101) \\ \tau(0110) \\ \tau(1001) \\ \tau(1010) \\ \tau(1100) \\ \tau(0001) \\ \tau(0010) \\ \tau(0100) \\ \tau(1000) \\ \tau(0000) \end{bmatrix}
\tag{2.5}
$$

where $\mathbf{F}$ is the (16×1) vector on the left-hand side of (2.5), $\mathbf{f}$ is the (16×1) vector on the right-hand side of (2.5), and D is the (16×16) symmetric matrix with elements $(+1)$ and (-1) occurring there. It will be noticed that any two rows of the matrix D are orthogonal to each other, and so are any two columns. The matrix equation (2.5) is equivalent to a set of simultaneous equations expressing the elements of the vector $\mathbf{F}$ in terms of the elements of the vector $\mathbf{f}$. Notice that each row of D corresponds to one of the factorial effects, i.e., an element of $\mathbf{F}$. Similarly, any column of D corresponds to one of the treatment effects, which is an element of $\mathbf{f}$. To compute any element of $\mathbf{F}$ in terms of the elements of $\mathbf{f}$, we simply take the row of D corresponding to this element of $\mathbf{F}$ and multiply it with the vector $\mathbf{f}$. Thus, for example, we have $F_{13} = \tau_1 - \tau_0 = 2\tau_1 - \mu$, where

$$
\begin{aligned}
\tau_1 = \tau(1111) + \tau(1011) + \tau(1110) + \tau(0101) \\
+ \tau(1010) + \tau(0001) + \tau(0100) + \tau(0000)
\end{aligned}
$$

Equations (2.5) can be solved for $\mathbf{f}$ in terms of $\mathbf{F}$; it can be shown that we have

$$
\mathbf{f} = (2^{-4})D\mathbf{F} \tag{2.7}
$$

Thus the elements of $\mathbf{f}$ are obtained from $\mathbf{F}$ in the same way as those of $\mathbf{F}$ are obtained from $\mathbf{f}$, except that in the former case we multiply by the constant 2^{-4}.

In a factorial experiment, interest usually lies in the estimation of the unknown factorial effects, namely μ, the main effects, the two-factor interactions, and other higher interactions. It is clear from the discussion in this section that given the set of true effects of the various treatments, we can compute the various factorial effects, μ, F_i, etc. However, in most scientific experiments, it is usually found that a large number of the higher order interactions are negligible. In fact, in a large number of situations, most of the 2-factor and higher order interactions are negligible. In other situations, some of the 2-factor interactions may not be negligible, and so on. Now the question is this. If it is known that, say, the 2-factor and higher order interactions are negligible, can we estimate μ and the main effects from a small set of treatments, or do we need the entire set of the 2^m treatment effects? In general, the following question arises. We are interested in estimating a certain set of factorial effects which may or may not include μ or main effects or other interactions. Can we estimate these effects from a small set of treatments? The answer to these questions is in the affirmative. Indeed, we can use a set of treatments for estimating a given set of effects (assuming that all of the remaining effects are negligible) such that the number of treatments that we need to take is equal to the number of effects that we wish to estimate (Srivastava, 1977).

In this chapter, we discuss 2^m factorial designs of resolution V which utilize N runs or assemblies. Designs are available for various practical values of m and N. These designs are of resolution V or VII. A design is said to be of resolution V (VII) if we can estimate μ, the main effects, and all interactions involving two (three) factors or a lesser number of factors assuming that all higher order interactions are negligible. Thus, in a 2^m factorial design which is of resolution V, the number of parameters is ν_m, where

$$\nu_m = 1 + \binom{m}{1} + \binom{m}{2} = 1 + m(m+1)/2 \tag{2.8}$$

Similarly, for a design of resolution VII, the number of parameters is ν_m where

$$\nu_m = 1 + \binom{m}{1} + \binom{m}{2} + \binom{m}{3} \tag{2.9}$$

For the resolution V case, designs are available for values of m in the range 4 to 11, and for resolution VII designs we have $m = 6, 7$, and 8. For each value of m, designs are available for each value of N starting from the value ν_m and ν_m', respectively, for the resolution V and resolution VII cases. It can be proved that if $N < \nu_m$, then we cannot estimate all of the parameters of interest in a resolution V design. The case for resolution VII is similar.

What is the need of giving designs for every value of N? In the classical theory developed during the 1940s and 1950s, designs were obtained where the estimates of any two parameters are uncorrelated with each other. Such designs are called orthogonal designs. However, for these designs N can only have certain values which are multiples of a power of 2. For example, when $m = 4$ or 5 in the classical theory, we need N to be a multiple of 2^4; for $m = 6$, N should be a multiple of 2^5; for $m = 7$, N should be a multiple of 2^6, etc., if the design is to be able to estimate all the ν_m parameters. These values of N are obviously too restrictive. Occasionally, one has only a certain fixed amount of funds for carrying out a given kind of experiment. This amount dictates the choice of N. For example, when $m = 7$, we have $\nu_m = 29$, but in the classical theory, we need N to be a multiple of 64. The number of treatments needed in the classical theory are excessive. It is possible that with the funds that one has, one can conduct an experiment with 42 treatments and no more. One would, therefore, like to have a design which involves 42 treatments. Having a design with less than 42 treatments may or may not be desirable. It may not be desirable if the person wants to have the maximum accuracy in the experiment given the amount of funds that one possesses. If accuracy is not of decisive importance, the person may want to do an experiment with fewer than 42 treatments. In this case, one's choice of the value of N (which should be between 29 and 42) will be dictated by other considerations. We discuss this in more detail in Section 13.

A relatively small value of N would be desirable in many other situations. One situation is where the error variance σ^2 is very small. In other words, the random fluctuation on any experimental unit or from unit to unit is quite small. Another situation where relatively small values of N would be very important are those in which the cost of each experimental unit and the cost of performing the experiment on each unit are quite large. An example of the latter can be given in the manufacture of airplane wings. Suppose there are seven different factors influencing the performance of an airplane wing. Suppose the performance is being measured in terms of a response variable y, and suppose that y is influenced by the set of seven different factors, each of which we wish to experiment on with two levels. To do the whole experiment, we need to try 128 level combinations, which means that we need to manufacture airplane wings of 128 different specifications. Clearly, in this case, the inclusion of any particular treatment into the design would be quite expensive. In this case, therefore, one may wish to use the smallest value of N that is possible and which at the same time could allow us to estimate all of the factorial effects of interest.

One important feature of the designs discussed here is that they are "balanced." By this we mean that the variances and covariances do not depend on factors. To elaborate, consider two factorial effects, say θ_1 and θ_2, where θ_1 and θ_2 may or may not be distinct. Let $\hat{\theta}_1$ and $\hat{\theta}_2$ be the corresponding estimators based on the given design (say T). Then the design T is said to be balanced if and only if for all factorial effects θ_1 and θ_2 which are of interest, and which we are going to estimate, the value of $\text{cov}(\hat{\theta}_1, \hat{\theta}_2)$ depends only on three numbers, namely (a) the number of factors in θ_1, (b) the number of factors in θ_2, and (c) the number of factors which are common to both θ_1 and θ_2. (In case one of θ_1 and θ_2 is μ, then the number of factors in μ is taken to be zero.) Thus, for example, if T is balanced, then $\text{cov}(\hat{F}_{12}, \hat{F}_{14}) = \text{cov}(\hat{F}_{23}, \hat{F}_{34}) = \text{cov}(\hat{F}_{25}, \hat{F}_{45})$, etc.,

$$\text{var}(\hat{F}_{12}) = \text{var}(\hat{F}_{13}) = \text{var}(\hat{F}_{34}),$$

etc. Note that we do not necessarily have $\text{var}(\hat{F}_1) = \text{var}(\hat{F}_{12})$, because the number of factors involved in (F_1) is one while the number of factors involved in (F_{12}) is two. Similarly, $\text{cov}(\hat{F}_{12}, \hat{F}_{14})$ and $\text{cov}(\hat{F}_{13}, \hat{F}_{24})$ are not necessarily equal, since in the first case F_{12} and F_{14} have one factor in common, whereas in the second case F_{13} and F_{24} do not have any factor in common.

The most important benefit from having a balanced design is that it lends ease in the analysis and interpretation of the results. For the balanced design, the computations for obtaining estimators and writing down their variances and covariances are quite simple. We will observe this fact more closely in Section 13.

One other, and probably the most important, feature of the designs presented is that they maximize the amount of information. This is a mathematically somewhat complicated concept, and we shall therefore not go into details here. However, roughly speaking, the designs that we discuss minimize the sum of the variances of the estimates of all the factorial effects that are of interest to us and that we are estimating. The information is thus measured in terms of the reciprocal of this sum of variances. It turns out that in those cases in which the value of N is such that orthogonal designs exist, a balanced design must necessarily maximize the amount of information in this sense. For most other values of N, it appears that the balanced designs either do maximize the amount of information or are quite close to them. However, there are cases for which this is perhaps not so. On the other hand, it may be remarked here that even though a balanced design may not quite maximize the information, it may be preferred to an unbalanced design if the difference in the amount of information between the two designs is not much. The reason, again, is that the use of an unbalanced design may require much more effort in the analysis and interpretation of results.

3. 2^m FACTORIAL DESIGNS

We continue the discussion of the preceding section in a more mathematical direction. First, we introduce additional notation. An r-factor interaction $F_{i_1 i_2 \cdots i_r}$ involving the factors number $i_1, \ldots, i_r$ will sometimes be denoted by $F_{i_1} F_{i_2} \cdots F_{i_r}$ or by $F_1^{k_1} \cdots F_u^{k_u}$, where $k_u = 1$ if $u = i_1, \ldots, i_r$, and $k_u = 0$ otherwise. Let $\mathbf{F}$ be the $(2^m \times 1)$ vector of interactions given by

$$\mathbf{F}' = \{M; F_1, F_2, \ldots, F_m; F_{12}, F_{13}, \ldots, F_{m-1,m}; F_{123}, \ldots; \ldots; F_{12\cdots m}\} \tag{3.1}$$

The treatment $(j_1, \ldots, j_m)$ will sometimes be denoted by $f_1^{j_1} \cdots f_m^{j_m}$, or by $f_{i_1} \cdots f_{i_r}$, if $j_u = 0$, for $u = i_1, \ldots, i_r$, and $j_u = 1$, when $u \neq i_1, \ldots, i_r$. Also, ϕ denotes the treatment $(1, 1, \ldots, 1)$. The $(2^m \times 1)$ vector of treatments is then

$$\mathbf{f} = \{\phi; f_1 f_2, \ldots, f_m; f_1 f_2, f_1 f_3, \ldots, f_{m-1} f_m; f_1 f_2 f_3, \ldots; \ldots; f_1 f_2 \cdots f_m\} \tag{3.2}$$

To illustrate this, notice that if $m = 6$, F_{134}, $F_1 F_3 F_4$, and $F_1{}^1 F_2{}^0 F_3{}^1 F_4{}^1 F_5{}^0 F_6{}^0$ denote the same interaction, and $(1, 0, 1, 1, 0, 0)$, $f_2 f_5 f_6$, and $f_1{}^1 f_2{}^0 f_3{}^1 f_4{}^1 f_5{}^0 f_6{}^0$ denote the same treatments. Finally, the true effect of the treatment $f_1^{j_1} \cdots f_m^{j_m}$ will be denoted by the same symbol.

Consider the interaction $F_1^{k_1} F_2^{k_2} \cdots F_m^{k_m}$; this is a linear combination of the true effects of the various level combinations:

$$F_1^{k_1} F_2^{k_2} \cdots F_m^{k_m} = \sum_{j_1, \ldots, j_m} d_{k_1}(j_1)\, d_{k_2}(j_2) \cdots d_{k_m}(j_m)(f_1^{j_1} \cdots f_m^{j_m}) \tag{3.3}$$

where

$$d_k(j) = 1 \quad \text{if} \quad k = 0; \qquad d_1(0) = -1, \quad d_1(1) = 1 \tag{3.4}$$

The preceding is well known. From this, the following result, which is also well known, can easily be derived.

Theorem 3.1. If the j_r $(r = 1, \ldots, m)$, which take values 0 and 1, are regarded as numbers over the real field, then we have

$$\begin{aligned} F_1^{k_1} F_2^{k_2} \cdots F_m^{k_m} &= \sum_{j_1, \ldots, j_m} (2j_1 - 1)^{k_1} \cdots (2j_m - 1)^{k_m} (f_1^{j_1} \cdots f_m^{j_m}) \\ &= \prod_{r=1}^{m} (f_r{}^1 + (-1)^{k_r} f_r{}^0) \end{aligned} \tag{3.5}$$

where the last expression on the right-hand side is to be expanded considering the f_r^j $(r = 1, \ldots, m;\ j = 0, 1)$ as pure symbols (indeterminates) which commute.

Equation (3.3) can be expressed as

$$\mathbf{F} = D\mathbf{f} \tag{3.6}$$

where D is a $(2^m \times 2^m)$ matrix. The rows and columns of D correspond, respectively, to the various elements of $\mathbf{F}$ and $\mathbf{f}$. Also, the element of D which lies at the intersection of the row corresponding to the interaction $F_1^{k_1} \cdots F_m^{k_m}$ and the column corresponding to the treatment $f_1^{j_1} \cdots f_m^{j_m}$ is $d_{k_1}(j_1)$ $d_{k_2}(j_2) \cdots d_{k_m}(j_m)$, where for all r, j_r, and k_r take the values 0 and 1. It is clear that each element of the matrix D equals $+1$ or -1. Thus, each row of D has a sum of squares which equals 2^m. Also, D is symmetric, an important (unpublished) fact recently studied by the author. Each element in the first row and the first column of D equals 1. Again, let $(k_1, \ldots, k_m)$ and $(k_1', \ldots, k_m')$ be two distinct vectors with the k taking values 0 and 1. Consider the two rows of D corresponding to the interactions $F_1^{k_1} \cdots F_m^{k_m}$ and $F_1^{k_1'} \cdots F_m^{k_m'}$. The sum of products of these two rows of D

$$\sum_{j_1, \ldots, j_m} d_{k_1}(j_1) \cdots d_{k_m}(j_m)\, d_{k_1'}(j_1) \cdots d_{k_m'}(j_m) = \prod_{r=1}^{m} \left[\sum_{j_r=0}^{1} d_{k_r}(j_r)\, d_{k_r'}(j_r) \right] \tag{3.7}$$

Since the vectors $(k_1, \ldots, k_m)$ and $(k_1', \ldots, k_m')$ are distinct, there exists a value of r such that $k_r \neq k_r'$. Then

$$\sum_{j_r=0}^{1} d_{k_r}(j_r)\, d_{k_r'}(j_r) = \sum_{j=0}^{1} d_1(j) = 0 \tag{3.8}$$

by virtue of (3.4). This shows that any two rows of D are mutually orthogonal. Thus $(2^{-m/2}D)$ is an orthogonal matrix. It therefore follows that

$$\mathbf{f} = 2^{-m} D\mathbf{F} \tag{3.9}$$

Now, consider a fractional factorial design or, briefly, a design T with N runs so that T can be written as an $(N \times m)$ matrix containing zeros and ones such that any row of T denotes a treatment combination, and the columns of T correspond to the set of m factors. Notice that any treatment combination can appear zero, one, or more times in T. Also, N may have any value. Let $\mathbf{y}$ be an $(N \times 1)$ vector whose elements correspond to the rows of T. Also, the elements of $\mathbf{y}$ are defined as follows. Consider an experiment with N experimental units, such that each of the N treatments in T is applied to one and only one distinct experimental unit. Thus, from these N units, we will have N observations, one observation corresponding to each treatment in T. These observations are the elements of $\mathbf{y}$, arranged in the same order as they occur in T. If in T a treatment occurs several times as a row, then the corresponding observations can be put in any arbitrary but fixed order in $\mathbf{y}$.

As usual, we assume that the observations from different experimental units are independent. The means and the variances of the observations are then given by

$$\text{Exp}(\mathbf{y}) = (E^{*\prime}\mathbf{F})2^{-m} \tag{3.10a}$$

$$\text{var}(\mathbf{y}) = \sigma^2 I_N \tag{3.10b}$$

where σ^2 is a known or unknown nonnegative number, and where $E^{*\prime}$ is a matrix derived from D' by omitting those rows of D' which correspond to treatments which do not occur in T. Also, if a treatment occurs more than once in T, then the corresponding row of D' occurs the same number of times in $E^{*\prime}$. Notice that the rows of $E^{*\prime}$ correspond to the treatments in T and are arranged in the same order.

Now suppose that $\mathbf{F}_0$ is a vector of size $(2^m - v) \times 1$, where $0 \leq v \leq 2^m$, such that the elements of $\mathbf{F}_0$ are a subset of the elements of $\mathbf{F}$, and furthermore the elements of $\mathbf{F}_0$ are assumed to be negligible. Let $\boldsymbol{\xi}$ be a $(v + 1)$ vector which contains the remaining elements of $\mathbf{F}$, i.e., elements which are not in $\mathbf{F}_0$. Let E' be a matrix which is obtained from the matrix $E^{*\prime}$ by omitting those columns of $E^{*\prime}$ which correspond to interactions that are included in the vector $\mathbf{F}_0$. Then Eq. (3.10a) becomes

$$\text{Exp}(\mathbf{y}) = (E'\boldsymbol{\xi})2^{-m} \tag{3.10c}$$

Notice that Eq. (3.10a) and (3.10c) reflect the fact that we are assuming that no nuisance factors are present. In other words, we are assuming that there are no block effects.

Now consider the problem of estimating the parameters in $\boldsymbol{\xi}$, from the linear model given by (3.10b) and (3.10c). As is well known, the normal equations are given by

$$(2^{-m})EE'\boldsymbol{\xi} = E\mathbf{y} \tag{3.11}$$

where EE' ($=M$, say) is the information matrix. The matrix M is important and is studied in detail in the following sections. Here we comment upon the computation of the right side of (3.11). Construct a vector $\mathbf{y}^*(2^m \times 1)$ defined in the following manner. The elements of $\mathbf{y}^*$ correspond one-to-one to the elements of $\mathbf{f}$. If $\pi \in \mathbf{f}$, then let $y^*(\pi)$ denote the corresponding element of $\mathbf{y}^*$. Further, let $\mathbf{y}^*(\pi) = 0$, if $\pi \notin T$, and let $\mathbf{y}^*(\pi)$ be the total yield of π (from all observations in $\mathbf{y}$ which correspond to π) if $\pi \in T$. Then, from the definition of E' and $E^{*\prime}$, it is easy to check that

$$E\mathbf{y} = D_0\mathbf{y}^* \tag{3.12}$$

where D_0 is a $(v \times 2^m)$ matrix which is obtained from D by omitting all rows of D which correspond to elements of $\mathbf{F}$ that are in $\mathbf{F}_0$. In other words, the vector $E\mathbf{y}$ is obtained the same way from $\mathbf{y}^*$ as the elements of ξ are obtained from the vector $\mathbf{f}$. There exist simple techniques (like Yates' method) of obtaining the elements of $\mathbf{F}$, and hence of ξ, from the vector $\mathbf{f}$. Thus, these simple techniques can be used to obtain $E\mathbf{y}$ after $\mathbf{y}^*$ has been computed. Since $\mathbf{y}^*$ is defined in a simple way, it is clear that the right-hand side of Eq. (3.11) is quite simple to compute *without having to write the matrix* E. Later, we show the same is true of M.

A design T is said to be *nonsingular* with respect to a set of parameters ξ if and only if the corresponding information matrix M is nonsingular. All the elements of the vector ξ are estimable if and only if the design T is nonsingular. If M is nonsingular, then the estimate of ξ is given by

$$\hat{\xi} = M^{-1}(E\mathbf{y}) \cdot 2^m \tag{3.13}$$

Also, the dispersion matrix of the estimator is given by

$$V(\hat{\xi}) = \sigma^2 2^{2m} M^{-1} = \sigma^2 2^{2m} V \qquad \text{(say)} \tag{3.14}$$

Thus, the variance is proportional to the inverse of the information matrix M. As indicated in Eq. (3.13), the matrix M^{-1} will be denoted by V and will be referred to as the variance–covariance matrix. Equation (3.14) explains why M is called the information matrix. Intuitively speaking, the "larger" the information matrix M, the "smaller" is the covariance matrix V.

One important problem in factorial experimentation, for a given vector of parameters ξ, is to choose a design T such that the corresponding variance–covariance matrix V is "small." This particular question is the main topic of this chapter.

4. THE COUNTING OPERATOR AND THE INFORMATION MATRIX

Consider the set of $2m + 2$ symbols f_r^j ($j = 0, 1$; $r = 1, \ldots, m$), and f_{00} and f_{01}. Throughout this chapter we shall occasionally, without any danger of confusion, treat these symbols as a set of indeterminates. We assume that these symbols are commutative and associative with respect to multiplication. Let $\mathscr{P}$ be the ring of polynomials in these symbols over the real field. We shall assume that f_{00} and f_{01} are the identity elements of $\mathscr{P}$ with respect to addition and multiplication, respectively. We shall further assume that each of these $2m + 2$ symbols is idempotent; in other words,

$$(f_r^j)^2 = f_r^j \qquad \text{for all permissible } r \text{ and } j \tag{4.1a}$$

Furthermore, we shall assume that

$$f_r^0 f_r^1 = f_{00}, \qquad r = 1, \ldots, m \tag{4.1b}$$

To avoid triviality, we assume that the symbols f_{00} and f_{01} are distinct from the symbols $f_r^{\,j}$ for all r and j. Using this fact, it can be easily checked that we have

$$f_r^{\,1} + f_r^{\,0} = f_{01} \qquad \text{for all} \quad r \tag{4.1c}$$

Because of this fact, every polynomial $P \in \mathscr{P}$ is expressible in terms of the symbols $f_{00}, f_{01}, f_1^{\,1}, \ldots, f_m^{\,1}$.

Now consider a design T, and let $(t_{j1}, \ldots, t_{jm})$, $j = 1, \ldots, N$, be the jth row of T. Then there is a unique polynomial in $\mathscr{P}$, denoted by P_T, which provides a representation for T and which is given by the equation

$$P_T = \sum_{j=1}^{N} (f_1^{t_{j1}} \cdots f_m^{t_{jm}}) \tag{4.2}$$

It can be easily checked that if P is any polynomial in $\mathscr{P}$, then it can be uniquely represented in terms of the symbols f_{00}, f_{01}, and the $f_r^{\,1}$ $(r = 1, \ldots, m)$. Such a representation of P is unique and may be called a 1-level representation of P. Similarly, P has a dual unique representation. This representation is obtained by expressing P in such a way that each term in the polynomial is a real number multiplied with a symbol of the form $f_1^{j_1} f_2^{j_2} \cdots f_m^{j_m}$, where $j_1, \ldots, j_m$ have values 0 or 1, and where no term in P equals a real number times any other term of P. Such a representation of P may be called a 2-level representation or a treatment representation of P. The representation of P_T in (4.2) is a 2-level representation.

We shall now define a weight function w over $\mathscr{P}$. Let $P \in \mathscr{P}$; then we let

$$w(P) = \text{sum of the coefficients in the treatment representation of } P \tag{4.3}$$

We now define the counting operator λ over $\mathscr{P} \times \mathscr{T}$, where $\mathscr{T}$ is the set of all possible designs T. Let $P \in \mathscr{P}$ and $T \in \mathscr{T}$. Then we define

$$\lambda(P, T) = w(PP_T) \tag{4.4}$$

From Eq. (4.4), it is easy to check that if $(j_1, j_2, \ldots, j_m)$ is any treatment, then

$$\lambda(f_1^{j_1} \cdots f_m^{j_m}, T) = \text{number of times the treatment } (j_1, \ldots, j_m) \text{ occurs as a row of } T \tag{4.5}$$

For this reason, the operator λ is sometimes called the *counting operator*.

Now let $T_1(N_1 \times m)$ and $T_2(N_2 \times m)$ be two designs. Then, we denote by $T_1 + T_2(N_1 + N_2 \times m)$ the design obtained by adjoining the matrices T_1 and T_2. It is then clear that

$$P_{T_1 + T_2} = P_{T_1} + P_{T_2} \tag{4.6}$$

$$\lambda(P, T_1 + T_2) = \lambda(P, T_1) + \lambda(P, T_2) \tag{4.7}$$

Since for each design T, P_T (its polynomial representation) is a polynomial in $\mathscr{P}$, one can generalize the definition of the λ operator in the following way. The λ operator can be defined over $\mathscr{P} \times \mathscr{P}$. Thus, let $P_1 \in \mathscr{P}$, $P_2 \in \mathscr{P}$. Then we define

$$\lambda(P_1, P_2) = w(P_1 P_2) \tag{4.8}$$

Further properties of the λ operator will be found in the paper by Bose and Srivastava (1964a). These properties particularly relate to situations where the design T is presented or obtained through various customary processes.

We now consider the evaluation of the information matrix M, for any general but fixed design T. Let r, u, v be nonnegative integers, such that $r + u + v \leq m$. Let $i_1, i_2, \ldots, i_{r+u+v}$ be distinct integers out of the set of integers $\{1, 2, \ldots, m\}$. Let π_1 and π_2 be two interactions belonging to the vector ξ, where

$$\pi_1 = F_{i_1} F_{i_2} \cdots F_{i_r} F_{i_{r+1}} \cdots F_{i_{r+u}}$$

and

$$\pi_2 = F_{i_1} \cdots F_{i_r} F_{i_{r+u+1}} \cdots F_{i_{r+u+v}}$$

Let $\varepsilon(\pi_1, \pi_2)$ denote the element of the matrix M standing in the row and column corresponding, respectively, to the interaction π_1 and π_2. It is clear that, since M is symmetric, we have $\varepsilon(\pi_1, \pi_2) = \varepsilon(\pi_2, \pi_1)$. The following theorem is a slight generalization of a special case of Theorem 3.1 in Bose and Srivastava (1964a), and can be easily proved along the same lines.

Theorem 4.1. Let $Q(\pi_1, \pi_2)$ be a polynomial in $\mathscr{P}$ given by

$$Q(\pi_1, \pi_2) = (f^1_{i_{r+1}} - f^0_{i_{r+1}})(f^1_{i_{r+2}} - f^0_{i_{r+2}}) \cdots (f^1_{i_{r+u+v}} - f^0_{i_{r+u+v}}) \tag{4.9}$$

Then, we have

$$\varepsilon(\pi_1, \pi_2) = \lambda(Q(\pi_1, \pi_2), T) \tag{4.10}$$

Corollary 4.1. Define the nonnegative integers μ as below for all distinct integers $i_1, \ldots, i_r$ belonging to the set $\{1, 2, \ldots, m\}$:

$$\begin{aligned}\mu(i_1, \ldots, i_r) &= \lambda(f^1_{i_1} \cdots f^1_{i_r}, T) \\ &= \text{number of treatments in } T \text{ in which factors} \\ &\quad \text{number } i_1, \ldots, i_r \textit{ all} \text{ occur at level 1}\end{aligned} \tag{4.11}$$

Then, we have

$$\varepsilon(\mu, F_{i_1}) = 2\mu(i_1) - N = \varepsilon(F_{i_2}, F_{i_1}F_{i_2}) \tag{4.12}$$

$$\begin{aligned}
\varepsilon(\mu, F_{i_1}F_{i_2}) &= 4\mu(i_1, i_2) - 2\mu(i_1) - 2\mu(i_2) + N \\
&= \varepsilon(F_{i_1}, F_{i_2}) = \varepsilon(F_{i_1}F_{i_3}, F_{i_2}F_{i_3}) \\
\varepsilon(\mu, F_{i_1}F_{i_2}F_{i_3}) &= \varepsilon(F_{i_1}, F_{i_2}F_{i_3}) \\
&= 8\mu(i_1, i_2, i_3) - 4\mu(i_1, i_2) - 4\mu(i_1, i_3) \\
&\quad - 4\mu(i_2, i_3) + 2\mu(i_1) + 2\mu(i_2) + 2\mu(i_3) - N \\
\varepsilon(F_{i_1}F_{i_2}, F_{i_3}F_{i_4}) &= 16\mu(i_1, i_2, i_3, i_4) - 8[\mu(i_1, i_2, i_3) + \mu(i_1, i_2, i_4) \\
&\quad + \mu(i_1, i_3, i_4) + \mu(i_2, i_3, i_4)] + 4[\mu(i_1, i_2) + \mu(i_1, i_3) \\
&\quad + \mu(i_1, i_4) + \mu(i_2, i_3) + \mu(i_2, i_4) + \mu(i_3, i_4)] \\
&\quad - 2[\mu(i_1) + \mu(i_2) + \mu(i_3) + \mu(i_4)] + N \\
&\;\;\vdots
\end{aligned}$$

Define

$$\begin{aligned}\xi_l = \{&\mu; F_1, \ldots, F_l; F_{12}, \ldots, F_{m-1,m}; F_{123}, \ldots; \ldots; \\ &F_{12\cdots l}, \ldots, F_{m-l+1,m-l+2,\ldots,m}\}\end{aligned} \tag{4.13}$$

Then, for a design of resolution $(2l + 1)$ the vector ξ equals ξ_l, and T is such that the matrix M is nonsingular. For a design of resolution $(2l + 1)$, the vector ξ has v elements where v is given by

$$v = 1 + \binom{m}{1} + \binom{m}{2} + \cdots + \binom{m}{l} \tag{4.14}$$

In this chapter we are concerned mostly with designs of resolution V. We shall present the theory for obtaining designs of this type which are optimal within the class of balanced designs.

5. ORTHOGONAL DESIGNS AND ORTHOGONAL ARRAYS

A design T is said to be *orthogonal* with respect to the parameter vector ξ if and only if the corresponding information matrix M is a diagonal matrix. For the case of the 2^m type of factorial designs, the following theorem can be easily established.

Theorem 5.1. Let ξ be any (arbitrary but fixed) parameter vector, and let $T(N \times m)$ be a design. If T is orthogonal (i.e., if the corresponding information matrix M is diagonal), then T is D-, A-, and E-optimal among all designs having N treatments.

The proof of this theorem follows by observing that whatever T and ξ may be, the matrix M has N along the diagonal.

From this theorem, the importance of orthogonal designs is clear. We now proceed to obtain necessary and sufficient conditions for a design T to be orthogonal when $\xi = \xi_l$. Consider M for this case. If M is diagonal, then from Theorem 4.1 and Corollary 4.1, it can be easily checked that we must have the following:

For all subsets of $2l$ distinct integers $i_1, \ldots, i_{2l}$ from the set $\{1, \ldots, m\}$, we have $N = 2\mu(i_1) = 2^2\mu(i_1, i_2) = 2^3\mu(i_1, i_2, i_3) = \cdots = 2^{2l}\mu(i_1, \ldots, i_{2l})$ (5.1)

The conditions (5.1) indicate that for the design T to be orthogonal, the integer N should be divisible by the number 2^{2l}. A set of equivalent conditions is given by

$\lambda\{f_{i_1}^{j_1} \cdots f_{i_{2l}}^{j_{2l}}, T\} = N \times 2^{-2l}$ for all $j_1, \ldots, j_{2l} = 0$ or 1, and for all subsets of $2l$ distinct integers $i_1, \ldots, i_{2l}$ from the set $\{1, \ldots, m\}$ (5.2)

We now define an orthogonal array with two symbols and parameters m, N, t, and μ. Such an array, often denoted by OA(m, N, t, μ), is an $m \times N$ matrix R whose elements are (without loss of generality) the symbols 0 and 1 and, furthermore, which is such that in every $t \times N$ submatrix of R, each of the 2^t possible distinct column vectors having zeros and/or ones occur exactly μ times. Thus, the conditions (5.1) and (5.2) are equivalent to saying that the design T is such that, written as a matrix, $T'(m \times N)$ is an orthogonal array of strength $2l$, in which case we have $\mu = N \times 2^{-2l}$.

Now, suppose that N is an integer such that there does exist an orthogonal array T' of size $m \times N$ and of strength $2l$. Then it can be easily checked that a necessary condition that a design T^* be A-, D-, and E-optimal is that $T^{*\prime}(m \times N)$ be an orthogonal array of strength $2l$. Notice that for any given N, there may exist more than one orthogonal array of strength $2l$. Thus an optimal design is not necessarily unique.

The preceding clearly indicates the importance of orthogonal arrays which were first defined by C. R. Rao. In the following we give the most important method known for obtaining such arrays. There are very few orthogonal arrays useful as factorial designs which have been obtained by methods other than this.

The matrix B is said to have the property Q_t if and only if every nonzero vector in the row space of B has at least $(t + 1)$ nonzero elements. We shall consider matrices B which are defined over a field and, in this chapter, we shall always assume that this field is the field GF(2). The following is the main result.

Theorem 5.2. Let $B(k \times m)$ be a matrix over GF(2) with rank k. Consider the equation

$$B\mathbf{x} = \mathbf{C} \tag{5.3}$$

where $\mathbf{C}(k \times 1)$ is any vector over GF(2), and $\mathbf{x}' = (x_1, \ldots, x_m)$ is a vector of variables. The number of distinct vectors $\mathbf{x}$ over GF(2) which satisfy (5.3) is 2^{m-k}. Furthermore, if B has the property Q_t, then if the distinct solutions $\mathbf{x}$ to (5.3) are written as columns of an $(m \times 2^{m-k})$ matrix R, then R is an OA$(m, 2^{m-k}, t, 2^{m-k-t})$.

Consider EG$(m, 2)$ the finite Euclidean geometry of m dimensions based on the finite field GF(2). It is clear that the solutions of (5.3) constitute an $(m - k)$-dimensional flat space of EG$(m, 2)$.

We now define the property P_t of a matrix. A matrix B^* (over any given field) is said to have the property P_t if and only if no set of t rows of B^* is linearly dependent (over the given field). The following result gives a simple connection between the properties Q_t and P_t.

Theorem 5.3. Let $B(k \times m)$ be a matrix of rank k over some field, and let $B^*(m \times \overline{m - k})$ be a matrix of rank $(m - k)$ over the same field such that

$$BB^* = O(k \times \overline{m - k}) \tag{5.4}$$

Then B has the property Q_t if and only if B^* has property P_t.

Corollary 5.1. If $U(k \times \overline{m - k})$ is a matrix over GF(2) such that the matrix $[I_k \vdots U]$, of size $(k \times m)$, has property Q_t over GF(2), then the matrix

$$\begin{bmatrix} U \\ \cdots \\ I_{m-k} \end{bmatrix}$$

of size $(m \times \overline{m - k})$ has property P_t over GF(2).

The representation of the matrix B occurring in Theorem 5.2 in the form $[I_k \vdots U]$ serves another important purpose. Suppose we partition the vector $\mathbf{x}$ as $\mathbf{x}' = [\mathbf{x}_1' \vdots \mathbf{x}_2']$, where $\mathbf{x}_1$ and $\mathbf{x}_2$ have k and $(m - k)$ elements, respectively. Then, the equations $B\mathbf{x} = \mathbf{C}$ are equivalent to

$$[I_k \vdots U][\mathbf{x}_1' \vdots \mathbf{x}_2']' = \mathbf{C} \tag{5.5}$$

But (5.5) is equivalent to

$$\mathbf{x}_1 = \mathbf{C} - U\mathbf{x}_2 \tag{5.6}$$

The importance of the foregoing equation lies in the fact that we can give all possible 2^{m-k} values to the $(m - k \times 1)$ vector $\mathbf{x}_2$. Also, for any given value of $\mathbf{x}_2$, $\mathbf{x}_1$ can be calculated from (5.6). Thus, for any $\mathbf{x}_2$, we know the corresponding $\mathbf{x}_1$ and hence we know all the possible 2^{m-k} solutions $\mathbf{x}$ of Eqs. (5.3) or (5.5). This gives us an easy method of obtaining all the solutions. Notice that Eqs. (5.3) can always be expressed in the form (5.5), if the rank of B is k. This is done by multiplying by the inverse of a nonsingular $k \times k$ submatrix of B on both sides of Eqs. (5.3) and rearranging the columns of B and the elements of $\mathbf{x}$, if necessary, to produce the form (5.5).

The general problem of obtaining a matrix B with r columns and having the property P_t and, furthermore, having the maximum number of rows that is possible (for this fixed value of r) is called *Bose's packing problem.* As we notice, this is a very important problem in the theory of factorial designs. In this chapter we shall not touch on the problem of confounding since we are not assuming the presence of blocks. However, there too a similar problem arises. The same problem also arises in the field of coding theory. Recall that the Hamming distance between two vectors $\mathbf{x}_{(1)}$ and $\mathbf{x}_{(2)}$ (of the same size) is the number of nonzero elements in the vector $\mathbf{x}_{(1)} - \mathbf{x}_{(2)}$. Then it is easy to check that if a matrix B has property Q_t, the row space of this matrix is such that if $\mathbf{x}_{(1)}$ and $\mathbf{x}_{(2)}$ are any two vectors belonging to the row space of B, then the Hamming distance between $\mathbf{x}_{(1)}$ and $\mathbf{x}_{(2)}$ is at least $2t + 1$. It is clear then that if the vectors in the row space of B are used as code words, then this code will be able to correct any set of t errors. In other words, if one of these vectors is sent and because of noise the elements in any t or lesser number of places in this vector get changed, then we shall be able to determine the vector that was sent correctly.

We now present, for the sake of completeness, some well-known results concerning properties P_t or Q_t for matrices over GF(2). Consider an $(m \times r)$ matrix B^* having property P_t for a fixed value of r. Let $m_t(r)$ denote the largest value of m that such a matrix B^* could possibly have. Then it is well known that

$$m_2(r) = 2^r - 1, \qquad m_3(r) = 2^{r-1}, \qquad m_r(r) = r + 1 \qquad (\text{for all} \quad r) \tag{5.7}$$

Some general methods of writing down matrices with the property P_t, for general t [especially over GF(2)] are the well-known matrices associated with BCH codes and Srivastava codes (e.g., see Berlekamp, 1968).

6. BALANCED DESIGNS AND BALANCED ARRAYS

It was seen in the preceding section that orthogonal arrays, and hence orthogonal designs, are rather scarce in the sense that they exist only for a very few values of N. Thus, for most values of N, the optimal designs are not orthogonal designs because the latter do not simply exist. For this reason, we need to go into a larger class of designs of which the orthogonal designs are a subclass. The bigger class that we shall investigate in this chapter is the class of balanced designs now defined.

Consider a vector of parameters $\xi(v \times 1)$. Let $T(N \times m)$ be a design, and let V_T be the variance matrix of the BLU (best linear unbiased) estimators $\hat{\xi}$ of ξ, when the design T is used. Then T is said to be balanced with respect to ξ if and only if we have

$$V_T = V_{T*} \tag{6.1}$$

for every $T^*(N \times m)$ obtainable from T by a permutation of the columns of T. We may express this last statement in a different way by saying that the design T is balanced if and only if the matrix V_T is invariant under a permutation of the factor symbols. Note that this definition is the same as that in Section 1.

Now, suppose that $\xi = \xi_l$ for some l. The matrix V for a balanced design has a certain pattern or structure. This pattern or structure can be described in terms of the linear associative algebra to which this matrix belongs. We now elaborate this algebra.

Consider the set of all l-factor interactions, and assume that the vector containing these $\binom{m}{l}$ interactions written in lexicographic order is denoted by $\phi_l(\binom{m}{l} \times 1)$, $l = 0, 1, \ldots, m$. We now define a multidimensional partially balanced (MDPB) association scheme within and between these sets ϕ_l for various values of l. Such schemes were first defined by Srivastava (1961) and an account will be found in the article by Bose and Srivastava (1964b). Some further studies in this direction have been made by Srivastava and Anderson (1971). The MDPB scheme which arises in the theory of balanced designs is usually called the factorial association scheme and is defined thus. Consider two interactions containing $r + u$ and $r + v$ factors, respectively, where r, u, v are nonnegative integers with $r + u + v \leq m$. Consider two interactions

$$\pi_1 = F_{i_1} F_{i_2} \cdots F_{i_r} F_{i_{r+1}} \cdots F_{i_{r+u}}$$

and

$$\pi_2 = F_{i_1} F_{i_2} \cdots F_{i_r} F_{i_{r+u+1}} \cdots F_{i_{r+u+v}}$$

where $i_1, i_2, \ldots, i_{r+u+v}$ are distinct integers belonging to the set $\{1, 2, \ldots, m\}$. Then the interactions π_1 and π_2 are said to be $(u + v)$th associates of each other. Thus, the two interactions are said to be αth associates if there are α factors each of which is involved in the symbol for one of the interactions but none of which occurs in the symbols for both of the interactions. Let $\varepsilon^{-}(\pi_1, \pi_2)$ denote the element of the matrix V which stands in the row corresponding to the interaction π_1 and the column corresponding to the interaction π_2. We show that the value of $\varepsilon^-(\pi_1, \pi_2)$ depends only on the values of r, u, and v and is otherwise independent of the integers $i_1, \ldots, i_{r+u+v}$. Let $j_1, \ldots, j_{r+u+v}$ be distinct integers out of the set $\{1, 2, \ldots, m\}$, where r, u, and v are as before.

Let $\pi_1^* = F_{j_1} \cdots F_{j_r} F_{j_{r+1}} \cdots F_{j_{r+u}}$ and $\pi_2^* = F_{j_1} \cdots F_{j_r} F_{j_{r+u+1}} \cdots F_{j_{r+u+v}}$. We show that $\varepsilon^-(\pi_1^*, \pi_2^*) = \varepsilon^-(\pi_1, \pi_2)$. Consider the design T, and let the design $T^*(N \times m)$ be obtained from T by permuting the columns of T as follows: (a) the columns $i_1, i_2, \ldots, i_{r+u+v}$ should now be numbered as the columns $j_1, \ldots, j_{r+u+v}$; and (b) the remaining columns of T should be given numbers from the set $\{1, 2, \ldots, m\} - \{j_1, \ldots, j_{r+u+v}\}$. Thus, T^* is obtained from T by a permutation of columns of T. Since the design T is balanced, we must have $V_T = V_{T*}$. However, the element $\varepsilon^-(\pi_1^*, \pi_2^*)$ in V_{T*} is the same as the element $\varepsilon^-(\pi_1, \pi_2)$ in V_T. This proves our assertion. Now let M_T denote the information matrix corresponding to T so that $M_T = V_T^{-1}$. From this it follows that, in an obvious notation, $M_T = M_{T*}$, so that $\varepsilon(\pi_1^*, \pi_2^*) = \varepsilon(\pi_1, \pi_2)$, where ε denotes the elements in the various cells of M_T. Now, from (4.9) and (4.10) it follows that

$$\varepsilon(\pi_1, \pi_2) = \lambda \left[\prod_{p=r+1}^{r+u+v} (f_{i_p}^1 - f_{i_p}^0), T \right] \tag{6.2}$$

In view of the foregoing discussion, the right-hand side of (6.2) depends only on r, u, and v and is independent of the subscripts $i_1, \ldots, i_{r+u+v}$. From this fact, after some computations, it can be shown that we have

$$\lambda[f_{j_1}^{g_1} f_{j_2}^{g_2} \cdots f_{j_{2l}}^{g_{2l}}, T] = \rho(g_1, \ldots, g_{2l}) \tag{6.3}$$

where the function ρ is independent of $j_1, j_2, \ldots, j_{2l}$ and depends only on $g_1, \ldots, g_{2l}$ where the g_i's take the values 0 and 1.

Now let us define the weight of a vector $(\theta_1, \ldots, \theta_m)$, written $w(\theta_1, \ldots, \theta_m)$, to be the number of nonzero elements in the vector. Then the right-hand side of (6.3) depends only on $w(g_1, \ldots, g_{2l})$.

We now define a balanced array, or briefly B-array. A B-array T^* of strength t, two symbols, and size $(m \times N)$ is an $(m \times N)$ matrix T^* such that if T_0^* is any $(t \times N)$ submatrix of T^*, then the number of times any t-vector [a (0, 1) vector with t elements] occurs in T_0^* depends only on the weight of the t-vector. In other words, there exist nonnegative integers μ_i $(i = 0, 1, \ldots, t)$ such that if $\mathbf{g}' = (g_1, \ldots, g_t)$ is any t-vector, then the number of times $\mathbf{g}$ occurs as a column of T_0^* is μ_i, where $i = w(\mathbf{g})$, and where μ_i are independent of T_0^* so long as T_0^* is a $(t \times N)$ submatrix of T^*. In the terminology of B-arrays, the condition (6.3) reduces to requiring that $T'(m \times N)$ be a B-array of strength $2l$.

It is easy to see that an orthogonal array of strength t is a special case of a B-array of the same strength where $\mu_i = N \cdot 2^{-t}$, for all i. Although, as mentioned earlier, orthogonal arrays exist only for very few values of N, it can be easily shown that B-arrays exist for all values of N and m. For those values of N for which orthogonal arrays do exist, the optimal design within the class of B-arrays (i.e., balanced designs) is automatically an orthogonal array. However, for other values of N for which orthogonal arrays do not exist, there exist good B-arrays which provide a very useful alternative.

From now on, this chapter shall be concerned with the problem of obtaining designs which are optimal within the class of balanced designs.

A design T for which the information matrix M is not diagonal is said to be a nonorthogonal design. Since orthogonal designs rarely exist, balanced designs are mostly nonorthogonal. However, balanced designs are preferable to designs which are nonorthogonal and also unbalanced for a variety of reasons, the most important being the ease in the analysis and interpretation of the results.

7. EXISTENCE CONDITIONS FOR BALANCED ARRAYS

In the preceding section, we saw the statistical importance of balanced arrays. The question now arises: When does a balanced array of a given strength and index set exist? In this section, we shall summarize some results regarding this from Srivastava (1972). In passing, we point out that Dembowski (1968) has called B-arrays of strength t designs of type $(0, t)$. For some other work on B-arrays and arrays in general, see Rafter (1971) and Rao (1973).

Consider a B-array of size $(m \times N)$, strength t, and index set $(\mu_0{}^t, \mu_1{}^t, \ldots, \mu_t{}^t)$. Let T denote such an array and let T_0 denote a subarray of size $(t \times N)$. Since each t-vector of weight i occurs $\mu_i{}^t$ times in T_0, it is clear that

$$N = \binom{t}{0}\mu_0{}^t + \binom{t}{1}\mu_1{}^t + \cdots + \binom{t}{t}\mu_t{}^t \tag{7.1}$$

Furthermore, the following results are easy to prove.

Theorem 7.1. Let $T(m \times N)$ be a B-array with two symbols, strength t, and index set $(\mu_0{}^t, \mu_1{}^t, \ldots, \mu_t{}^t)$. Then the following results hold:

(a) T is also of strength $(t-1)$ with index set $(\mu_0^{t-1}, \ldots, \mu_{t-1}^{t-1})$, where

$$\mu_j^{t-1} = \mu_j{}^t + \mu_{j+1}^t, \qquad j = 0, 1, \ldots, t-1 \tag{7.2}$$

(b) If $T_0(m_0 \times N)$ is a subarray of T, then T_0 is also a B-array of strength t and with the same index set as T.

(c) Let $T_1(m \times N_1)$ be a B-array of strength t which is continued in T and which has index set $(\pi_0, \pi_1, \ldots, \pi_t)$. Let $(T - T_1)$ denote the $(m \times N - N_1)$ array obtained by deleting those columns of T which are in

T_1. Then $(T - T_1)$ is also a B-array of strength t and has index set $(\mu_0{}^t - \pi_0, \ldots, \mu_t{}^t - \pi_t)$. In particular, let T^* be the m-rowed array obtained from T by deleting all columns of T which have weight 0 or m. Then T^* is a B-array. (The array T^* is sometimes called a *trim* array.)

(d) Let $T_2(m \times N_2)$ be a B-array of strength t and index set $(\mu_{20}, \ldots, \mu_{2t})$. Let $T + T_2$ be the $(m \times \overline{N + N_2})$ array obtained by adjoining T and T_2. Then $T + T_2$ is also of strength t and has index set $(\mu_0{}^t + \mu_{20}, \ldots, \mu_t{}^t + \mu_{2t})$.

(e) If $T_p(m \times N)$ is obtained from T by permuting rows and/or columns of T, then T_p is a B-array of strength t with the same index set as T.

(f) The array T is said to be of full strength if $t = m$. Let Ω_{mi} $(i = 0, 1, \ldots, m)$ be the set of all the $\binom{m}{i}$ m-vectors (with elements 0 or 1) of weight i, written as an $(m \times \binom{m}{i})$ matrix. Then Ω_{mi} is a B-array of full strength and index set $(W_0{}^m, W_1{}^m, \ldots, W_m{}^m)$, where $W_i = \binom{m}{i}$ and $W_j = 0$, $j \neq i$. Furthermore, if $0 \leq t \leq m$, then Ω_{mi} is also a B-array of strength t with index set $(W_0{}^t, \ldots, W_t{}^t)$, where for $j = 0, 1, \ldots, t$, we have $W_j^t = \binom{m-t}{i-j}$. [*Note:* Throughout this chapter, we use the usual convention that $\binom{a}{b} = 0$ if $a < b$.]

(g) If N satisfies (7.1), then a B-array of size $(t \times N)$ with index set $(\mu_0{}^t, \ldots, \mu_t{}^t)$ always exists, and is obtained by writing down, for each i, all t-vectors of weight i $(i = 0, 1, \ldots, t)$.

Consider now the existence of the B-array t when $m = t + 1$; call this array T^1. For any positive integer u, let Ξ_u denote the set of integers $\{1, 2, \ldots, u\}$. As before, we shall use the counting operator λ so that if $\boldsymbol{\theta}(u \times 1)$ is a vector and Θ is a u-rowed matrix, then $\lambda(\boldsymbol{\theta}, \Theta)$ will denote the number of times $\boldsymbol{\theta}$ occurs as a column of Θ. Also, for any positive integers l and m, J_{lm} and O_{lm} will denote matrices whose elements are all ones or all zeros, respectively. Also, let $\boldsymbol{\omega}_1(t + 1 \times 1)$ denote a vector which has zeros exactly at the coordinates $j_1, \ldots, j_k$ $(\in \Xi_{t+1})$ and 1 elsewhere. Let $\lambda(J_{t+1,1}, T^1) = d$ and $\lambda(\boldsymbol{\omega}_1, T^1) = v_1(j_1, \ldots, j_k)$. Then it can easily be shown that

$$v_1(j_1, \ldots, j_k) = \mu_{t-k+1}^t - \mu_{t-k+2}^t + \mu_{t-k+3}^t - \cdots + (-1)^{k+1}\mu_t{}^t + (-1)^k d \tag{7.3}$$

Since the numbers d and v are nonnegative for all values of k, the necessity part of the following theorem is established.

Theorem 7.2. A necessary and sufficient condition that a $(t + 1)$-rowed B-array with index set $\boldsymbol{\mu}'$ exists is that there exists an integer d (≥ 0) such that

$$d \leq \min_{0 \leq 2r \leq t} (\mu_t{}^t - \mu_{t-1}^t + \mu_{t-2}^t - \cdots + \mu_{t-2r}^t) = \psi_{12} \quad \text{(say)} \tag{7.4a}$$

$$d \geq \max_{0 \leq 2r-1 \leq t} (\mu_t{}^t - \mu_{t-1}^t + \cdots - \mu_{t-2r+1}^t) = \psi_{11} \quad \text{(say)} \tag{7.4b}$$

If (7.4) are satisfied, there exist $(1 + \psi_{12} - \psi_{11})$ nonisomorphic $(t + 1)$-rowed arrays with index set $\boldsymbol{\mu}'$.

The proof of the sufficiency part of the theorem follows from the fact that if (7.4) are satisfied, then we can construct T_1 by writing various columns, with a column of the type $\boldsymbol{\omega}_1$ described earlier occurring $v_1(i_1, \ldots, i_k)$ times, and this being done for all possible $\boldsymbol{\omega}_1$.

Next, consider a $(t + 2)$-rowed array T^2 with the same index set as T^1. Let T^{1i} $(i = 1, \ldots, t + 2)$ denote the $(t + 1)$-rowed array obtained from T^2 by omitting the ith row. Let $\boldsymbol{\omega}_2$ denote a $(t + 2 \times 1)$ vector which has 0 at positions $(j_1, \ldots, j_k)$ and 1 elsewhere. Let $d_0 = \lambda(J_{t+2,1}, T^2)$; $d_i = \lambda(J_{t+1,1}, T^{1i})$; $i = 1, \ldots, t + 2$; and $v_2(j_1, \ldots, j_k) = \lambda(\boldsymbol{\omega}_2, T^2)$. Then it is easy to check that

$$v_2(j_1) = d_{j_1} - d_0$$

$$v_2(j_1, j_2) = \mu_t{}^t - (d_{j_1} + d_{j_2}) + d_0$$

$$\begin{aligned} &v_2(j_3, j_4, \ldots, j_{t-k+2}) + v_2(j_1, j_3, j_4, \ldots, j_{t-k+2}) \\ &\quad + v_2(j_2, j_3, \ldots, j_{t-k+2}) + v_2(j_1, j_2, j_3, \ldots, j_{t-k+2}) = \mu_k{}^t \end{aligned} \tag{7.5}$$

for all permissible k. From Eqs. (7.5), it can be shown that

$$\pi(j_1, \ldots, j_{2r-1}) = -d_0 + \sum_{a=1}^{2r-1} d_{j_a} + \sum_{q=0}^{2r-2} (-1)^q \mu_{t-q}^t (q - 2r + 2) \qquad \text{if} \quad (2r - 1) \in \Xi_{t+2} \tag{7.6a}$$

$$\pi(j_1, j_2, \ldots, j_{2r}) = d_0 - \sum_{a=1}^{2r} d_{j_a} + \sum_{q=0}^{2r-2} \mu_{t-q}^t (-1)^q (2r - 1 - q) \qquad \text{if} \quad 2r \in \Xi_{t+2} \tag{7.6b}$$

Using Eqs. (7.6a) and (7.6b), the following result can be established.

Theorem 7.3. A necessary and sufficient condition for the existence of a $(t + 2)$-rowed B-array T^2 of strength t and index set $\boldsymbol{\mu}'$ is that $\psi_{12} \geq \psi_{11}$, and that there exist constants $d_0, d_1, \ldots, d_{t+2}$ such that

$$\psi_{12} \geq d_1 \geq \cdots \geq d_{t+2} \geq \psi_{11} \tag{7.7}$$

and

$$d_0 \leq \min_r \left\{ \sum_{i=0}^{2r} d_{t+2-i} + \sum_{q=0}^{2r} (-1)^q (q - 2r) \mu_{t-q}^t \right\} = \psi_{22} \quad \text{(say)} \tag{7.8a}$$

$$d_0 \geq \max_r \left\{ \sum_{i=1}^{2r} d_i + \sum_{q=0}^{2r-2} (-1)^{q+1} (2r - q - 1) \mu_{t-q}^t \right\} = \psi_{21} \quad \text{(say)} \tag{7.8b}$$

where the maximum and minimum are taken over all permissible r. [This obviously leads to the conclusion that, for every distinct set of d_i, $i = 0, 1, \ldots, t+2$, satisfying (7.8), there exists a distinct (nonisomorphic) array with the prescribed parameters.]

The following corollary is important for the later sections of this chapter.

Corollary 7.1. For $t = 4$, the following results hold:

I. A set of necessary and sufficient conditions for the existence of a five-rowed array T of strength 4, and index set $\boldsymbol{\mu}' = (\mu_0, \ldots, \mu_4)$, is that there exists an integer d such that

$$d \geq \psi_{11}(\boldsymbol{\mu}) \equiv \max(0, \mu_4 - \mu_3, \mu_4 - \mu_3 + \mu_2 - \mu_1)$$
$$d \leq \psi_{12}(\boldsymbol{\mu}) \equiv \min(\mu_4, \mu_2 - \mu_3 + \mu_4, \mu_0 - \mu_1 + \mu_2 - \mu_3 + \mu_4) \qquad (7.9)$$

II. A set of necessary and sufficient conditions for the existence of a six-rowed array T of strength 4 and index set $\boldsymbol{\mu}'$ is that there exists an integer d_0 such that

$$d_0 \geq \psi_{21}(\boldsymbol{\mu}) \equiv \max(0, d_{1\cdot 2} + \theta_1, d_{1\cdot 4} + \theta_2, d_{1\cdot 6} + \theta_3)$$
$$d_0 \leq \psi_{22}(\boldsymbol{\mu}) \equiv \min(d_6, d_{4\cdot 6} - \theta_4, d_{2\cdot 6} - \theta_5) \qquad (7.10)$$

where $d_{i\cdot j} = d_i + d_{i+1} + \cdots + d_j$, $j \geq i$; for each i, $d = d_i$ satisfies (7.9); and $d_1 \geq d_2 \geq \cdots \geq d_6$. Also the θ's in (7.10) are given in terms of the μ's by

$$\theta_1 = -\mu_4, \qquad \theta_2 = \mu_2 + 2\mu_3 - 3\mu_4$$
$$\theta_3 = -\mu_0 + 2\mu_1 - 3\mu_2 + 4\mu_3 - 5\mu_4, \qquad \theta_4 = -\mu_3 + 2\mu_4 \qquad (7.11)$$
$$\theta_5 = -\mu_1 + 2\mu_2 - 3\mu_3 + 4\mu_4$$

Equivalent conditions (7.10) can be expressed as

$$\begin{array}{l} (a) \\ (b) \\ (c) \\ (d) \\ (e) \\ (f) \\ (g) \\ (h) \\ (i) \\ (j) \\ (k) \\ (l) \end{array} \begin{bmatrix} d_6 \\ -d_1 - d_2 + d_6 \\ -d_1 - d_2 - d_3 - d_4 + d_6 \\ -d_1 - d_2 - d_3 - d_4 - d_5 \\ d_4 + d_5 + d_6 \\ -d_1 - d_2 + d_4 + d_5 + d_6 \\ -d_1 - d_2 - d_3 + d_5 + d_6 \\ -d_1 - d_2 - d_3 \\ d_2 + d_3 + d_4 + d_5 + d_6 \\ -d_1 + d_3 + d_4 + d_5 + d_6 \\ -d_1 + d_5 + d_6 \\ -d_1 \end{bmatrix} \geq \begin{bmatrix} 0 \\ -\mu_4 \\ -\mu_2 + 2\mu_3 - 3\mu_4 \\ -\mu_0 + 2\mu_1 - 3\mu_2 + 4\mu_3 - 5\mu_4 \\ -\mu_3 + 2\mu_4 \\ \mu_4 - \mu_3 \\ -\mu_2 + \mu_3 - \mu_4 \\ -\mu_0 + 2\mu_1 - 3\mu_2 + 3\mu_3 - 3\mu_4 \\ -\mu_1 + 2\mu_2 - 3\mu_3 + 4\mu_4 \\ -\mu_1 + 2\mu_2 - 3\mu_3 + 3\mu_4 \\ -\mu_1 + \mu_2 - \mu_3 + \mu_4 \\ -\mu_0 + \mu_1 - \mu_2 + \mu_3 - \mu_4 \end{bmatrix} \qquad (7.12)$$

Theorems 7.2 and 7.3 provide necessary and sufficient conditions for the existence of m-rowed arrays of strength t if $m \leq t + 2$. In what follows, we present some necessary conditions for larger values of m. Consider the B-array T as before. Suppose T contains X_q columns which are of weight q each, $0 \leq q \leq m$. Let $N(t, j)$ denote the total number of t-vectors in T which are of weight j each. Then it can be shown that $N(t, j)$ equals both the right-hand and the left-hand sides of

$$\sum_{q=0}^{m} \binom{q}{j}\binom{m-q}{t-j} X_q = \binom{m}{t}\binom{t}{j} \mu_j^t, \qquad j = 0, 1, \ldots, t \tag{7.13}$$

where, as before, $\binom{u}{v} = 0$, if $u < v$. Equations (7.13) are called single diophantine equations (SDE). The reason is that they involve only one set of unknown variables, namely the X_q. These equations provide necessary conditions for the existence of T, since if T exists, they must have nonnegative integer solutions for the X_q.

Now, consider the $(t + 2)$-rowed arrays contained in T. If T_0 is such an array, then corresponding to T_0 there must exist integers $d_{00}, d_{10}, \ldots, d_{t+2,0}$ which satisfy Eqs. (7.8a) and (7.8b) for the d_i ($i = 0, 1, \ldots, t + 2$). Now, suppose that for a fixed value of m, t, and the index set $\boldsymbol{\mu}' = (\mu_0{}^t, \ldots, \mu_t{}^t)$, there exist g different solutions, say $(d_{0v}, d_{1v}, \ldots, d_{t+2,v})$, $v = 1, \ldots, g$, for the d_i in Eqs. (7.8). Also, let ζ_v ($v = 1, \ldots, g$) be the number of subarrays T_v which correspond to the solution $(d_{0v}, \ldots, d_{t+2,v})$. Then, it is easy to check that

$$\sum_{v=1}^{g} \zeta_v N_v(t+2, j) = \sum_{q=0}^{m} \binom{q}{j}\binom{m-q}{t-j} x_q \tag{7.14}$$

where $N_v(t + 2, j)$ denotes the number of columns of the array T_v, each of which have weight j. Now, using Eqs. (7.6a) and (7.6b) one can prove that

$$N_v(t+2, j)_{t-j \text{ odd}} = \binom{t+2}{j}\Bigg[-d_{0v} + (t+2-j)\bar{d}_v + \sum_{q=0}^{t-j+1} (-1)^q \mu_{t-q}^t (q+j-t-1)\Bigg] \tag{7.15a}$$

$$N_v(t+2, j)_{t-j \text{ even}} = \binom{t+2}{j}\Bigg[d_{0v} - (t+2-j)\bar{d}_v + \sum_{q=0}^{t-j} (-1)^{q+1} \mu_{t-q}^t (q+j-t-1)\Bigg] \tag{7.15b}$$

where

$$\bar{d}_v = (t+2)^{-1}\left(\sum_{i=1}^{t+2} d_{iv}\right) \tag{7.16}$$

Using the value of $N_v(t+2, j)$ from Eqs. (7.15a) and (7.15b) and substituting this into Eq. (7.14), we get a set of equations which involve three sets of nonnegative integral unknowns, namely the ζ's, X's, and d's. These equations (7.14) are therefore called the triple diophantine equations (TDE). When $g = 1$, we call these the double diophantine equations (DDE) since we always have

$$\sum_{1}^{g} \zeta_v = \binom{m}{t+2} \tag{7.17}$$

which for the case of $g = 1$ gives us the value of ζ_1, so that there are only two sets of unknowns left, namely the X's and the d's. These results are summarized in the following:

Theorem 7.4. A necessary condition for the existence of the array $T(m \times N)$ of strength t and index set $\boldsymbol{\mu}' = (\mu_0{}^t, \ldots, \mu_t{}^t)$ to exist is that Eqs. (7.13) and (7.14) have nonnegative integral solutions in the X's, ζ's and d's.

Let us now consider the DDE in a little more detail. These correspond to the case when Eqs. (7.8) have only one solution, say $(d_{00}, \ldots, d_{t+2,0})$. Let $N_0(t+2, j)$ be the value of the expression in Eqs. (7.15) when the values d_{j0} $(j = 0, \ldots, t+2)$ are substituted in place of the d_{jv}. Then, we get the following result:

Corollary 7.2. The DDE are given by

$$\binom{0}{j}\binom{m}{t-j} X_0 + \binom{1}{j}\binom{m-1}{t-j} X_1 + \cdots + \binom{m}{j}\binom{m-m}{t-j} X_m$$
$$= \binom{m}{t+2} N_0(t+2, j), \quad j = 0, 1, \ldots, t+2 \tag{7.18}$$

Corollary 7.3. The DDE (7.18) cannot hold if there exists a value of j such that

$$N_0(t+2, j) = N_0(t+2, j+2) = 0, \qquad \text{and} \qquad N_0(t+2, j+1) \neq 0 \tag{7.19}$$

Corollary 7.3 is proved by observing that in Eq. (7.18) if for some j, $N_0(t+2, j) = 0$, then all the X's occurring with nonzero coefficients in that equation must be 0.

For ease of reference in later sections, we present the special case of the DDE when $t = 4$:

Corollary 7.4. Let

$$\begin{aligned}
\delta_0 &= (\mu_0 - 2\mu_1 + 3\mu_2 - 4\mu_3 + 5\mu_4) - 6\bar{d} + d_0 = -\theta_3 - 6\bar{d} + d_0 \\
\delta_1 &= (\mu_1 - 2\mu_2 + 3\mu_3 - 4\mu_4) + 5\bar{d} - d_0 = -\theta_0 + 5\bar{d} - d_0 \\
\delta_2 &= (\mu_2 - 2\mu_3 + 3\mu_4) - 4\bar{d} + d_0 = -\theta_2 - 4\bar{d} + d_0 \\
\delta_3 &= (\mu_3 - 2\mu_4) + 3\bar{d} - d_0 = -\theta_4 + 3\bar{d} - d_0 \qquad (7.20)\\
\delta_4 &= \mu_4 - 2\bar{d} + d_0 = -\theta_1 - 2\bar{d} + d_0 \\
\delta_5 &= \bar{d} - d_0 \\
\delta_6 &= d_0
\end{aligned}$$

Then, for $t = 4$, the DDE reduce to

$$\begin{aligned}
\binom{m}{6} X_0 + \binom{m-1}{6} X_1 + \cdots + \binom{6}{6} X_{m-6} &= \binom{m}{6}\binom{6}{0}\delta_0 \\
\binom{1}{1}\binom{m-1}{5} X_1 + \binom{2}{1}\binom{m-2}{5} X_2 + \cdots + \binom{m-5}{1}\binom{5}{5} X_{m-5} &= \binom{m}{6}\binom{6}{1}\delta_1 \\
&\vdots \\
\binom{6}{6}\binom{m-6}{0} X_6 + \binom{7}{6}\binom{m-7}{0} X_7 + \cdots + \binom{m}{6}\binom{0}{0} X_m &= \binom{m}{6}\binom{6}{6}\delta_6
\end{aligned} \qquad (7.21)$$

In the following, we give a set of inequalities which have been found very useful in investigating the existence of seven-rowed balanced arrays.

Theorem 7.5. Consider a B-array $T(7 \times N)$ with index set $\boldsymbol{\mu}'$. Also, let T^* be a $(6 \times N)$ array with the same index set $\boldsymbol{\mu}'$, and further suppose that T^* is a subarray of T. Corresponding to T^*, let a value of $(d_0, \mathbf{d}')$ which satisfies (7.9)–(7.10) be denoted by $(d_0^*, d_1^*, \ldots, d_6^*)$. Let $\boldsymbol{\delta}^{*\prime}$ be obtained using (7.20), with $d_i = d_i^*$, and let $\pi_i^* = \binom{6}{i}\delta_i$, for $i = 0, \ldots, 6$. Then the following inequalities must be satisfied:

$$\begin{aligned}
&(1)\quad \pi_{i-1}^* + \pi_i^* \geq x_i, \qquad 1 \leq i \leq 6 \\
&(2)\quad \pi_{i-1}^* + \pi_i^* + \pi_{i+1}^* \geq x_i + x_{i+1}, \qquad 1 \leq i \leq 5 \\
&(3)\quad \pi_{i-1}^* + \pi_i^* + \pi_{i+1}^* + \pi_{i+2}^* \geq x_i + x_{i+1} + x_{i+2}, \qquad 1 \leq i \leq 4 \\
&(4)\quad \pi_{i-1}^* + \pi_i^* + \pi_{i+1}^* + \pi_{i+2}^* + \pi_{i+3}^* \qquad (7.22)\\
&\qquad\qquad \geq x_i + x_{i+1} + x_{i+2} + x_{i+3}, \qquad 1 \leq i \leq 3 \\
&(5)\quad \pi_{i-1}^* + \pi_i^* + \pi_{i+1}^* + \pi_{i+2}^* + \pi_{i+3}^* + \pi_{i+4}^* \\
&\qquad\qquad \geq x_i + x_{i+1} + x_{i+2} + x_{i+3} + x_{i+4}, \qquad 1 \leq i \leq 2
\end{aligned}$$

This theorem is easily proved by noticing that in any inequality, the left-hand side denotes the number of column vectors in T^* whose weights equal the suffixes of π^*. Also, the corresponding right-hand side denotes the total number of column vectors in T which have weights equal to the suffixes of the X's.

8. THE CHARACTERISTIC ROOTS OF THE INFORMATION MATRIX FOR BALANCED DESIGNS

We shall first consider the designs of resolution V. The information matrix for such designs has five distinct elements γ_i $(i = 1, \ldots, 5)$:

$$\begin{aligned}
&\gamma_1 = N = \varepsilon(\mu, \mu) = \varepsilon(A_i, A_i) = \varepsilon(A_{ij}, A_{ij}), &&\gamma_2 = \varepsilon(\mu, A_i) = \varepsilon(A_j, A_{ij})\\
&\gamma_3 = \varepsilon(\mu, A_{ij}) = \varepsilon(A_i, A_j) = (A_{ik}, A_{jk}), &&\gamma_4 = \varepsilon(A_i, A_{jk})\\
&\gamma_5 = \varepsilon(A_{ij}, A_{kl})
\end{aligned} \tag{8.1}$$

where the suffixes i, j, k, l are all distinct and belong to Ξ_m. Let the matrix M correspond to a balanced design T which is of strength 4 and has index set $\boldsymbol{\mu}' = (\mu_0, \ldots, \mu_4)$. Then it is easy to check that the γ_i are given by

$$\begin{aligned}
&\gamma_1 = N = \mu_0 + 4\mu_1 + 6\mu_2 + 4\mu_3 + \mu_4, &&\gamma_5 = \mu_0 - 4\mu_1 + 6\mu_2 - 4\mu_3 + \mu_4\\
&\gamma_2 = (\mu_4 - \mu_0) + 2(\mu_3 - \mu_1), &&\gamma_4 = (\mu_4 - \mu_0) - 2(\mu_3 - \mu_1)\\
&\gamma_3 = \mu_4 - 2\mu_2 + \mu_0
\end{aligned} \tag{8.2}$$

The rows and columns of M correspond to the parameters $\boldsymbol{\xi}$ which in this case can be divided into three sets $\{\mu\}$, $\{F_i\}$, $\{F_{ij}\}$ where these sets stand for the general mean, the main effect, and the two-factor interactions, respectively, and have 1, $\binom{m}{1}$, and $\binom{m}{2}$ elements. Let M be partitioned according to these three sets as

$$M = \begin{array}{c} \\ \\ \\ \end{array}\begin{array}{ccc|c} \{\mu\} & \{F_i\} & \{F_{ij}\} & \\ \hline M_{00} & M_{01} & M_{02} & \{\mu\} \\ M_{10} & M_{11} & M_{12} & \{F_i\} \\ M_{20} & M_{21} & M_{22} & \{F_{ij}\} \end{array} \tag{8.3}$$

Thus, M_{01} is $1 \times \binom{m}{1}$, M_{11} is $\binom{m}{1} \times \binom{m}{1}$, M_{12} is $\binom{m}{1} \times \binom{m}{2}$, and so on. Let

$$M_0 = \begin{bmatrix} M_{01} \\ M_{21} \end{bmatrix}, \qquad M^* = \begin{bmatrix} M_{00} & M_{02} \\ M_{20} & M_{22} \end{bmatrix} \tag{8.4}$$

$$[M^* - \delta I]^{-1} = \begin{bmatrix} Q_{00} & Q_{02} \\ Q_{20} & Q_{22} \end{bmatrix} \tag{8.5}$$

Then it is easy to check that (using the usual formula for the inverse of partitioned matrices)

$$Q_{02} = -(M_{00} - \delta I)^{-1} M_{02} Q_{22} \tag{8.6a}$$

$$Q_{22} = [(M_{22} - \delta I) - M_{20}(M_{00} - \delta I)^{-1} M_{02}]^{-1} \tag{8.6b}$$

$$Q_{00} = (M_{00} - \delta I)^{-1} + (M_{00} - \delta I)^{-1} M_{02} Q_{22} M_{20} (M_{00} - \delta I)^{-1} \tag{8.6c}$$

$$\begin{aligned} |M - \delta I| = |M^* - \delta I| \, |(M_{11} - \delta I) - (M_{10} Q_{00} M_{01} + M_{12} Q_{20} M_{01} \\ + M_{10} Q_{02} M_{21} + M_{12} Q_{22} M_{21})| \end{aligned} \tag{8.7}$$

Now, M_{02} is a vector of size $1 \times \binom{m}{2}$ which has the element γ_3 everywhere. Also, M_{22} is a square matrix with $\binom{m}{2}$ rows which belongs to the linear associative algebra L_5 generated by the association matrices of the general triangular association scheme. The notation L_5 was introduced by Bose and Srivastava (1964b), where the characteristic roots of the association matrices of this algebra (which is commutative) are presented. Using these characteristic roots and the fact that the algebra L_5 is commutative, it can be easily checked that the (possibly) distinct roots of Q_{22} are $(\pi_1 - \delta)^{-1}$, $(\pi_2 - \delta)^{-1}$, $(\pi_3 - \delta)^{-1}$ with respective multiplicities 1, $m - 1$, and m', where

$$\begin{aligned} \pi_1 &= \gamma_1 + 2(m-2)\gamma_3 + m''\gamma_5 \\ \pi_2 &= \gamma_1 + (m-4)\gamma_3 - (m-3)\gamma_5 \\ \pi_3 &= \gamma_1 - 2\gamma_3 + \gamma_5 \\ m' &= m(m-3)/2 \\ m'' &= (m-2)(m-3)/2 \end{aligned} \tag{8.8}$$

Now, it is well known that since Q_{22}^{-1} belongs to the linear associative algebra generated by the association matrices of a partially balanced association scheme, Q_{22} can be calculated knowing the characteristic roots of the association matrices. In this manner, the matrix Q_{22} is obtained by Srivastava and Chopra (1971b). Having obtained Q_{22}, one can calculate Q_{02} and Q_{00} using (8.6a) and (8.6c). Finally, one can substitute the Q so obtained in the second term on the right-hand side of (8.7), which then reduces to the determinant of a matrix of the form $\sigma_1 I + \sigma_2 J$ where σ_1 and σ_2 are scalars, and J is a matrix having 1 everywhere. The fact that the second term on the right-hand side of (8.7) reduces to this form is not an accident; it follows from the fact that the matrix M belongs to the linear associative algebra generated by the association matrices of the multidimensional partially balanced association scheme which corresponds to the factorial association scheme defined earlier in this chapter. This fact is proved by Bose and

Srivastava (1964b). Also, $M^* - \delta I = (\gamma_1 - \gamma_3)I + \gamma_3 J$. Thus, the right-hand side of (8.7) and hence the characteristic polynomial of M can be computed. Thus, after a series of computations one finds that the characteristic polynomial of M^{-1} is given by

$$|M^{-1} - \delta I| = (c_3\delta^3 - c_2\delta^2 + c_1\delta - 1)(\delta c_6 - 1)^{m'}(c_5\delta^2 - c_4\delta + 1)^{m-1} \tag{8.9}$$

where

$$c_1 = 3\gamma_1 + (3m - 5)\gamma_3 + m''\gamma_5 \tag{8.10a}$$

$$c_2 = 3\gamma_1{}^2 + 2\gamma_1\gamma_3(3m - 5) + 2m''\gamma_1\gamma_5 + \frac{(m-1)(3m-8)}{2}\gamma_3{}^2 + m''(m-1)\gamma_3\gamma_5 - m\gamma_2{}^2 - \frac{m-1}{2}(2\gamma_2 + (m-2)\gamma_4)^2 \tag{8.10b}$$

$$c_3 = \gamma_1{}^3 - \binom{m}{2}(m-1)\gamma_3{}^3 + \frac{(m-1)(3m-8)}{2}\gamma_1\gamma_3{}^2 + (3m-5)\gamma_1{}^2\gamma_3 + m''\gamma_1{}^2\gamma_5 + m''(m-1)\gamma_1\gamma_3\gamma_5 - m\gamma_1\gamma_2{}^2 - mm''\gamma_2{}^2\gamma_5 + 2m\gamma_2{}^2\gamma_3 + m(m-1)(m-2)\gamma_2\gamma_3\gamma_4 - \frac{m-1}{2}\gamma_1(2\gamma_2 + (m-2)\gamma_4)^2 \tag{8.10c}$$

$$c_4 = 2\gamma_1 + (m - 5)\gamma_3 - (m - 3)\gamma_5 \tag{8.10d}$$

$$c_5 = (\gamma_1 - \gamma_3)(\gamma_1 - (m-3)\gamma_5 + (m-4)\gamma_3) - (\gamma_2 - \gamma_4)^2(m-2) \tag{8.10e}$$

$$c_6 = 16\mu_2 \tag{8.10f}$$

Since the matrix M is positive semidefinite, being equal to EE', it must have nonnegative roots. Thus, this computation of the characteristic polynomial of M leads to the following result regarding the existence of B-arrays:

Theorem 8.1. A set of necessary conditions that a B-array of size $(m \times N)$

with two symbols, strength 4, and index set $\boldsymbol{\mu}' = (\mu_0, \ldots, \mu_4)$ exists is

$$\mu_2 \geq 0 \qquad \text{(which is always true)} \tag{8.11a}$$

$$(m-1)(\mu_1 + \mu_3) \geq 2(m-5)\mu_2 \tag{8.11b}$$

$$(m-4)\mu_2{}^2 \leq \mu_2(\mu_1 + \mu_3) + (m-2)\mu_1\mu_3 \tag{8.11c}$$

$$c_1 \geq 0, \qquad c_2 \geq 0, \qquad c_3 \geq 0 \tag{8.11d}$$

Furthermore, M is nonsingular if and only if each of these six inequalities is strict.

The following result is important in the theory of optimal designs.

Theorem 8.2. The matrix M (and hence M^{-1}) cannot have more than six distinct characteristic roots. Let the roots of M^{-1} be ζ_i ($i = 1, \ldots, 6$). From (8.9), let the multiplicities be 1, 1, 1, $(m-1)$, $(m-1)$, and m, respectively. Then

$$\begin{aligned} &\zeta_1 + \zeta_2 + \zeta_3 = c_2/c_3, \qquad \zeta_1^{-1} + \zeta_2^{-1} + \zeta_3^{-1} = c_1, \qquad \zeta_1\zeta_2\zeta_3 = c_3^{-1} \\ &\zeta_4 + \zeta_5 = c_4/c_5, \qquad \zeta_4\zeta_5 = c_5^{-1}, \qquad \zeta_6 = 1/16\mu_2 \end{aligned} \tag{8.12}$$

$$\begin{aligned} \operatorname{tr} M^{-1} &= (c_2/c_3) + (m-1)(c_4/c_5) + m'/16\mu_2 \\ |M^{-1}| &= (16\mu_2)^{-m'} c_5^{-(m-1)} c_3^{-1} \\ Ch_{\max} M^{-1} &= \max(\zeta_1, \ldots, \zeta_6) \end{aligned} \tag{8.13}$$

Finally, the following result is needed since we need a nonempty class of designs before we can search for an optimal one within it.

Theorem 8.3. For every $N\ [\geq v = 1 + \binom{m}{1} + \binom{m}{2}]$, the class of nonsingular balanced 2^m factorial designs with N treatment combinations is nonempty.

To prove Theorem 8.3, it is clearly sufficient to show that the class is nonempty for the case in which $N = v$. However, for this case, a nonsingular balanced design is obtained by taking all treatment combinations which have weights 0, 1, or $(m-2)$.

Using ideal theory, Yamamoto, Shirakura, and Kuwada (1976) have generalized the above work of Srivastava and Chopra (1971) to the case of designs of resolution $2l + 1$. We present the case $l = 3$.

Theorem 8.4. The characteristic polynomial $\psi(\lambda)$ of the information matrix M of a balanced fractional 2^m factorial design T of resolution VII (considered as a B-array of strength 6, with index set $\{\mu_0, \mu_1, \ldots, \mu_6\}$) is given by

$$\psi(\lambda) = [\phi_1(\lambda)]^{m(m-1)(m-5)/6}[\phi_2(\lambda)]^{m(m-3)/2}[\phi_3(\lambda)]^{m-1}\phi_4(\lambda) \tag{8.14}$$

where

$$\phi_1(\lambda) = (\gamma_0 - 3\gamma_2 + 3\gamma_4 - \gamma_6 - \lambda)^{m(m-1)(m-5)/6} \tag{8.15a}$$

$$\phi_2(\lambda) = \begin{vmatrix} \gamma_0 - 2\gamma_2 + \gamma_4 - \lambda & (m-4)^{1/2}(\gamma_1 - 2\gamma_3 + \gamma_5) \\ \text{(sym)} & \begin{array}{l}\gamma_0 + (m-7)\gamma_2 \\ \quad - (2m-11)\gamma_4 + (m-5)\gamma_6 - \lambda\end{array} \end{vmatrix}^{m(m-3)/2} \tag{8.15b}$$

$$\phi_3(\lambda) = \begin{vmatrix} \gamma_0 - \gamma_2 - \lambda & (m-2)^{1/2}(\gamma_1 - \gamma_3) & \binom{m-2}{2}^{1/2}(\gamma_2 - \gamma_4) \\ & \begin{array}{l}\gamma_0 + (m-4)\gamma_2 \\ \quad - (m-3)\gamma_4 - \lambda\end{array} & \begin{array}{l}\dfrac{(m-3)^{1/2}}{2}\{2\gamma_1 + (m-6)\gamma_3 \\ \quad - (m-4)\gamma_5\}\end{array} \\ \text{(sym)} & & \begin{array}{l}\gamma_0 + (2m-9)\gamma_2 \\ \quad + \dfrac{(m-4)(m-9)}{2}\gamma_4 \\ \quad - \dbinom{m-4}{2}\gamma_6 - \lambda\end{array} \end{vmatrix}^{m-1} \tag{8.15c}$$

$$\phi_4(\lambda) = \begin{vmatrix} \gamma_0 - \lambda & (m\gamma_1)^{1/2} & \binom{m^{1/2}}{2}\gamma_2 & \binom{m^{1/2}}{3}\gamma_3 \\ & \gamma_0 + (m-1)\gamma_2 - \lambda & \begin{array}{l}\dfrac{(m-1)^{1/2}}{\sqrt{2}}\{2\gamma_1 \\ \quad + (m-2)\gamma_3\}\end{array} & \begin{array}{l}\dbinom{m-1}{2}^{1/2}/3\{3\gamma_2 \\ \quad + (m-3)\gamma_4\}\end{array} \\ & & \begin{array}{l}\gamma_0 + 2(m-2)\gamma_2 \\ \quad + \dbinom{m-2}{2}\gamma_4 - \lambda\end{array} & \begin{array}{l}\dfrac{(m-2)^{1/2}}{\sqrt{3}} \\ \times\left\{3\gamma_1 + 3(m-3)\gamma_3 \right. \\ \quad \left. + \dbinom{m-3}{2}\gamma_5\right\}\end{array} \\ \text{(sym)} & & & \begin{array}{l}\gamma_0 + 3(m-3)\gamma_2 \\ \quad + 3\dbinom{m-3}{2}\gamma_4 \\ \quad + \dbinom{m-3}{3}\gamma_6 - \lambda\end{array} \end{vmatrix} \tag{8.15d}$$

where, for $i = 0, 1, \ldots, 2l$, we have

$$\gamma_i = \sum_{j=0}^{2l} \sum_{p=0}^{i} (-1)^p \binom{i}{p} \binom{2l-i}{j-i+p} \mu_j \tag{8.16}$$

9. OPTIMAL BALANCED 2^m FACTORIAL DESIGNS OF RESOLUTION V, $m = 4, 5, 6$

This section is concerned with optimal balanced designs which are obtainable as a direct consequence of the work presented in the preceding two sections. Before discussing the derivation of optimal designs, it is important to notice that the problem of optimal balanced designs is broadly composed of two distinct parts: (a) the analytical problem, and (b) the combinatorial problem. Consider any fixed value of m and N. The analytical problem consists of determining which index set $\boldsymbol{\mu}' = (\mu_0, \ldots, \mu_4)$ gives rise to the smallest value of the optimality criterion. Thus, suppose the optimality criterion is tr M^{-1}. Then the question is which value of $\boldsymbol{\mu}'$ would minimize this quantity. In the preceding section, we presented an expression for tr M^{-1} in terms of the elements of $\boldsymbol{\mu}'$. Thus, using this expression, one can compute the value of tr M^{-1} for various values of $\boldsymbol{\mu}'$ with relative ease. Notice that the importance of the work in Section 8 lies in the fact that without it we would have had to evaluate M for every value of $\boldsymbol{\mu}'$ and then invert M and finally obtain tr M^{-1}. This would be a long and tedious process. Thus, the results of Section 8 help tremendously in the analytical problem.

Now suppose that a certain value of $\boldsymbol{\mu}'$, say $\boldsymbol{\mu}^{*\prime}$, minimizes the optimality criterion, say tr M^{-1}. The combinatorial problem is this: Does there exist a B-array of size $m \times N$ and of strength 4 whose index set is $\boldsymbol{\mu}^{*\prime}$? From the preceding section we know the necessary and sufficient conditions for the existence of B-arrays of strength t when $m \leq t + 2$. Thus, for $t = 4$, we need $m \leq 6$. Similarly, $t = 6$, which corresponds to resolution VII, requires $m \leq 8$. These are the two cases dealt with by Srivastava and Chopra (1971c) and Shirakura (1976), respectively. We briefly describe the computation of optimal designs for $t = 4$.

Notice that the relation (7.1) between N and the index set $\boldsymbol{\mu}'$ holds, so that $N = \mu_0 + 4\mu_1 + 6\mu_2 + 4\mu_3 + \mu_4$. Thus, when looking at the analytical problem, we restrict ourselves to values of $\boldsymbol{\mu}'$ which satisfy (7.1) for the given N.

Consider $m = 4$. In this case there is no combinatorial problem at all since, as mentioned earlier, for any $\boldsymbol{\mu}'$, we can write down the corresponding array. Thus, we look only at the analytical problem and find that $\boldsymbol{\mu}'$ for which the optimality criterion is minimized. Again, we restrict ourselves to the values of $\boldsymbol{\mu}'$ satisfying (7.1). As mentioned earlier, if an orthogonal array of strength 4 exists for the given value of N, then the optimal design is this orthogonal array, in which case all the μ_i ($i = 0, \ldots, 4$) are equal. Thus, it is easy to see that the value of the optimality criteria like trace would be minimized for values of the index set for which the μ_i are about equal. This helps us in looking at various values of $\boldsymbol{\mu}'$ which are optimal or close to the optimal. Thus, for each value of N in a certain practical range, the best value of $\boldsymbol{\mu}'$ was obtained and tabulated.

For $m = 5$, a similar procedure is adopted. First, the value of $\boldsymbol{\mu}'$ which minimizes the value of the criterion for a given value of N is found. For this value of $\boldsymbol{\mu}'$, check the conditions in Theorem 7.2 for the case $m = 5$. If the value of $\boldsymbol{\mu}'$ under consideration satisfies this condition, then we know that the array exists and can be written down, and our problem is solved. If it does not satisfy this condition, then we know that the array does not exist. In this case, we look at the analytical problem again, and find the next best value of $\boldsymbol{\mu}'$, and proceed as before.

For $m = 6$, the same procedure is followed as in the preceding case except that the conditions of Theorem 7.3 are also checked as a part of the combinatorial problem.

10. OPTIMAL BALANCED DESIGNS OF RESOLUTION V OF THE 2^7 SERIES, $N \leq 42$

For resolution V designs, the case $m = 7$ corresponds to the smallest value of m for which we have only necessary conditions for existence of arrays. The method of obtaining optimal designs in this case is as follows. We first study the combinatorial problem of existence of arrays and show what class of arrays actually does exist. Then within this class, we obtain the value of the criterion for each value of the index set $\boldsymbol{\mu}'$ and, thus, obtain the optimal design. We therefore proceed to study the combinatorial problem. The method of the study is to use all the different necessary conditions that we have in such a way as to be able to obtain new results regarding existence. Since the various conditions are to be used in a detailed manner, we present special cases of Theorem 7.4 and its corollaries for $m = 7$.

Throughout this section and the next, we shall be considering B-arrays $T(m \times N)$, $m = 7$, and of strength 4, with index set $\boldsymbol{\mu}' = (\mu_0, \ldots, \mu_4)$.

Theorem 10.1. For $m = 7$, the SDE reduce to

$$\begin{aligned}
&\text{(a)} \quad 35X_0 + 15X_1 + 5X_2 + X_3 = 35\mu_0 \\
&\text{(b)} \quad 5X_1 + 5X_2 + 3X_3 + X_4 = 35\mu_1 \\
&\text{(c)} \quad 5X_2 + 9X_3 + 9X_4 + 5X_5 = 105\mu_2 \\
&\text{(d)} \quad X_3 + 3X_4 + 5X_5 + 5X_6 = 35\mu_3 \\
&\text{(e)} \quad X_4 + 5X_5 + 15X_6 + 35X_7 = 35\mu_4
\end{aligned} \tag{10.1}$$

Theorem 10.2. For $m = 7$, the TDE reduce to

$$\begin{array}{llll} \text{(a)} & 7X_0 + X_1 = \pi_0 & \text{(b)} & 6X_1 + 2X_2 = 6\pi_1 \\ \text{(c)} & 5X_2 + 3X_3 = 15\pi_2 & \text{(d)} & 4X_3 + 4X_4 = 20\pi_3 \\ \text{(e)} & 3X_4 + 5X_5 = 15\pi_4 & \text{(f)} & 2X_5 + 6X_6 = 6\pi_5 \\ \text{(g)} & X_6 + 7X_7 = \pi_6 & & \end{array} \tag{10.2}$$

where, for $j = 0, 1, \ldots, 6$, we have $\pi_j = \sum_{r=1}^{g} \delta_{rj} y_r$, and the y_r are nonnegative integers with $\sum_1^g y_r = 7$. Also, the δ_{rj} are obtainable using (7.20).

Corollary 10.1. The DDE correspond to $g = 1$, so that $\pi_j = 7\delta_{ij} = 7\delta_j$, say.

We shall consider trim designs for the purpose of the existence of arrays. Thus we shall assume $X_0 = X_7 = 0$. Define

$$\begin{array}{lll} u = X_1 + X_6, & v = X_2 + X_5, & w = X_3 + X_4 \\ u' = X_1 - X_6, & v' = X_2 - X_5, & w' = X_3 - X_4 \\ \mu' = \mu_1 + \mu_3, & \mu_0' = \mu_1 - \mu_3, & \mu'' = \mu_0 + \mu_4 \\ & \mu_0'' = \mu_0 - \mu_4 & \end{array} \tag{10.3}$$

The SDE imply

$$\begin{array}{llll} \text{(a)} & 15u + 5v + w = 35\mu'', & \text{(b)} & 15u' + 5v' + w' = 35\mu_0'' \\ \text{(c)} & 5u + 5v + 4w = 35\mu', & \text{(d)} & 5u' + 5v' + 2w' = 35\mu_0' \\ \text{(e)} & 5v + 9w = 105\mu_2 & & \end{array} \tag{10.4}$$

$$u = 9\mu_2 + 5\mu'' - 8\mu' \geq 0, \qquad \text{i.e.,} \quad \mu'' \geq \tfrac{8}{5}\mu' - \tfrac{9}{5}\mu_2 \tag{10.5a}$$

$$v = 3[9\mu' - 11\mu_2 - 3\mu''] \geq 0, \qquad \text{i.e.,} \quad \mu'' \leq 3\mu' - \tfrac{11}{3}\mu_2 \tag{10.5b}$$

$$w = 5[6\mu_2 - 3\mu' + \mu''] \geq 0, \qquad \text{i.e.,} \quad \mu'' \geq 3\mu' - 6\mu_2 \tag{10.5c}$$

$$10u' - w' = 35(\mu_0'' - \mu_0'), \qquad \text{i.e.,} \quad 35 \mid (10u' - w') \tag{10.5d}$$

$$2v' + w' = 7(3\mu_0' - \mu_0''), \qquad \text{i.e.,} \quad 7 \mid (2v' + w') \tag{10.5e}$$

$$\begin{array}{ll} \text{(i)} & (\mu' \pm \mu_0'), \ (\mu'' \pm \mu_0''), \ (u \pm u'), \ (v \pm v'), \ (w \pm w') \ \text{ are even} \\ \text{(ii)} & |u'| \leq u, \ |v'| \leq v, \ |w'| \leq w, \ |\mu_0'| \leq \mu', \ |\mu_0''| \leq \mu'' \end{array} \tag{10.6}$$

$$\mu' \geq \tfrac{4}{3}\mu_2, \qquad \mu'' \geq \tfrac{1}{3}\mu_2 \tag{10.7}$$

Using these results, the following three theorems can be easily deduced.

Theorem 10.3. The number of assemblies N in a balanced trim design with $m \geq 7$ and with index set $\boldsymbol{\mu}'$ must satisfy

$$\max(\tfrac{21}{5}\mu_2 + \tfrac{28}{5}\mu', 7\mu') \leq N \leq \min(14\mu_2 + \tfrac{7}{3}\mu'', \tfrac{7}{3}\mu_2 + 7\mu') \tag{10.8}$$

Theorem 10.4. If T is a trim design with $m \geq 7$, then $\mu_2 = 3$ and 4 implies $N \geq 35$ and 51, respectively. Also $\mu_2 \geq 5$ implies $N \geq 60$.

Theorem 10.5. (a) There does not exist any trim design with $m = 7$, $\mu_2 = 1$ and the following set of values of N: (i) $24 \leq N \leq 27$, (ii) $31 \leq N \leq 34$, (iii) $38 \leq N \leq 41$, and (iv) $45 \leq N \leq 48$. (b) For the remaining values of N in the range $27 \leq N \leq 50$ (i.e., when $N = 28, 29, 30; 35, 36, 37; 42, 43, 44; 49, 50$), the values of (μ', μ'') corresponding to which designs might possibly exist are (4, 6), (4, 7), (4, 8), (5, 9), (5, 10), (5, 11), (6, 12), (6, 13), (6, 14), (7, 15), and (7, 16).

We shall now present a series of results for seven-rowed arrays. These results are needed for finding out the nature of the index sets $\boldsymbol{\mu}'$ corresponding to which arrays may exist. The proofs of many of the theorems presented here are quite similar. Therefore, we do not present the proofs or even their sketches, except in a few cases to give a flavor of the subject and also to enable the reader to understand the structure of the theory.

We begin with some results concerning the case when $\mu_2 = 1$.

Theorem 10.6. Let there exist a trim design T with $m \geq 7$, index set $\boldsymbol{\mu}'$, and $\mu_2 = 1$. Then $\mu_0 \geq \mu_1$ and $\mu_4 \geq \mu_3$.

Proof. Put $x = \mu_1 - \mu_0$, $y = \mu_3 - \mu_4$. From the five-row conditions of (7.4), we obtain $\psi_{11} = \max(0, -y, 1 - y - \mu_1) \leq \psi_{12} = \min(\mu_4, 1 - y, 1 - y - x)$, so that $1 - y - x \geq 0$, $y \leq 1$, $x \leq 1$. Hence $x + y > 0$ implies $x + y = 1$, so that $(x, y) = (0, 1)$ or $(1, 0)$. We show $(x, y) \neq (0, 1)$; the other case is similar. When $(x, y) = (0, 1)$, we get $d_1 = d_6 = 0$, and hence (7.12c)–(7.12e) show that the index set $\boldsymbol{\mu}'$ must be (0, 0, 1, 2, 1); however, this is rejected by (10.8). Hence $x + y \leq 0$. Since $x, y \leq 1$, we get $x \leq 0$, $y \leq 0$. This completes the proof.

Theorem 10.7. For a trim design T with $m = 7$ and $\mu_2 = 1$, one of the following sets of conditions must be satisfied: (i) $\{\mu_0 = \mu_1 = 0,\ \mu_3 \geq 3,\ 3\mu_3 = \mu_4 + 6\}$; (ii) $\{\mu_0 = 3\mu_1,\ \mu_3 \geq 3$, and $3\mu_3 = \mu_4 + 6\}$, or $\{3\mu_1 = \mu_0 + 6,\ \mu_1 \geq 3,\ \mu_4 = 3\mu_3\}$; (iii) $\{\mu_1 \geq 3,\ \mu_3 = \mu_4 = 0$, and $3\mu_1 = \mu_0 + 6\}$.

Corollary 10.2. For a trim design T with $m = 7$ and $\mu_2 = 1$, μ'' is divisible by 3.

Corollary 10.3. If a trim design with $m = 7$ and $\mu_2 = 1$ exists, then (μ', μ'') must take one of the values (4, 6), (5, 9), (6, 12), and (7, 15), with $N = 28, 35, 42$, and 49, respectively. This shows that N must be a multiple of 7, and that corresponding to a given such N, the index set $\boldsymbol{\mu}'$ must be of the form $(3\rho, \rho, 1, N/7 - \rho, 3N/7 - 3\rho - 6)$, where the integer ρ satisfies $N/7 - 3 \geq \rho \geq 0$.

Note that if an array with index set $(\mu_0^*, \mu_1^*, \mu_2^*, \mu_3^*, \mu_4^*)$ exists, then one with the "reverse" index set $(\mu_4^*, \mu_3^*, \mu_2^*, \mu_1^*, \mu_0^*)$ can be obtained from it by interchanging 0 and 1. Throughout this chapter, we therefore presented one index set in each case and ignored the reverse one.

Corollary 10.3 is easily obtainable from Theorem 10.5b and Theorem 10.7.

We now consider the case when $\mu_2 = 2$. The following result is easily obtainable from (10.7) and (10.8).

Theorem 10.8. Let T be a trim design with $m = 7$, $\mu_2 = 2$. Then (a) $\mu' > 3$; (b) when $\mu' = 4$, we must have $N = 31, 32$, and for $\mu' = 5$, $N = 37, 38$, or 39; (c) $\mu' \geq 6$ implies $N \geq 42$.

Theorem 10.9. Let T be a trim design with $m = 7$ and $\mu_2 = 2$. Then $\mu'' \geq \mu'$.

Theorem 10.10. Let T be an array having $\mu_2 = 2$, $m \geq 7$, $\mu_0 + \mu_4 = \mu_1 + \mu_3$. Then $\mu_0 = \mu_1$, and $\mu_3 = \mu_4$.

Theorems 10.9 and 10.10 lead to the following result:

Theorem 10.11. If T is a trim design with $m = 7$ and $\mu_2 = 2$, then $\mu' \neq 4$.

Theorem 10.12. If there exists a trim $(m \times N)$ B-array with $m = 7$, and $\mu_2 = 2$, then $N \neq 37, 38$, or 39.

Proof. We illustrate the use of the TDE by giving the proof for $N = 38$. From Theorem 10.8, we get $\mu' = 5$. Since $N = 4\mu_2 + 6\mu' + \mu''$, we get $\mu'' = 6$. The SDE in the form (10.4a), (10.4c), (10.4e) give $u = 8$, $v = w = 15$. Also, (10.1a), (10.1b), (10.1e), and (10.4b), respectively, imply that $5 | x_3$ and $7 | (x_1 - 2x_2 + x_3)$, $5 | x_4$, $7 | (x_4 - 2x_5 + x_6)$, and $7 | (2v' + w')$, where the vertical line means "divides." Solve the SDE at (10.1) using these facts; the solutions $\mathbf{X}' = (X_1, \ldots, X_6)$, recalling that $X_0 = X_7 = 0$ because the B-array is trim, turn out to be (6, 8, 10, 5, 7, 2), (5, 13, 0, 15, 2, 3), (2, 14, 5, 10, 1, 6), and (3, 9, 15, 0, 6, 5). Substituting back in the SDE, we find that the first two and the last two values of $\mathbf{X}'$ correspond, respectively, to $\boldsymbol{\mu}' = (4, 3, 2, 2, 2)$ and (3, 3, 2, 2, 3); these therefore are the only possible values of $\boldsymbol{\mu}'$ corresponding to which arrays may exist.

We prove the nonexistence of the array for $\boldsymbol{\mu}' = (4, 3, 2, 2, 2)$; the other case is similar. Using the five- and six-row conditions of (7.4) and (7.8), the procedure for which is fully illustrated in the proof of Theorem 11.7, it is found that the vector $\mathbf{d}' = (d_0, d_1, \ldots, d_6)$ can have four possible values, namely, (0, 1, 1, 1, 1, 1, 0), (1, 2, 1, 1, 1, 1, 1), (0, 1, 1, 1, 1, 1, 1), and (1, 1, 1, 1, 1, 1, 1). Thus, for the TDE, we get $g = 4$. Also, these four values of $\mathbf{d}'$ give $(\delta_0, \delta_3, \delta_6) = (1, \frac{1}{2}, 0)$, $(0, \frac{1}{2}, 1)$, $(0, 1, 0)$, and $(1, 0, 1)$, respectively. Thus, for

$\mathbf{X}' = (5, 13, 0, 15, 2, 3)$, the TDE give $y_1 + y_4 = 5$, $y_1 + y_2 + y_3 = 6$, $y_2 + y_4 = 3$, along with $y_1 + y_2 + y_3 + y_4 = 7$. Thus, $y_2 = 2$, which means there must be two six-rowed arrays (within the seven-rowed array under consideration) which correspond to the solution $\mathbf{d}' = (1, 2, 1, 1, 1, 1, 1)$. But, using the theory of Section 7, it is easy to check (e.g., by writing down the array) that a six-rowed array with this value of $\mathbf{d}'$ is such that the number of columns in it which have weight 5 or more is two. Thus, in the seven-rowed array under consideration, at most two columns could have weight 6 or more. But this contradicts the fact that $X_6 = 3$. Hence, the value of $\mathbf{X}'$ under consideration is rejected. The other value of $\mathbf{X}'$ is also similarly rejected. This completes the proof.

From Theorems 10.8, 10.11, and 10.12, we obtain the following:

Theorem 10.13. If T is a trim design with $m = 7$ and $\mu_2 = 2$, then $N \geq 42$.

We now present results for the case when $\mu_2 = 3$, and $N \leq 42$.

Theorem 10.14. For $\mu_2 = 3$ and $\mu' = 4$, the only possible value of $\boldsymbol{\mu}'$ for which trim designs may possibly exist is (1, 3, 3, 1, 0), apart from (0, 1) symmetry.

Theorem 10.15. If a design T with $m = 7$, $\mu_2 = 3$, and $\mu' = 5$ exists, then $N = 42$. Also, the corresponding values of $\boldsymbol{\mu}'$, apart from an interchange of 0 and 1, are (1, 3, 3, 2, 3) and (4, 4, 3, 1, 0).

Theorem 10.16. If T is a trim design with $m = 7$, and $27 \leq N \leq 41$, then the only values [apart from (0, 1) symmetry] of $\boldsymbol{\mu}'$ for which designs do exist are (i) (1, 3, 3, 1, 0) with $N = 35$; (ii) (0, 0, 1, 4, 6), (3, 1, 1, 3, 3) with $N = 28$; and (iii) (0, 0, 1, 5, 9), (3, 1, 1, 4, 6), (6, 2, 1, 3, 3) with $N = 35$.

Theorem 10.17. Let T be a trim design with $m = 7$ and $N = 42$. Then (a) apart from an interchange of 0 and 1, the possible values of $\boldsymbol{\mu}'$ are (0, 0, 1, 6, 12), (3, 1, 1, 5, 9), (6, 2, 1, 4, 6), (9, 3, 1, 3, 3), (1, 3, 3, 2, 3), and (4, 4, 3, 1, 0); and (b) the last two values of $\boldsymbol{\mu}'$ give rise to tr $M^{-1} = \infty$.

The proof of these four theorems is relatively simple. In particular, Theorem 10.16 is obtained from Theorems 10.13–10.15 and 10.7. Also, it turns out that corresponding to each index set arrived at in Theorem 10.16, there exists an array which is not only of strength 4 but is indeed of full strength. Now we present the results on optimal designs corresponding to $m = 7$ and $N \leq 42$.

Theorem 10.18. Let α and β denote nonnegative integers. Let T be an optimal balanced design of resolution V with N (≥ 29) runs and index set $\boldsymbol{\mu}'$.

(a) Then $\boldsymbol{\mu}'$ is not of the form $(1+\alpha, 3, 3, 1, \beta)$, $(\alpha, 0, 1, 4, 6+\beta)$, or $(\alpha, 0, 1, 5, 9+\beta)$.

(b) If $N \leq 41$, then apart from (0, 1) symmetry, $\boldsymbol{\mu}'$ must be of one of the forms $(\alpha+3, 1, 1, 3, \beta+3)$, $(\alpha+3, 1, 1, 4, \beta+6)$, or $(\alpha+6, 2, 1, 3, \beta+3)$.

(c) For $N = 42$, the values in (b) as well as the first four values in Theorem 10.17 are possible.

Parts (a) and (b) of Theorem 10.18 are easily proved by considering the expression for tr M^{-1} and comparing the value of the trace corresponding to the index set given with the value of the trace corresponding to a design T_0 obtained as follows. Take the optimal design with $N = 29$ and add to it $N - 29$ treatments, each of which is $(0, 0, \ldots, 0)$. The value of tr M^{-1} can be calculated for T_0, and it is shown that the value of the trace for the index sets shown is bounded below by the value of tr M^{-1} corresponding to T_0. Finally, using the SDE, it can be shown that the optimal designs are unique, apart from an interchange of 0 and 1.

11. OPTIMAL BALANCED 2^7 FACTORIAL DESIGNS WITH $43 \leq N \leq 68$

In this section we shall very briefly summarize the main results of the author obtained in three joint papers with Chopra, namely those with (i) $43 \leq N \leq 48$, (ii) $49 \leq N \leq 55$, and (iii) $56 \leq N \leq 68$. There are a large number of theorems proved in these papers and in the five papers dealt with in the following section. Even a bare statement of these theorems would cover about one-fifth of this chapter. The proof of the results in these papers is broadly similar to the ones discussed in the preceding section, except for some cases. However, the proofs are not straightforward, and in general more complex. The interested reader should look into the original papers to get an idea of these.

In view of this, we present only the most important results. Proofs are given only in two cases, and there too just to illustrate two basic results which should be included in this chapter because of their generality and importance.

Theorem 11.1. (a) Let $T(7 \times N)$ be a trim design with $\mu_2 \geq 4$. Then $N \geq 51$. (b) Let $T(7 \times N)$ be a trim design with $43 \leq N \leq 48$. Then $\mu_2 \neq 3$.

Theorem 11.2. There does not exist any seven-rowed trim design with $\mu_2 = 2$, and $27 \leq N \leq 48$.

Theorem 11.3. (a) Let $T(7 \times N)$ be a trim B-array with index set $\boldsymbol{\mu}'$, $\mu_2 = 3$, and $27 \leq N \leq 48$. Then, apart from an interchange of 0 and 1 in the array, the possible values of $\boldsymbol{\mu}'$ are (i) (1, 3, 3, 1, 0) with $N = 35$; and (ii) (1, 3, 3, 2, 3) and (4, 4, 3, 1, 0) with $N = 42$.

(b) Let $T'(7 \times N)$ be an optimal design with index set $\boldsymbol{\mu}'$, and $43 \leq N \leq 48$. Then, apart from duality, $\boldsymbol{\mu}'$ is of one of the following three figures, where $\alpha, \beta \geq 0$: $(1 + \alpha, 3, 3, 1, \beta)$, $(1 + \alpha, 3, 3, 2, 3 + \beta)$, and $(4 + \alpha, 4, 3, 1, \beta)$.

Theorem 11.4. (a) There do exist seven-rowed trim B-arrays with index set (i) (6, 4, 2, 3, 3), (ii) (9, 7, 2, 0, 0), and (iii) (3, 1, 2, 6, 6).

(b) If $49 \leq N \leq 55$, there does not exist any $(7 \times N)$ trim B-array with $\mu_2 = 2$, except for the three cases indicated in (a).

Theorem 11.5. Consider a $(7 \times N)$ trim B-array with index set $\boldsymbol{\mu}'$, and $49 \leq N \leq 55$. Then (a) $\mu_2 \leq 3$; (b) if $\mu_2 = 3$, then $N = 49$, and the possible values of $\boldsymbol{\mu}'$ are (i) (4, 4, 3, 2, 3), (ii) (7, 5, 3, 1, 0), and (iii) (6, 3, 3, 3, 1), apart from duality. Furthermore, arrays do exist corresponding to each of these values of $\boldsymbol{\mu}'$.

Theorem 11.6. Consider a B-array T of size $(7 \times N)$ with the index set $\boldsymbol{\mu}'$ and with $\mu_2 \geq 7$. Then $N \geq 78$.

By the time work was started on the case $56 \leq N \leq 68$, the diophantine analysis which was developed earlier for $29 \leq N \leq 55$ was systematized. This enabled the development of a computer program through which the conditions (8.11), (7.9), (7.10), (7.22), along with the conditions implied by the SDE and the TDE, could be fully checked. Using this program, we looked at trim B-arrays with $56 \leq N \leq 68$, and obtained (for each N) every index set which satisfied *all* the aforementioned conditions. Such index sets are presented by Srivastava and Chopra (1974, Table I, p. 273). Corresponding to one of these index sets, namely (2, 4, 4, 3, 2), it was shown that no arrays exist. Corresponding to most of the remaining index sets whether or not an array exists is not known.

However, for the purpose of obtaining optimal designs, the following procedure was followed. We compare the value of the optimality criterion for a given value of N, for index sets of the form $(\mu_0{}^*, \mu_1, \mu_2, \mu_3, \mu_4{}^*)$ where $\mu_0{}^* \geq \mu_0$ and $\mu_4{}^* \geq \mu_4$, and where $(\mu_0, \mu_1, \mu_2, \mu_3, \mu_4)$ is one of the index sets corresponding to this value of N listed in Table 13.1. After the computations were made, it turned out that the index sets for which optimal designs did exist were found to correspond to index sets of trim arrays which also could be shown to exist. Indeed, all such arrays are of full strength.

For $43 \leq N \leq 55$, optimal designs were obtained as in the preceding section, and using Theorems 11.3b, 11.4a, and 11.5b.

We present the following theorem and its proof to illustrate a simple application of the π^* of (7.22).

Theorem 11.7. There does not exist any trim seven-rowed B-array with index set (4, 3, 2, 3, 4).

Proof. Here $\boldsymbol{\theta}' = (-4, -8, -12, 5, 8)$, and $1 = \psi_{11} \leq d_6 \leq d_5 \leq \cdots \leq d_1 \leq \psi_{12} = 3$. The six-row conditions (7.12) give $d_{4 \cdot 6} \leq 2$, $d_{1 \cdot 3} \leq 7$, and $d_{5 \cdot 6} \geq 3$. Hence, $\theta_4 \geq 2$, $d_6 \leq d_3 \leq 2$, so that $d_4 = d_5 = 2$, $d_6 = 1$ or 2. For $d_6 = 1$, the inequality $d_{1 \cdot 2} \leq -1 + d_{4 \cdot 6}$ gives $d_{1 \cdot 2} \leq 4$, implying $d_1 = d_2 = 2$. Similarly, for $d_6 = 2$, we get $d_2 = 2, d_1 = 2$ or 3. Computing the ψ_{21} and ψ_{22}, we find that $d_1 = d_6 = 2$ gives $d_0 = 0$ or 1; $d_1 = d_5 = 2, d_6 = 1$ gives $d_0 = 0$; and $d_1 = 3$, $d_2 = d_6 = 1$ gives $d_0 = 1$. Thus, we have four values of $\mathbf{d}' = (d_0, \ldots, d_6)$, and hence four values of $\boldsymbol{\pi}^{*\prime}$ in (7.22), namely (0, 12, 0, 20, 0, 12, 0), (1, 6, 15, 0, 15, 6, 1), (1, 7, 10, 10, 5, 11, 0), and (0, 11, 5, 10, 10, 7, 1). Using (10.5)–(10.7), and (10.1), we can show that corresponding to the index set under consideration, the solution vector to the SDE is $(x_1, \ldots, x_6) = (5, 12, 5, 5, 12, 5)$. Using (7.22) (ii) with $i = 1$ and 5, all the values obtained for $\boldsymbol{\pi}^{*\prime}$ get rejected, except the second one. This value, however, is rejected by the DDE. This completes the proof.

12. THE CASE $m = 8$

We start with a general result (from Srivastava and Chopra, 1971a) on the comparison of index sets, which help in narrowing down the class of candidates for optimal designs.

Theorem 12.1. If T_1, T_2, and T_3 are, respectively, three balanced resolution V designs of the 2^8 series, with index sets $(\mu_0, \mu_1, \mu_2, \mu_1, \mu_0)$, $(\mu_0 - z, \mu_1, \mu_2, \mu_1, \mu_0 + z)$, and $(\mu_0, \mu_1 - z, \mu_2, \mu_1 + z, \mu_0)$, where z is a nonzero integer, then we have

$$\operatorname{tr} V_{T_2} > \operatorname{tr} V_{T_1} < \operatorname{tr} V_{T_3} \tag{12.1}$$

Next, we present combinatorial results from Srivastava and Chopra (1973), Chopra and Srivastava (1974), and Chopra (1975a, b).

Theorem 12.2. For a trim B-array with two symbols, $t = 4$ and $m \geq 8$, we have $N \geq \frac{35}{3}\mu_2$. (The bound is attainable for $\mu_2 = 6$.)

Theorem 12.3. Let T be a trim B-array with $\mu_2 > 0$. Then μ_1 and μ_3 are not both zero, and μ_0 and μ_4 are not both zero.

Theorem 12.4. There exist no trim $(8 \times N)$ B-arrays of strength 4, and $N \leq 59$, except (apart from duality) for the index sets $\boldsymbol{\mu}' = (3, 3, 1, 0, 0)$; (6, 4,

1, 0, 0); (6, 4, 1, 1, 4), (10, 5, 1, 0, 0); (10, 5, 1, 1, 4), (14, 6, 1, 0, 0), (6, 4, 1, 2, 8); and (6, 4, 1, 3, 12), (10, 5, 1, 2, 8), (14, 6, 1, 1, 4), and (18, 7, 1, 0, 0). These arrays correspond to values of N equal to 21, 28, 36, and 52, and are all of full strength.

Theorem 12.5 (Intermediate Diophantine Equations). Consider an $(m \times N)$ B-array T of two symbols, strength t, and index set $(\mu_0{}^t, \mu_1{}^t, \ldots, \mu_t{}^t)$. Suppose that $(t+1)$-rowed arrays with this index set do exist, so that [from (7.4a) and (7.4b)] there exists an integer d satisfying $\psi_{11} \le d \le \psi_{12}$. Let X_q $(q = 0, 1, \ldots, m)$ be the number of columns of T whose weight is q, and let

$$\phi_{t+1,d} = d$$

$$\phi_{id} = \mu_i^t - \mu_{i+1}^t + \mu_{i+2}^t - \cdots + (-1)^{t-i}\mu_t^t + (-1)^{t-i+1}d, \qquad \text{for} \quad 0 \le i \le t \tag{12.2}$$

For any d such that $\psi_{11} \le d_0 \le \psi_{12}$, let there be Z_d subarrays in T [of size $(t+1) \times N$ each]. Then, for $0 \le i \le t+1$, we have

$$\sum_{q=0}^{m} \binom{q}{i}\binom{m-q}{t+1-i} X_q = \binom{t+1}{i}\left[\sum_d Z_d \phi_{id}\right], \qquad \sum_d Z_d = \binom{m}{t+1} \tag{12.3}$$

where $\sum_d$ runs over all the $(1 + \psi_{12} - \psi_{11})$ values of d satisfying $\psi_{11} \le d \le \psi_{12}$; hence a necessary condition that T exists is that the system of $(t+2)$ equations (12.3) has a solution in nonnegative integers X's and Z's.

Equations (12.3) are usually referred to as the intermediate diophantine equations (IDE). These are obtained in a way similar to the SDE and TDE. In the following theorem, we illustrate their application, using the notation

$$\alpha^* = \mu_1 - \mu_0, \; \beta^* = \mu_3 - \mu_4.$$

Theorem 12.6. Consider a B-array T with parameters $(m, N, t; \boldsymbol{\mu}_0')$, where $m \ge 8$, $t = 4$, $\mu_2 = 3$, and $\alpha^* + \beta^* = 0$. Then $\alpha^* = -\beta^* = 0$, 1, or -1.

Proof. Suppose $\beta^* = -\alpha^* > 1$. The condition $\psi_{11} \le \psi_{12}$ leads to (i) $\beta^* = 3, \psi_{11} = \psi_{12} = 0, d_1 = d_6 = 0$; and (ii) $\beta^* = 2, \psi_{11} = 0, \psi_{12} = 1$. Consider (i). The six-row conditions (b)–(d), (e), and (i) of (7.12) give $\psi_{21} = \psi_{22} = 0$, $d_0 = \bar{d} = 0$, $\delta_3 = \delta_5 = 0$, and $\delta_4 = \mu_4 = 3$, contradicting the DDE. For (ii), the index d in (12.3) takes two values 0 and 1. For $d = 0$, (12.2) gives $\phi_{50} = 0$, $\phi_{40} = \mu_4$, $\phi_{30} = \beta^* = 2$, $\phi_{20} = 1$. Similarly, $d = 1$ implies $\phi_{51} = 1$, $\phi_{41} = \mu_4 - 1$, $\phi_{31} = 3$, $\phi_{21} = 0$. Adding up the IDE for $m = 8$, $t = 4$, with $i = 2$ and 5, we get $\binom{8}{5} = 56 = Z_0 + Z_1 = 2X_2 + 3X_3 + (2.4)X_4 + 2X_5 + 6X_6 + 21X_7$. On the other hand, the SDE (for $m = 8$, $t = 4$, and $q = 3, 4$) give $280\beta = 560 = 5X_3 + 12X_4 + 10X_5 - 20X_6 - 105X_7 - 280X_8$. Subtracting the second equation from 10 times the first, we obtain $0 = 20X_2 + 25X_3 + 12X_4 + 10X_5 + 80X_6 + 315X_7 + 280X_8$.

Since the X's are nonnegative, we get $X_2 = \cdots = X_8 = 0$. But then in the IDE for $m = 8$, $t = 4$, $i = 3$, we get $0 = 10(2Z_0 + 3Z_1) = 10Z_1 + 20 \times 56$, giving a contradiction since $Z_1 \geq 0$. This completes the proof.

The following results can be proved on the lines of those in the preceding two sections by using the IDE, TDE, and so on.

Theorem 12.7. Let T be a possible B-array with $\mu_2 = 3, m \geq 8, \alpha^* = 0$, and $\beta^* > 0$. Then $\beta^* = 1$. (Similarly, $\beta^* = 0$, $\alpha^* \geq 0$ implies $\alpha^* = 1$.)

Theorem 12.8. Let T be a B-array of strength 4 with $m \geq 8$, $\mu_2 = 3$. Then α^* and β^* cannot be both positive. If $\alpha^* + \beta^* > 0$, then either $\alpha^* = 0$, $\beta^* > 0$, or $\alpha^* > 0$, $\beta^* = 0$.

Theorem 12.9. There does not exist any $(8 \times N)$ trim B-array of strength 4 with $\mu_2 = 3$, and $N \leq 65$.

Theorem 12.10. For a B-array T of strength 4, with $m \geq 8$ and $\mu_2 = 5$, we have $N \geq 66$. (However, it is not known whether a B-array of strength 4 with $m = 8$, $\mu_2 = 5$, and $N = 66$ exists.)

Theorem 12.11. Let T be an $(8 \times N)$ trim B-array with $\mu_2 = 2$ and $37 \leq N \leq 65$. Then T exists only when $N = 56$ or 64, the index sets $\boldsymbol{\mu}'$ being (6, 4, 2, 4, 6) or (0, 0, 2, 8, 12), and (10, 5, 2, 4, 6).

Theorem 12.12. Consider a B-array $T(8 \times N)$ with $\mu_2 = 4$ and $N \leq 65$. Then $N = 56$ or 64, with $\boldsymbol{\mu}' = (4, 6, 4, 1, 0)$, and (8, 7, 4, 1, 0) or (4, 6, 4, 2, 4). Arrays exist in each of these cases and are of full strength.

In view of these theorems, the optimal designs up to the case $N \leq 56$ correspond to the value of $\mu_2 = 1$. When $N \geq 57$, the value $\mu_2 = 2$ takes over. For the range $60 \leq N \leq 65$, if $T(8 \times N)$ has $\mu_2 = 1$, then tr $V_T \geq 1.2500$. Also, $\mu_2 \geq 5$ implies $N \geq 66$. Hence, we are restricted to the cases $\mu_2 = 2$ and 4.

13. TABLES OF OPTIMAL DESIGNS AND HOW TO USE THEM

We now consider optimal balanced designs, how to write them down using the tables, and how to compute estimates, etc. Tables 13.1 and 13.2 give information on the source of the tables. Here, we present a full discussion of their use, which will enable the reader to go straight to the tables without having to read their explanations in the body of the text in which the table is contained, where such explanations are usually intermixed with other mathematical discussions.

TABLE 13.1

Papers Containing Tables of Designs of Resolutions V and VII

m	N	Reference
4	11–28	Srivastava and Chopra (1971c, p. 261)
5	16–32	Srivastava and Chopra (1971c, pp. 261, 266)
6	22–40	Srivastava and Chopra (1971c, pp. 262, 266)
4	29–64	Chopra (1975c, p. 15)
5	33–64	Chopra (1975c, pp. 16, 17)
6	41–64	Chopra (1975c, p. 18)
7	29–42	Chopra and Srivastava (1971a, pp. 591–592)
7	43–48	Chopra and Srivastava (1977)
7	49–55	Chopra and Srivastava (1973b)
7	56–68	Srivastava and Chopra (1974, pp. 276–277)
8	37–51	Chopra and Srivastava (1974, p. 50)
8	52–65	Chopra (1975a,b, pp. 99; 168)
9–11	—	Chopra (1972, abstract)
6[a]	42–64	Shirakura (1976, pp. 522–523)
7[a]	64–90	Shirakura (1976, pp. 524–525, 528)
8*	93–128	Shirakura (1976, pp. 526–529)

[a] Resolution VII.

Note: (1) In the paper of Shirakura, the factors are denoted by θ rather than F; in other papers they are denoted by A.

(2) In some papers, the value of λ' and of the covariance for different N are given in separate tables. Cases in which λ' is not given at all are covered in Table 13.2

(3) In many papers, the value of λ' is given for the dual design.

(4) For some cases, particularly $m = 5$, two values of λ' are given for certain values of N. As explained in the text, any of these values of λ' can be chosen.

(5) The figures given under the columns of "Var" and "Cov" in these tables are $(\sigma^{-2}2^{-2m})$ times the actual values of the corresponding variance or covariance.

For any given value of m and N, we first find the value of the vector $\boldsymbol{\lambda}' = (\lambda_0, \lambda_1, \ldots, \lambda_m)$. Note that for $m = 4$, we have $\lambda_0 = \mu_0, \lambda_1 = \mu_1, \lambda_2 = \mu_2$, $\lambda_3 = \mu_3$, and $\lambda_4 = \mu_4$, so that the λ are not separately given. For any given value of N, the value of the vector $\boldsymbol{\mu}' = (\mu_0, \mu_1, \mu_2, \mu_3, \mu_4)$ is given for each design; this is of interest to the more mathematically oriented readers, and refers to the development in various sections. However, the value of $\boldsymbol{\mu}'$ is not needed for writing down the treatments of a design, except for the case $m = 4$. We illustrate the method using the case $m = 7$, $N = 30$; all the other cases are exactly the same. From the corresponding tables, we find that $\boldsymbol{\lambda}' = (\lambda_0, \lambda_1, \ldots, \lambda_7) = (2, 0, 1, 0, 0, 0, 1, 0)$, so that $\lambda_0 = 2$, $\lambda_2 = \lambda_6 = 1$, and $\lambda_1 = \lambda_3 = \lambda_4 = \lambda_5 = \lambda_7 = 0$. Now, to write down the required design T, we simply write down each vector of weight i exactly λ_i times, where $i = 0, 1, \ldots, 7$.

TABLE 13.2

Values of λ′ for Various Values of N for Optimal Balanced 2^8 Factorial Designs[a]

	N	$\mu' = (\mu_0, \mu_1, \mu_2, \mu_3, \mu_4)$	$\lambda' = (\lambda_0, \lambda_1, \lambda_2, \lambda_3, \lambda_4, \lambda_5, \lambda_6, \lambda_7, \lambda_8)$
$m = 8$			
	37	(7, 4, 1, 1, 4)	(1, 0, 2, 0, 0, 0, 0, 1, 0)
	38	(8, 4, 1, 1, 4)	(2, 0, 2, 0, 0, 0, 0, 1, 0)
	39	(9, 4, 1, 1, 4)	(3, 0, 2, 0, 0, 0, 0, 1, 0)
	40	(10, 4, 1, 1, 4)	(4, 0, 2, 0, 0, 0, 0, 1, 0)
	41	(11, 4, 1, 1, 4)	(5, 0, 2, 0, 0, 0, 0, 1, 0)
	42	(12, 4, 1, 1, 4)	(6, 0, 2, 0, 0, 0, 0, 1, 0)
	43	(13, 4, 1, 1, 4)	(7, 0, 2, 0, 0, 0, 0, 1, 0)
	44	(14, 4, 1, 1, 4)	(8, 0, 2, 0, 0, 0, 0, 1, 0)
	45	(7, 4, 1, 2, 8)	(1, 0, 2, 0, 0, 0, 0, 2, 0)
	46	(8, 4, 1, 2, 8)	(2, 0, 2, 0, 0, 0, 0, 2, 0)
	47	(9, 4, 1, 2, 8)	(3, 0, 2, 0, 0, 0, 0, 2, 0)
	48	(10, 4, 1, 2, 8)	(4, 0, 2, 0, 0, 0, 0, 2, 0)
	49	(11, 4, 1, 2, 8)	(5, 0, 2, 0, 0, 0, 0, 2, 0)
	50	(12, 4, 1, 2, 8)	(6, 0, 2, 0, 0, 0, 0, 2, 0)
	51	(13, 4, 1, 2, 8)	(7, 0, 2, 0, 0, 0, 0, 2, 0)
	52	(10, 5, 1, 2, 8)	(0, 1, 2, 0, 0, 0, 0, 2, 0)
	53	(11, 5, 1, 2, 8)	(1, 1, 2, 0, 0, 0, 0, 2, 0)
	54	(8, 4, 1, 3, 12)	(2, 0, 2, 0, 0, 0, 0, 3, 0)
	55	(9, 4, 1, 3, 12)	(3, 0, 2, 0, 0, 0, 0, 3, 0)
	56	(10, 4, 1, 3, 12)	(4, 0, 2, 0, 0, 0, 0, 3, 0)
	57	(6, 4, 2, 4, 7)	(0, 0, 2, 0, 0, 0, 2, 0, 1)
	58	(7, 4, 2, 4, 7)	(1, 0, 2, 0, 0, 0, 2, 0, 1)
	59	(8, 4, 2, 4, 7)	(2, 0, 2, 0, 0, 0, 2, 0, 1)
	60	(8, 4, 2, 4, 8)	(2, 0, 2, 0, 0, 0, 2, 0, 2)
	61	(8, 4, 2, 4, 9)	(2, 0, 2, 0, 0, 0, 2, 0, 3)
	62	(9, 4, 2, 4, 9)	(3, 0, 2, 0, 0, 0, 2, 0, 3)
	63[b]	(3, 4, 4, 4, 4)	
	64[c]	(4, 4, 4, 4, 4)	
	65[b]	(5, 4, 4, 4, 4)	
$m = 7$			
	43	(2, 3, 3, 2, 3)	(1, 0, 0, 1, 0, 0, 1, 0)
	44	(3, 3, 3, 2, 3)	(2, 0, 0, 1, 0, 0, 1, 0)
	45	(4, 3, 3, 2, 3)	(3, 0, 0, 1, 0, 0, 1, 0)
	46	(4, 3, 3, 2, 4)	(3, 0, 0, 1, 0, 0, 1, 1)
	47	(5, 3, 3, 2, 4)	(4, 0, 0, 1, 0, 0, 1, 1)
	48	(5, 3, 3, 2, 5)	(4, 0, 0, 1, 0, 0, 1, 2)

[a] These values are not presented explicitly in the corresponding papers.

[b] A design can be obtained for $N = 63$ or 65 from the suggested one for $N = 64$, by deleting or adding the treatment (0, 0, 0, 0, 0, 0, 0, 0).

[c] This is a classical orthogonal case. A design can be obtained by taking all treatments $(x_1, x_2, \ldots, x_8)$ which are solutions of the equations $x_1 + x_2 + x_3 + x_4 + x_5 = x_1 + x_2 + x_6 + x_7 + x_8 = 0 \pmod 2$, where the x take the value 0 or 1. Computation (mod 2) means that every odd integer is replaced by 1, and even by 0; for example, $1 + 1 + 1 + 1 + 1 = 1$, while $0 + 1 + 0 + 0 + 1 = 0 \pmod 2$.

Thus, in the present case, we must write down the vector (0, 0, 0, 0, 0, 0, 0) twice and the vectors of weight 2 like (0, 1, 0, 1, 0, 0, 0) and also the vectors of weight 6 like (1, 1, 0, 1, 1, 1, 1), each exactly once. Since the other λ are zero, vectors of other weights do not occur in the present design. Thus, the present design, written in transposed form, is given by

$$T' = \begin{bmatrix} 0\,0\,1\,1\,1\,1\,1\,1\,0\,0\,0\,0\,0\,0\,0\,0\,0\,0\,0\,0\,0\,0\,0\,0\,1\,1\,1\,1\,1\,1 \\ 0\,0\,1\,0\,0\,0\,0\,0\,1\,1\,1\,1\,1\,0\,0\,0\,0\,0\,0\,0\,0\,0\,0\,1\,0\,1\,1\,1\,1\,1 \\ 0\,0\,0\,1\,0\,0\,0\,0\,1\,0\,0\,0\,0\,1\,1\,1\,1\,0\,0\,0\,0\,0\,0\,1\,1\,0\,1\,1\,1\,1 \\ 0\,0\,0\,0\,1\,0\,0\,0\,0\,1\,0\,0\,0\,1\,0\,0\,0\,1\,1\,1\,0\,0\,0\,1\,1\,1\,0\,1\,1\,1 \\ 0\,0\,0\,0\,0\,1\,0\,0\,0\,0\,1\,0\,0\,0\,1\,0\,0\,1\,0\,0\,1\,1\,0\,1\,1\,1\,1\,0\,1\,1 \\ 0\,0\,0\,0\,0\,0\,1\,0\,0\,0\,0\,1\,0\,0\,0\,1\,0\,0\,1\,0\,1\,0\,1\,1\,1\,1\,1\,1\,0\,1 \\ 0\,0\,0\,0\,0\,0\,0\,1\,0\,0\,0\,0\,1\,0\,0\,0\,1\,0\,0\,1\,0\,1\,1\,1\,1\,1\,1\,1\,1\,0 \end{bmatrix} \tag{13.1}$$

Notice that the vectors of any particular weight, say weight 2, have been written in a systematic way in (13.1); this particular scheme is evident from (13.1) by inspection. Although it does not matter in what order one writes the treatments in the treatment matrix T, a systematic way of writing is generally more convenient and reduces the incidence of error in the analysis and presentation of the data.

Notice that for any design obtained by the methods described herein, we can obtain a dual design by interchanging the symbols 0 and 1 in the original design. Thus, in the design dual to the one in (13.1), the vectors of weight 7 will occur twice, those of weight 1 or 5 once each, and other vectors do not occur. It can be proved that the statistical properties of any design and its dual are identical. Thus, from the statistical point of view, it does not matter whether one uses a particular design or its dual. In many cases, one could randomly choose between the two. In other cases, cost considerations might indicate the choice of one over another. This would happen, for example, when for certain factors the cost for applying, say, the level 1 is higher than that of using the level 0. Also, it should be borne in mind that the levels 0 and 1 of any factor are indicated here merely as symbols; they can be assigned, in any manner one likes, to the actual physical levels of a particular factor. For example, suppose nitrogen fertilizer is one of the factors, and suppose it is to be used at the levels 30 and 60 lb/acre. Then, the level 30 lb/acre could be represented either by 0 or by 1, and similarly for the level 60 lb/acre. Cost factors may be used to make decisions regarding this.

Again, for many values of N, there are two designs which are both optimal. This happens particularly for the case $m = 5$. Since we give the value of $\boldsymbol{\mu}'$, readers familiar with Section 7 can easily write down all the possible designs. However, the variances and covariances depend only on $\boldsymbol{\mu}'$ and are as indicated.

Next, we discuss the variances and covariances of the various estimates.

Since the designs are balanced, we need to give only 10 numbers for each case. These are, respectively, $V(\hat{\mu})$, $V(\hat{F}_i)$, $V(\hat{F}_{ij})$, $C(\mu, \hat{F}_i)$, $C(\mu, \hat{F}_{ij})$, $C(\hat{F}_i, \hat{F}_j)$, $C(\hat{F}_i, \hat{F}_{ij})$, $C(\hat{F}_i, \hat{F}_{jk})$, $C(\hat{F}_{ij}, \hat{F}_{ik})$, and $C(\hat{F}_{ij}, \hat{F}_{kl})$, where V denotes variance, C covariance, the caret over a parameter denotes the (best linear unbiased) estimate of that parameter, and i, j, k, and l are any distinct integers belonging to the set $\{1, 2, \ldots, m\}$. Because of balance, these numbers give all the variances and covariances of the estimates of parameters in a resolution V design. Thus, for example, for the case $m = 7$, $V(\hat{F}_2) = V(\hat{F}_4) = V(\hat{F}_i)$, $V(\hat{F}_{12}) = V(\hat{F}_{ij})$, $C(\hat{F}_{12}, \hat{F}_{25}) = C(\hat{F}_{34}, \hat{F}_{37}) = C(\hat{F}_{ij}, \hat{F}_{ik})$, etc. In the tables for each design, all the variances and covariances are presented. This should be especially convenient for the users of the designs, since now they do not need to compute these quantities.

The next question is how to calculate the various estimators $(\hat{F}_i)$. In the following, we present formulas for their computation. The estimates are given by

$$
\begin{aligned}
\hat{\mu} &= [\zeta_{00}\, V(\hat{\mu}) + \zeta_{01}\, C(\hat{\mu}, \hat{F}_i) + \zeta_{02}\, C(\hat{\mu}, \hat{F}_{ij})]2^m \\
\hat{F}_i &= [\zeta_{10,i}\, C(\hat{\mu}, \hat{F}_i) + \zeta_{11,i}\, V(\hat{F}_i) + \zeta_{12,i}\, C(\hat{F}_i, \hat{F}_j) + \zeta_{13,i}\, C(\hat{F}_i, \hat{F}_{ij}) \\
&\quad + \zeta_{14,i}\, C(\hat{F}_i, \hat{F}_{jk})]2^m \qquad (13.2)\\
\hat{F}_{ij} &= [\zeta_{20,ij}\, C(\hat{\mu}, \hat{F}_{ij}) + \zeta_{21,ij}\, C(\hat{F}_i, \hat{F}_{ij}) + \zeta_{22,ij}\, V(\hat{F}_{ij}) + \zeta_{23,ij}\, C(\hat{F}_i, \hat{F}_{jk}) \\
&\quad + \zeta_{24,ij}\, C(\hat{F}_{ij}, \hat{F}_{ik}) + \zeta_{25,ij}\, C(\hat{F}_{ij}, \hat{F}_{kl})]2^m
\end{aligned}
$$

where the ζ are real numbers whose computation we now discuss. Notice that in formula (13.2), the symbols i, j, k, l denote any set of four distinct integers chosen out of the set $\{1, 2, \ldots, m\}$. Also, the ζ are numbers whose computation is performed using the following steps:

(1) First, for any treatment $(j_1, \ldots, j_m)$, we calculate a quantity denoted by $y_0(j_1, \ldots, j_m)$, called *total yield*, as follows. If the treatment $(j_1, \ldots, j_m)$ does not occur in the design T which we have used for our experiment, then we take $y_0(j_1, \ldots, j_m) = 0$. On the other hand, if the treatment $(j_1, \ldots, j_m)$ does occur in T, then we take $y_0(j_1, \ldots, j_m)$ to equal the total yield from all experimental units on which the treatment $(j_1, \ldots, j_m)$ has been applied.

(2) Next, we compute what we call the *intermediate estimates* of the various factorial effects of interest. For designs of resolution V, these intermediate estimates will be denoted by $\hat{\hat{\mu}}$, $\hat{\hat{F}}_1, \ldots, \hat{\hat{F}}_m$, $\hat{\hat{F}}_{12}, \ldots, \hat{\hat{F}}_{m-1,m}$. The intermediate estimate of any factorial effect is computed from the total yield of the various treatment combinations in the same way as the corresponding factorial effect is computed from the true effect of the various treatment combinations. Thus, for example, $\hat{\hat{\mu}}$ will be simply the sum of the quantities $y_0(j_1, \ldots, j_m)$ over all treatments $(j_1, \ldots, j_m)$. Similarly, $\hat{\hat{F}}_1$ will be the sum of the total yields $y_0(j_1, \ldots, j_m)$ for all treatments in which $j_1 = 1$ minus the sum

of the total yields of all treatments in which $j_1 = 0$, and similarly for the other parameters.

(3) The quantities ζ are now computed from the intermediate estimate by the following formula. Let

$$\phi_1 = \sum_{i=1}^{m} \hat{\hat{F}}_i$$

$$= \text{sum of the intermediate estimates of the main effect}$$

$$\phi_2 = \sum_{\substack{i,j=1 \\ i<j}}^{m} \hat{\hat{F}}_{ij}$$

$$= \text{sum of the intermediate estimates of all two-factor interactions} \tag{13.3}$$

$$\phi_{2i} = \hat{\hat{F}}_{1i} + \hat{\hat{F}}_{2i} + \cdots + \hat{\hat{F}}_{i-1,i} + \hat{\hat{F}}_{i,i+1} + \cdots + \hat{\hat{F}}_{im}$$

$$= \text{sum of the intermediate estimates of all two-factor interactions which involve the } i\text{th factor}; \quad i = 1, \ldots, m$$

Then, for all integers i and j in the set $\{1, \ldots, m\}$, with $i < j$, the ζ are given by

$$\begin{aligned} &\zeta_{00} = \hat{\hat{\mu}}, \quad \zeta_{01} = \phi_1, \quad \zeta_{02} = \phi_2, \quad \zeta_{10,i} = \hat{\hat{\mu}}, \\ &\zeta_{11,i} = \hat{\hat{F}}_i, \quad \zeta_{12,i} = \phi_1 - \hat{\hat{F}}_i, \quad \zeta_{13,i} = \phi_{2i}, \quad \zeta_{14,i} = \phi_2 - \phi_{2i} \\ &\zeta_{20,ij} = \hat{\hat{\mu}}, \quad \zeta_{21,ij} = \hat{\hat{F}}_i + \hat{\hat{F}}_j, \quad \zeta_{22,ij} = \hat{\hat{F}}_{ij}, \quad \zeta_{23,ij} = \phi_1 - \hat{\hat{F}}_i - \hat{\hat{F}}_j \\ &\zeta_{24,ij} = \phi_{2i} + \phi_{2j} - 2F_{ij}, \quad \zeta_{25,ij} = \phi_2 - \phi_{2i} - \phi_{2j} + \hat{\hat{F}}_{ij} \end{aligned} \tag{13.4}$$

The ζ's obtained from (13.4) are finally substituted into (13.3) to obtain the estimates of the factorial effects of resolution V designs.

Similar formulas can be written down for the resolution VII case, but are not being presented here for lack of space.

The next question of interest is what are the expected values of the estimates $\hat{\mu}$, $\hat{F}_i$, $\hat{F}_{ij}$ for different values of i and j, if the assumption that the three-factor and higher order effects are zero is not correct? Well, the expected value of the estimators can be found in a straightforward manner. Since the optimal designs presented here turn out to have a very elegant structure, it is possible to simplify the formula for these expected values considerably. These simplified formulas are contained in a paper by the author, yet unpublished.

For a resolution V design, the degree of freedom due to error is given by n_e where

$$n_e = N - 1 - m(m + 1)/2 \tag{13.5}$$

Also, the error sum of squares S_e^2 is given by

$$S_e^2 = S_1 - S_0 \tag{13.6}$$

where

$$\begin{aligned} S_1 &= \text{the sum of squares of all the } N \text{ observations} \\ &\quad \text{obtained from the } N \text{ units} \\ S_0 &= \hat{\mu}\hat{\hat{\mu}} + \sum_{i=1}^{m} \hat{F}_i\hat{\hat{F}}_i + \sum_{\substack{i,j=1 \\ i<j}}^{m} \hat{F}_{ij}\hat{\hat{F}}_{ij} \end{aligned} \tag{13.7}$$

What about confidence intervals on the various parameters? We illustrate using F_{12}. Under the usual assumption that the observations on the N experimental units are independent and normally distributed with the same variance σ^2, we have $\text{var}(\hat{F}_{12}) = \sigma^2 2^{2m}[\sigma^{-2}2^{-2m} \text{ var}(\hat{F}_{12})]$, where the quantity $[\sigma^{-2}2^{-2m} \text{ var}(\hat{F}_{12})]$ is given in the tables. (For example, for $m = 7$, $N = 44$, we have $\sigma^{-2}2^{-2m} \text{ var}(\hat{F}_{12}) = 0.0253$.) Thus, an estimate of $\text{var}(\hat{F}_{12})$ is $2^{2m}[2^{-2m}\sigma^{-2} \text{ var}(\hat{F}_{12})]S_e^2/n_e = \hat{\sigma}^2(\hat{F}_{12})$, say. Thus, a confidence interval on F_{12} with a confidence coefficient $(1 - \alpha)$ is given by

$$\hat{F}_{12} - t_{\alpha/2}\hat{\sigma}(\hat{F}_{12}) \leq F_{12} \leq \hat{F}_{12} + t_{\alpha/2}\hat{\sigma}(\hat{F}_{12}) \tag{13.8}$$

where $t_{\alpha/2}$ is such that the number $(1 - t_{\alpha/2})$ is the upper $(1 - \alpha/2)$ percentage point of the t-distribution. In other words, we have $\text{Prob}(t \geq t_{\alpha/2}) = \alpha/2$, where t has the t-distribution.

What is the procedure for the estimation of the true effect of any treatment combination? To answer this, recall from Section 2 that if we are given the value of the various factorial effects, namely μ, the main effects F_i, and the two-factor and higher order interactions, then we can compute the treatment effects from these. Indeed, as indicated in Eq. (2.7), the procedure for computing the treatment effects from the factorial effects is about the same as that of computing the factorial effects from the treatment effects. Now, to calculate the estimates of the treatment effects, we proceed in the same way as earlier except that for the factorial effects, we take the estimates $\hat{\mu}$, $\hat{F}_i$, $\hat{F}_{ij}$ for the parameters in the resolution V design, and for the three-factor and higher order interactions, we assume the value 0. Starting from these estimates and following the same procedure, we compute the estimates of the treatment effects. It can be shown that these estimates are best linear unbiased estimates.

Now, we come to a more basic question. How to choose N. As mentioned earlier, this will depend on the considerations of cost and of the accuracy needed in the estimates of the various parameters. However, occasionally there is some flexibility, and one wants to choose a value of N within a certain relatively small range of values. In such a situation, it will be profitable to look at the column entitled E, which denotes the (absolute) efficiency of the corresponding design. We look at the column of E in the range of values of N under consideration, and choose that value of N at which the value of E takes a big jump from the value of E for the design corresponding to the value of N just preceding it. For example, suppose $m = 5$ and we are interested in the range $23 \leq N \leq 28$. Then it might be most preferable to choose the value $N = 26$ because the jump in the value of E relative to $N = 25$ is from 0.82 to 0.90.

The absolute efficiency E is a very crude measure of the efficiency of the given design. Since for different values of N, the optimal design (which is optimal in the class of all designs, both balanced and unbalanced) is usually not known, we cannot compute the actual efficiency of the designs given. The absolute efficiency of a design is its efficiency compared to an orthogonal design with the same value of N if the latter existed. However, as remarked in Section 2, such designs exist very, very rarely. Thus, it is conjectured that the actual efficiency for the majority of the designs presented here is close to 100. On the other hand, there are some known cases of unbalanced designs which are somewhat superior to the design given here. The efficiency E is computed by the formula

$$
\begin{aligned}
E &= \frac{(\mathrm{tr}\ V) \text{ for an orthogonal design}}{(\mathrm{tr}\ V) \text{ for the given design}} \\
&= \frac{1 + m(m+1)/2}{N[(\mathrm{tr}\ V) \text{ for the given designs}]}
\end{aligned}
\tag{13.9}
$$

For mathematically oriented readers, we may remark that the value of E is in a sense a measure of the "nonorthogonality" inherent in the pair (m, N).

Before concluding this section, we raise a more fundamental question. Consider a situation in which we are planning to use a resolution V design. We are obviously under the belief that most of the three-factor and higher order interactions are negligible. But how can we be sure that these factorial effects are actually small? It is not difficult to imagine that perhaps a few of the factorial effects which we have assumed negligible are actually not so. This is also observed from practical experience. What can be done about this? A new formulation of the whole subject of designs has been made recently by the author. This is the theory of *search linear models* and *search designs*. There we reformulate our basic problem by saying that we are interested in the estimation of μ, the main effects, and the two-factor interactions; however, we also want to *search* the nonzero interactions which we have had to assume to be negligible. Notice that the concept of *search* arises here since we do not know which interactions are actually negligible. Interested readers are referred to Section 15 for a few more details.

14. COMPARISON OF OPTIMALITY CRITERIA

Consider a statistical problem, where we are estimating two parameters, say θ_1 and θ_2, with respective estimates $\hat{\theta}_1$ and $\hat{\theta}_2$. Suppose $\text{var}(\hat{\theta}_1) = \sigma_1{}^2$, $\text{var}(\hat{\theta}_2) = \sigma_2{}^2$, and $\text{cov}(\hat{\theta}_1, \hat{\theta}_2) = \rho\sigma_1\sigma_2$. Then if V is the variance–covariance matrix of $(\hat{\theta}_1, \hat{\theta}_2)$, we have

$$|V| = \sigma_1{}^2\sigma_2{}^2(1 - \rho^2) \tag{14.1}$$

This shows that $|V|$ may be small not only because the variances $\sigma_1{}^2$ and $\sigma_2{}^2$ are small, but also because the correlation ρ may be large. This fact is also supported by certain situations with more than two parameters; see, for example, the work of Srivastava and Anderson (1974) on resolution IV designs. Thus, the determinant criterion for optimal designs has this feature, which makes it somewhat less satisfying in many situations. For this and certain other reasons, the author prefers the trace criterion in situations involving one response, and also in multiresponse problems where the different responses denote essentially similar physical variables. But where the responses are physically dissimilar, for example "amount of sunshine" and "amount of cow manure," the determinant criterion is preferred.

As is shown by the work of Kiefer and in the discrete case by that of Srivastava and Anderson (1970), the three criteria of D-, A-, and E-optimality usually imply each other. However, there are many exceptions. To compare designs with respect to the criteria, a "partial" study was undertaken by the author for a large number of values of m and N. For each index set $\boldsymbol{\mu}'$ which corresponds to an A-optimal design, adjacent index sets were compared. [Two index sets $(\mu_0, \ldots, \mu_4)$ and $(\mu_0{}^*, \ldots, \mu_4{}^*)$ were considered adjacent if $|\mu_i - \mu_i{}^*| \leq 1$, for $i = 1, 2, 3$.] It was felt that index sets not adjacent to the one corresponding to an A-optimal design would not be optimal with respect to other criteria as well, but there is no proof of this. In some cases, it did turn out that of two index sets $\boldsymbol{\mu}$ and $\boldsymbol{\mu}^*$, one was A-optimal and the other D-optimal; however, in most such cases, one or both of the two index sets did correspond to a combinatorially nonexistent design. This shows that so far as discrete designs are concerned, optimality cannot be decided merely by analytical methods; combinatorial existence studies are also necessary.

Out of the cases considered, very few values of m and N are such that there exist two different designs optimal under two different criteria. We present one example here. The index sets (4, 2, 1, 1, 4) and (5, 2, 1, 1, 3), with $m = 6$, $N = 26$, both correspond to existent designs (say T_1 and T_2). We find that both T_1 and T_2 are E-optimal, while T_1 is D-optimal, and T_2 is A-optimal; the arithmetic and geometric means of the reciprocal of the roots of the information matrix M corresponding to T_1 are 0.05005871 and 0.4515952, and those for T_2 are 0.05005490 and 0.04520806. This shows that T_1 and T_2 are not really too different either with respect to A- or D-optimality.

15. SEARCH DESIGNS

As we indicated at the end of Section 5, a basic difficulty with the theory presented here and, indeed, with all the theory of optimal design developed so far, is that we cannot be sure that the parameters that we have assumed negligible are actually so. To meet this situation, a new concept called search linear models has been introduced which we now briefly discuss. Let $\mathbf{y}(N \times 1)$ be a vector of observations, such that

$$\operatorname{Exp}(\mathbf{y}) = A_1\xi_1 + A_2\xi_2, \qquad \operatorname{var}(\mathbf{y}) = \sigma^2 I_N \tag{15.1}$$

where $A_1(N \times v_1)$ and $A_2(N \times v_2)$ are known matrices, $\xi_1(v_1 \times 1)$ and $\xi_2(v_2 \times 1)$ are vectors of parameters of which ξ_1 is completely unknown, and σ^2 is the (usually unknown) error variance. We have partial information about ξ_2. It is known that at most k elements of ξ_2 are nonzero, where k is a known or unknown integer, usually much smaller than v_2 in applications. The problem is this: We want to estimate the parameters in ξ_1 and also search the nonnegligible elements of ξ_2 and estimate them. Of course, besides estimation, other inference problems could be considered. Notice that the model (15.1) is basically different from an ordinary linear model. If we know which elements of ξ_2 are negligible, we could include the rest of the elements in ξ_1 and would then have an ordinary linear model. However, since we do not know which elements of ξ_2 are nonzero, we do not know exactly which linear model holds. Forcing this situation into a linear model would require us to estimate all the parameters in ξ_1 and ξ_2, thus necessitating a rather large experiment in which we must have $N \geq v_1 + v_2$. Since there is a large amount of information available on ξ_2, namely that a large number of its elements are negligible, we do not need to run a very large experiment. This situation, which involves search, entails both design and inference problems. A beginning on such problems has been made in the last few years in the papers of the author and his associates. For lack of space, we cannot discuss these problems in detail here. However, a very brief introduction to the subject would be very useful to the readers of this chapter.

The case in which $\sigma^2 = 0$ is unrealistic. Still, however, it is very important in the search linear model defined in (15.1). The reason is that if the structure of the observations $\mathbf{y}$ (and hence the matrices A_1 and A_2) is such that we cannot solve the search and estimation problem when there is no noise (i.e., $\sigma^2 = 0$), we cannot hope to solve this problem when noise is present. In other words, the structure of the observations, and hence the design aspect of the problem, is largely concerned with the case in which $\sigma^2 = 0$. When $\sigma^2 > 0$, the structure of the observations (in other words, the design) needed for "search" usually remains the same; only the inference aspect gets influenced. If the design is such that in the noiseless case we can do the search correctly, and estimate all the nonzero parameters in ξ_2 and also the elements of ξ_1 accurately (i.e., with variance 0), then such a design is called a *search design.* When $\sigma^2 > 0$, then even though a search design is used, the probability of correct search would in general be expected to be less than unity, and as usual, various estimates of parameters will have a nonzero variance. The following result is basic.

Theorem 15.1. Consider the noiseless case, and assume k is known. A necessary and sufficient condition that the observation $\mathbf{y}$ have the structure of a search design is that we have

$$\operatorname{rank}(A_1 \vdots A_{20}) = v_1 + 2k, \ldots \tag{15.2}$$

for *every* submatrix $A_{20}(N \times 2k)$ of A_2.

The concept of search linear models was introduced by Srivastava (1975), wherein many of the basic ideas and results, including the preceding theorem, were presented. A related concept, the weakly resolvable models, and certain sequential ideas were discussed by Srivastava (1976). An important application of this concept is to factorial experiments. Thus, consider the situation where we normally use a resolution V design: here we identify ξ_1 with the general mean μ, the main effects F_i, and the two-factor interactions F_{ij}. Then the vector ξ_2 is identified with the three-factor and higher order effects. Thus, we have $v_1 = 1 + \binom{m}{1} + \binom{m}{2}$, and $v_2 = 2^m - v_1$. A design is said to be of resolution 5.k if in the noiseless case, we can correctly search the nonnegligible parameters in ξ_2 (it being known that at most k of these are nonnegligible), and can precisely estimate these and also the elements of ξ_1. Designs of resolution 5.1 have been obtained by Srivastava and Ghosh (1977). Notice that although it is true that the value of k may not be known, it is intuitively clear that in general the design which we know is of resolution 5.1 would be better than a design of resolution 5, since we would be able to remove some of the bias in the corresponding linear model, by identifying at least one of the nonnegligible elements of ξ_2.

Some Monte Carlo and other studies have been done on the statistical inference problem concerning search, and the probability of correct search under a given method of search. It has been found that if the nonnegligible element of ξ_2 (assuming $k = 1$) equals 2σ, then there is a method of search under which the probability of correct search is 89%; when this value becomes 3σ, the probability jumps to 99%. This work is reported by Srivastava and Mallenby (1978).

Notice that to obtain a design of resolution 5.k, we would need a vector of observations $\mathbf{y}$, such that the corresponding matrices A_1 and A_2 are such that Eq. (15.2) is satisfied.

It is clear that if we are making an optimal design assuming $\xi_2 = 0$ when in fact some elements of ξ_2 are nonnegligible, then our so-called optimal design is not really optimal, because it is not bias-free. This led the author to introduce the concept of bias-free optimality theory (Srivastava, 1977). Needless to say, all the classical optimality theory suffers from this difficulty due to bias, including the cases where the factors are discrete or continuous.

However, from the experience the author has had so far with optimal search design theory, it seems that the theory of the classical optimal designs is still quite valuable. Not only may it be a natural stepping stone toward bias-free optimality theory, but it seems that many of the classical optimal designs may also turn out to be optimal in a certain sense with respect to the bias-free optimality criteria.

ACKNOWLEDGMENT

I am thankful to Pamela Eicher for her excellent typing of the manuscript.

REFERENCES

Atwood, C. L. (1976). Convergent design sequences for sufficiently regular optimality criteria, *Ann. Statist.* **4**, 1124–1138.

Berlekamp, E. R. (1968). "Algebraic Coding Theory." McGraw-Hill, New York.

Bose, R. C., and Srivastava, J. N. (1964a). Analysis of irregular factorial fractions, *Sankhyā Ser. A* **26**, 117–144.

Bose, R. C., and Srivastava, J. N. (1964b). Multidimensional partially balanced designs and their analysis with applications to partially balanced factorial fractions, *Sankhyā Ser. A* **26**, 145–168.

Box, M. J., and Draper, N. R. (1971). Factorial designs, the $|X'X|$ criterion, and some related matters, *Technometrics* **13**, 731–743.

Chopra, D. V. (1972). Trace-optimal balanced resolution V designs of 2^m series with $m = 9, 10, 11$, IMS Bull. 1 (Abstract).

Chopra, D. V. (1975a). Balanced optimal 2^8 fractional factorial designs of resolution V, $52 \leq N \leq 59$, "A Survey of Statistical Designs and Linear Models" (J. N. Srivastava, ed.), pp. 91–99. North-Holland Publ., Amsterdam.

Chopra, D. V. (1975b). Optimal balanced 2^8 fractional factorial designs of resolution V, with 60 to 65 runs, *Proc. Int. Statist. Inst.* (*Warsaw*) 164–168.

Chopra, D. V. (1975c). Some investigations on trace optimal balanced designs of resolution V and balanced arrays. Witchita State Univ. Bull. LI-4, Univ. Stud. 105.

Chopra, D. V., and Srivastava, J. N. (1973a). Optimal balanced 2^7 fractional factorial designs of resolution V with $N \leq 42$, *Amer. Inst. Statist. Math.* **25-3**, 587–604.

Chopra, D. V., and Srivastava, J. N. (1973b). Optimal balanced 2^7 fractional factorial designs of resolution V, $49 \leq N \leq 55$, *J. Comm. Statist.* **2-1**, 59–84.

Chopra, D. V., and Srivastava, J. N. (1974). Optimal balanced 2^8 fractional factorial designs of resolution V, $37 \leq N \leq 51$, *Sankhyā Ser. A* **36**, 41–52.

Chopra, D. V., and Srivastava, J. N. (1977). Optimal balanced 2^7 fractional factorial designs of resolution V, $43 \leq N \leq 48$, *Sankhyā* (to be published).

Cochran, W. G., and Cox, G. M. (1970). "Experimental Designs." Wiley, New York.

Dembowski, P. (1968). "Finite Geometries." Springer-Verlag, Berlin and New York.

Eccleston, J. A., and Hedayat, Y. (1974). On the theory of connected designs: Characterization and optimality, *Ann. Statist.* **2**, 1238–1255.

Federer, W. T. (1955). "Experimental Design: Theory and Application." Macmillan, New York.

Fedorov, V. V. (1972). "Theory of Optimal Experiments." Academic Press, New York.

Harville, D. A. (1975). Computing optimum designs for covariance models, "A Survey of Statistical Design and Linear Models" (J. N. Srivastava, ed.), pp. 209–228. North-Holland Publ., Amsterdam.

Herzberg, A. M., and Andrews, D. F. (1976). Some considerations in the optimal design of experiments in non-optimal situations, *J. Roy. Soc. Statist.* **38**, 284–289.

Hoke, A. T. (1975). The characteristic polynomial of the information matrix for second order models, *Ann. Statist.* **3**, 780–786.

Karlin, S., and Studden, W. J. (1966). Optimal experimental designs, *Ann. Math. Statist.* **37**, 1439–1888.

Kempthorne, O. (1952). "The Design and Analysis of Experiments." Wiley, New York.

Kempthorne, O. (1971). "The Design and Analysis of Experiments." Wiley, New York.

Kiefer, J. C. (1959). Optimum experimental designs, *J. Roy. Statist. Soc. Ser. B* **21**, 273–319.

Kiefer, J. (1961a). Optimum designs in regression problems II, *Ann. Math. Statist.* **32**, 298–325.

Kiefer, J. (1961b). Optimum experimental designs V, with applications to symmetric rotatable designs, *Proc. Berkeley Symp. Math. Statist. Probability, 4th* **1**, 381–405. Univ. of California Press, Berkeley, California.

Kiefer, J. (1962). Two more criteria equivalent to D-optimality of designs, *Ann. Math. Statist.* **33**, 792–796.

Kiefer, J. (1974). General equivalence theory for optimum designs (approximate theory), *Ann. Statist.* **2**, 849–879.

Kiefer, J. (1975). Construction and optimality of generalized Youden designs, "A Survey of Statistical Design and Linear Models" (J. N. Srivastava, ed.), pp. 333–353. North-Holland Publ., Amsterdam.

Kiefer, J., and Wolfowitz, J. (1959). Optimum designs in regression problems, *Ann. Math. Statist.* **30**, 271–294.

Lai, T. (1973). Optimal stopping and sequential tests which minimize the maximum expected sample size, *Ann. Math. Statist.* **1**, 659–673.

Mitchell, T. J. (1974). An algorithm for the construction of "D-optimal" experimental designs, *Technometrics* **16**, 203–211.

Nalimov, V. V., Golikova, T. I., and Nikeshina, N. G. (1970). On practical use of the concept of D-optimality, *Technometrics* **48**, 799–812.

Pearce, S. C. (1974). Optimality of design in plot experiments, *Proc. Int. Biometric Conf., 8th* (L. C. A. Corsten and T. Postelnicu, eds.), pp. 23–30. Editura Academiei Republich Socialiste Romania.

Puri, M. L., and Sen, P. K. (1971). "Nonparametric Methods in Multivariate Analysis." Wiley, New York.

Rafter, J. A. (1971). Contributions to the Theory and Construction of Partially Balanced Arrays, Ph.D. dissertation, Michigan State Univ.

Raghavarao, D. (1971). "Constructions and Combinatorial Problems in Designs of Experiments." Wiley, New York.

Raktoe, B. L. (1974). On classes of equi-information factorial arrangements, *Proc. Int. Biometric Conf., 8th* (L. C. A. Corsten and T. Postelnicu, eds.), pp. 31–44. Editura Academiei Republich Socialiste Romania.

Rao, C. R. (1973). Some combinatorial problems of arrays and applications to design of experiments, "A Survey of Combinatorial Theory" (J. N. Srivastava, ed.), Chapter 29. North-Holland Publ., Amsterdam.

Roy, S. N., Gnanadesikan, R., and Srivastava, J. N. (1970). "Analysis and Design of Certain Quantitative Multi-Response Experiments." Pergamon, Oxford.

Ruiz, F., and Seiden, E. (1974). On construction of some families of generalized Youden designs, *Ann. Statist.* **2**, 503–519.

Sacks, J., and Ylvisaker, D. (1970a). Designs for regression problems with correlated errors III, *Ann. Math. Statist.* **41**, 2057–2074.

Seiden, E., and Zemach, R. (1966). On orthogonal arrays, *Ann. Math. Statist.* **27**, 1355–1370.

Shirakura, T. (1976). Optimal balanced fractional 2^m factorial designs of resolution VII, $6 \leq m \leq 8$, *Ann. Statist.* **4**, 515–531.

Silvey, S. D., and Titterington, D. M. (1973). A geometric approach to optimal design theory, *Biometrika* **60**, 21–32.

Srivastava, J. N. (1961). Contributions to the Construction and Analysis of Designs, Univ. North Carolina, Inst. of Statist., Mimeo Ser., No. 301.

Srivastava, J. N. (1970a). Optimal balanced 2^m fractional factorial designs, S. N. Roy Memorial Volume, Univ. of North Carolina and Indian Statist. Inst., pp. 689–706.

Srivastava, J. N. (1972). Some general existence conditions for balanced arrays of strength t and 2 symbols, *J. Comb. Theory* **13**, 198–206.

Srivastava, J. N. (1975). Designs for searching non-negligible effects, "A Survey of Statistical Designs and Linear Models" (J. N. Srivastava, ed.), pp. 507–519. North-Holland Publ., Amsterdam.

Srivastava, J. N. (1976). Some further theory of search linear models, "Contributions to Applied Statistics," pp. 249–256. Swiss–Australian Region of Biometry Society.

Srivastava, J. N. (1977). Optimal search designs or designs optimal under bias-free optimality criteria, "Statistical Decision Theory and Related Topics" (S. S. Gupta and D. S. Moore, eds.), Vol. II, pp. 375–409.

Srivastava, J. N., and Anderson, D. A. (1970). Optimal fractional factorial plans for main effects orthogonal to two factor interactions 2^m series, *J. Amer. Statist. Assoc.* **65**, 828–843.

Srivastava, J. N., and Anderson, D. A. (1971). Factorial sub-assembly association scheme and multidimensional partially balanced designs, *Ann. Math. Statist.* **42**, 1167–1181.

Srivastava, J. N., and Anderson, D. A. (1974). A comparison of the determinant, trace, and largest root optimality criteria, *Comm. Statist.* **3**, 933–940.

Srivastava, J. N., and Chopra, D. V. (1971a). On the comparison of certain classes of balanced 2^8 fractional factorial designs of resolution V, with respect to the trace criterion, *J. Ind. Soc. Agric. Statist.* **23**, 124–131.

Srivastava, J. N., and Chopra, D. V. (1971b). On the characteristic roots of the information matrix of 2^m balanced factorial designs of resolution V with applications, *Ann. Math. Statist.* **42**, 722–734.

Srivastava, J. N., and Chopra, D. V. (1971c). Balanced optimal 2^m fractional factorial designs of resolution V, $m \leq 6$, *Technometrics* **13**, 257–269.

Srivastava, J. N., and Chopra, D. V. (1973). Balanced arrays and orthogonal arrays, "A Survey of Combinatorial Theory" (J. N. Srivastava, ed.), pp. 411–428. North-Holland Publ., Amsterdam.

Srivastava, J. N., and Chopra, D. V. (1974). Balanced trace optimal 2^7 fractional factorial designs of resolution V with 56 to 68 runs, *Utilitas Math.* **5**, 263–279.

Srivastava, J. N., and Ghosh, S. (1977). Balanced 2^m factorial designs of resolution V which allow search and estimation of one extra unknown effect, $4 \leq m \leq 8$, *Comm. Statist.—Theor. Methods* **A6(2)**, 141–166.

Srivastava, J. N., and Mallenby, D. M. (1978), Some studies on a new method of search in search linear models, submitted for publication.

Wahba, G. (1971). On the regression design problem of Sacks and Ylvisaker, *Ann. Math. Statist.* **42**, 1035–1053.

Webb, S. R. (1965). Expansible and contractible factorial designs and the application of linear programming to combinatorial problems, Aerospace Res. Lab. Tech. Rep. 65-116, Part I. Wright-Patterson Air Force Base, Ohio.

Whittle, P. (1973). Some general points in the theory of optimal experimental design, *J. Roy. Statist. Soc. Ser. B* **35**, 123–130.

Whitwell, T. D., and Morbey, G. K. (1961). Reduced designs of resolution V, *Technometrics* **3**, 459–477.

Wynn, H. P. (1975). Simple conditions for optimum design algorithms, "A Survey of Statistical Design and Linear Models" (J. N. Srivastava, ed.), pp. 571–579. North-Holland Publ., Amsterdam.

Yamamoto, S., Shirakura, T., and Kuwada, M. (1976). Characteristic polynomials of the information matrices of balanced fractional 2^m factorial designs of higher $(2l + 1)$ resolution, *Essays Probability Statist.* 73–94.

Zacks, S., and Eichhorn, B. H. (1975). Sequential search of optimal dosages: The linear regression case, "A Survey of Statistical Design and Linear Models" (J. N. Srivastava, ed.), pp. 609–628. North-Holland Publ., Amsterdam.

Author Index

Numbers in parentheses are reference numbers and indicate that an author's work is referred to although his name may not be cited in the text. Numbers in italics show the page on which the complete reference is listed.

T

U

V

W

Y

Z

Subject Index

T

W

Y

Z

A
B
C 8
D 9
E 0
F 1
G 2
H 3
I 4
J 5